Information Systems Engineering and Management

Volume 23

The book series "Information Systems Engineering and Management" (ISEM) publishes innovative and original works in the various areas of planning, development, implementation, and management of information systems and technologies by enterprises, citizens, and society for the improvement of the socio-economic environment.

The series is multidisciplinary, focusing on technological, organizational, and social domains of information systems engineering and management. Manuscripts published in this book series focus on relevant problems and research in the planning, analysis, design, implementation, exploration, and management of all types of information systems and technologies. The series contains monographs, lecture notes, edited volumes, pedagogical and technical books as well as proceedings volumes.

Some topics/keywords to be considered in the ISEM book series are, but not limited to: Information Systems Planning; Information Systems Development; Exploration of Information Systems; Management of Information Systems; Blockchain Technology; Cloud Computing; Artificial Intelligence (AI) and Machine Learning; Big Data Analytics; Multimedia Systems; Computer Networks, Mobility and Pervasive Systems; IT Security, Ethics and Privacy; Cybersecurity; Digital Platforms and Services; Requirements Engineering; Software Engineering; Process and Knowledge Engineering; Security and Privacy Engineering, Autonomous Robotics; Human-Computer Interaction; Marketing and Information; Tourism and Information; Finance and Value; Decisions and Risk; Innovation and Projects; Strategy and People.

Indexed by Google Scholar. All books published in the series are submitted for consideration in the Web of Science.

For book or proceedings proposals please contact Alvaro Rocha (amrrocha@gmail.com).

Brojo Kishore Mishra · Alvaro Rocha
Editors

Future of AI, IoT, and Sustainability

Springer

Editors
Brojo Kishore Mishra
Department of Computer Science
and Engineering
NIST University
Brahmapur, Odisha, India

Alvaro Rocha
School of Economics and Management
ISEG
University of Lisbon
Lisbon, Portugal

ISSN 3004-958X ISSN 3004-9598 (electronic)
Information Systems Engineering and Management
ISBN 978-3-031-76285-7 ISBN 978-3-031-76286-4 (eBook)
https://doi.org/10.1007/978-3-031-76286-4

This Springer imprint is published by the registered company Springer Nature Switzerland AG
The registered company address is: Gewerbestrasse 11, 6330 Cham, Switzerland

Contents

High-Tech Agriculture using AI and IOT

Tejinder Kaur and Amrit Preet Singh

Abstract The agricultural industry has always been an essential sector in society, and its development is crucial for the future of food security. With a growing global population, the demand for high-quality, nutritious, and sustainable food is increasing, and the traditional farming methods are no longer sufficient to meet these demands. In recent years, high tech agriculture has emerged as a promising solution, integrating advanced technologies like Artificial Intelligence (AI) and the Internet of Things (IOT) to optimize farming practices and enhance crop yields. Our aim is to explore the applications of AI and IOT in high tech agriculture and their impact on improving the efficiency and productivity of farming operations. We investigate the various ways that AI and IOT technologies are being used in agriculture, such as precision farming, smart irrigation, automated pest detection, and livestock monitoring. We highlight the benefits of precision farming, which involves using AI and IOT technologies to analyze data from sensors, drones, and satellite imagery to determine the optimal time for planting, fertilizing, and harvesting crops. This data-driven approach can help reduce costs, increase yields, and minimize environmental impact by reducing the use of fertilizers, pesticides, and water. We also discuss the use of IOT sensors in smart irrigation, which can monitor soil moisture levels and weather conditions to optimize irrigation schedules, reducing water waste and increasing crop yields. The use of automated pest detection using AI algorithms can help farmers detect and respond to pest infestations more quickly, reducing crop losses and minimizing the use of pesticides. Additionally, we explore the use of AI and IOT in livestock monitoring, which can help farmers track animal health and behavior, identify potential issues early, and optimize feed and nutrient intake to enhance animal welfare and productivity. Finally, we conclude that AI and IOT technologies have enormous potential in high tech agriculture, and their implementation can significantly improve farming practices and help address the challenges

T. Kaur (✉)
Department of MMICTBM, Maharishi Markandeshwar (Deemed To Be University), Mullana, Ambala-Haryana 133207, India
e-mail: teji7890@gmail.com

A. P. Singh
HIMT-Hoshiarpur Institute of Management and Technology, Hoshiarpur, Punjab, India

B. K. Mishra and A. Rocha (eds.), *Future of AI, IoT, and Sustainability*, Information Systems Engineering and Management 23,
https://doi.org/10.1007/978-3-031-76286-4_1

of food security and sustainability. By optimizing resource use, reducing environmental impact, and increasing crop yields, these technologies can pave the way for a more efficient and sustainable future in agriculture. However, the implementation of these technologies requires significant investment and infrastructure development, and more research is needed to fully realize their potential.

Keywords Agriculture · Sustainable · Farming · AI · IOT · Food · Planting · Fertilizing · And harvesting crops

1 Introduction

Agriculture is a crucial sector in the global economy, providing food and raw materials for various industries. In recent years, the agricultural industry has faced various challenges, including the growing demand for food, climate change, and limited resources. These challenges have prompted the development of innovative technologies that can address these issues and optimize farming practices [1]. The integration of Artificial Intelligence (AI) and the Internet of Things (IOT) into agriculture, referred to as high tech agriculture, has emerged as a promising solution that can enhance efficiency, reduce costs, and increase yields. AI is a branch of computer science that involves the development of algorithms and models that enable machines to learn and perform tasks that typically require human intelligence, such as decision-making and problem-solving. IOT, on the other hand, involves connecting devices and sensors to the internet to collect and exchange data, enabling them to interact with each other and the physical world. The combination of these two technologies in agriculture has revolutionized the industry, enabling farmers to make data-driven decisions, automate farming practices, and optimize resource use [2]. The application of AI and IOT in agriculture has numerous benefits, including improved crop yields, reduced resource use, and enhanced environmental sustainability. Precision farming is one of the most significant applications of AI and IOT in agriculture, involving the use of data collected from sensors, drones, and satellite imagery to optimize farming practices [1]. This data-driven approach allows farmers to determine the optimal time for planting, fertilizing, and harvesting crops, reducing costs and increasing yields [2]. Additionally, the use of IOT sensors in smart irrigation can help reduce water waste and increase crop yields, while automated pest detection using AI algorithms can reduce crop losses and minimize the use of pesticides. Livestock monitoring is another area where AI and IOT are being applied, enabling farmers to track animal health and behavior, identify potential issues early, and optimize feed and nutrient intake to enhance animal welfare and productivity [2]. Despite the potential benefits of high-tech agriculture, its implementation requires significant investment and infrastructure development. There are also concerns regarding the potential social and ethical implications of relying on technology to make farming decisions, particularly in developing countries where access to technology is limited. Our aim is to provide a comprehensive analysis of the applications of AI and IOT in high tech agriculture and their impact on farming practices [2]. The matter aims to explore the various ways that AI and IOT technologies are being used in agriculture and their potential to address the challenges of food security, environmental sustainability, and economic growth [2]. The text also seeks to identify the barriers and challenges to the adoption of these technologies in agriculture and the potential social and ethical implications of their implementation. High tech agriculture using AI and IOT has enormous potential to transform the agricultural industry and address the challenges of food security, sustainability, and economic growth [2]. The integration of these technologies into agriculture enables farmers to make data-driven decisions, automate farming practices, and optimize resource use. However, the implementation of these technologies requires significant investment and infrastructure development, and more research is needed to fully realize their potential [2].

2 High-Tech Agriculture Using AI and IOT

Over the past few decades, advancements in artificial intelligence (AI) have revolutionized many industries, including healthcare, finance, and retail [3]. In recent years, AI has also been increasingly used in agriculture, where it has the potential to transform the industry by making farming more Efficient, sustainable, and profitable [3] we will discuss how AI can be used for high tech agriculture [3]. One of the primary applications of AI in agriculture is precision agriculture, which refers to the use of technology to optimize crop yields and reduce waste. Precision agriculture involves the use of sensors, drones, and other technology to collect data on soil moisture, temperature, nutrient levels, and other factors that affect crop growth. This data is then used to make precise decisions about irrigation, fertilizer application, and other farming practices [3]. AI can enhance the capabilities of precision agriculture by using machine learning algorithms to analyze large amounts of data and make accurate predictions about crop growth and yield. For example, AI can be used to analyze satellite imagery to identify areas of a farm that are underperforming and suggest specific interventions to improve crop growth [3].

2.1 Autonomous Farming

Another application of AI in agriculture is autonomous farming, which refers to the use of robots and other autonomous systems to perform farming tasks. Autonomous farming has the potential to reduce labor costs, increase efficiency, and improve worker safety [4]. Autonomous farming systems can perform a wide range of tasks, such as planting, weeding, and harvesting crops [4]. AI can enhance the capabilities of autonomous farming systems by providing them with advanced computer vision and decision-making capabilities [4]. For example, AI-powered drones can be used to scan fields and identify areas that need irrigation or other interventions. Similarly, autonomous robots can use AI algorithms to navigate fields and identify and remove weeds [4].

2.2 Livestock Monitoring

AI can also be used to monitor the health and behavior of livestock, which can improve animal welfare and increase productivity [4]. AI-powered sensors can be implanted in animals to monitor their vital signs, such as heart rate and body temperature [4]. This data can be analyzed in real-time using machine learning algorithms to detect early signs of disease or other health issues. Similarly, AI can be used to monitor the behavior of livestock, such as their movement patterns and social interactions [4]. This can help farmers identify potential problems, such as overcrowding or aggressive behavior, and take steps to address those [4].

2.3 Predictive Analytics

AI can also be used to analyze large amounts of data to make accurate predictions about crop yields, weather patterns, and other factors that affect farming [5]. Predictive analytics can help farmers make informed decisions about when to plant and harvest crops, when to apply fertilizer and other inputs, and how to manage water resources [5]. For example, AI can be used to analyze weather patterns to predict when a drought or other weather event is likely to occur. This information can be used to adjust planting schedules or irrigation practices to minimize the impact of the weather event on crop yields [5].

2.4 Food Safety

AI can also be used to improve food safety by identifying potential food borne illnesses before they become widespread [6]. AI can analyze data from sensors and other sources to detect patterns that may indicate the presence of a food borne illness. This information can be used to trace the origin of the illness and take steps to prevent its spread [6]. AI can also be used to monitor food processing and distribution systems to ensure that food is handled and stored in a safe and sanitary manner [6]. AI-powered sensors can monitor the temperature and humidity of food storage areas to ensure that food is kept at the appropriate temperature to prevent spoilage and bacterial growth [6].

2.5 *Soil Health*

Finally, AI can be used to monitor and improve soil health. Healthy soil is essential for growing healthy crops, and AI can be used to identify and address issues that may be affecting soil health [7]. AI can play a significant role in crop genetic engineering by enabling faster and more accurate identification of gene sequences and their functions [7]. Crop genetic engineering is the process of modifying a plant's genetic makeup to produce desirable traits, such as increased yield, disease resistance, and improved nutritional value. Traditionally, this has been a time-consuming and labor-intensive process that involves the trial and error of different gene combinations. However, with the use of AI, the process can be made more efficient and accurate [7].

2.6 *Identification of Relevant Genes*

The first step in crop genetic engineering is identifying relevant genes [7]. AI algorithms can be used to scan massive databases of gene sequences to identify specific genes that are related to the traits of interest. This process can be much faster and more accurate than traditional methods of gene discovery [7].

2.7 *Gene Function Predictions*

Once the relevant genes have been identified, the next step is to determine their function. This is where AI can be particularly useful. Machine learning algorithms can analyze large datasets of gene expression data to identify the roles that specific genes play in various biological processes [7]. By combining this information with existing knowledge about gene function, AI can predict the functions of previously unknown genes [7].

2.8 *Gene Editing*

The final step in crop genetic engineering is gene editing. Once the relevant genes have been identified and their functions determined, CRISPR-Cas9 technology can be used to edit the genes to produce the desired traits [7]. AI can be used to optimize the CRISPR-Cas9 system for each specific gene, predicting the potential effects of different editing strategies [7].

2.9 Predictive Modeling

AI can also be used to build predictive models to predict the performance of the engineered crops in different environments [7]. These models can take into account factors such as soil type, climate, and pest and disease pressure. This can help to ensure that the crops perform as expected when grown in different regions and under different conditions [7].

2.10 Gene Expression Analysis

Once the engineered crops have been grown, AI can be used to analyze their gene expression to ensure that the intended genes have been properly edited and to monitor their performance [8]. This can help to identify any unintended effects of gene editing and ensure that the crops are performing as expected [8].

2.11 Taut Optimization

Finally, AI can be used to optimize the traits of the engineered crops. By monitoring the performance of the crops and analyzing their gene expression data, AI can identify potential areas for further improvement. This can help to ensure that the crops are continually improved and that their performance is optimized [8]. Overall, AI has the potential to revolutionize crop genetic engineering by enabling faster and more accurate identification of relevant genes, predicting gene function, optimizing gene editing strategies, building predictive models, analyzing gene expression data, and optimizing crop performance. By combining AI with traditional genetic engineering methods, it is possible to produce crops with improved yield, disease resistance, and nutritional value, helping to ensure food security for the growing global population. [8].

3 Role of IOT

The Internet of Things (IOT) has been a game changer in many industries, including agriculture. By using IOT devices and sensors, farmers can collect and analyze data to improve crop yields, optimize water and fertilizer usage, and reduce labor costs. In this essay, I will discuss how to use IOT in the field of agriculture. One of the most significant challenges facing farmers is the ability to monitor the health and growth of their crops in real-time. With IOT devices and sensors, farmers can remotely monitor and control irrigation, fertilization, and other factors that affect crop growth [8]. These devices can be installed in the soil, on plants, or even on drones to provide real-time data on soil moisture levels, temperature, humidity, and other environmental factors that affect plant growth [8]. By analyzing this data, farmers can make better decisions about when to water, fertilize, or spray their crops, leading to healthier plants and higher yields [8]. Another way IOT can be used in agriculture is by automating processes such as planting, harvesting, and sorting [8]. For example, self-driving tractors can be equipped with sensors that detect soil moisture and automatically adjust planting depths and spacing to optimize crop growth. Similarly, harvesting robots can be programmed to identify and pick only the ripest fruits and vegetables, reducing waste and increasing efficiency. Sorting and packaging can also be automated using IOT technology, allowing farmers to quickly and accurately sort and package their products for distribution. IOT can also be used to reduce labor costs and increase worker safety [8]. For example, drones equipped with cameras and sensors can be used to monitor crops, detect pests and diseases, and apply treatments [8]. This not only reduces the need for manual labor but also eliminates the need for workers to come into contact with potentially hazardous chemicals. Similarly, IOT devices can be used to monitor livestock health and behavior, reducing the need for manual inspections and increasing the safety of workers. IOT can be used to improve supply chain efficiency and reduce waste. By tracking the location and condition of products throughout the supply chain, farmers can optimize storage, transportation, and distribution, reducing spoilage and waste [8]. Additionally, IOT devices can be used to monitor and control the storage and processing of crops, ensuring that they are kept at optimal temperatures and humidity levels, extending their shelf life and reducing waste. IOT technology has the potential to revolutionize the agriculture industry. By using sensors, drones, and other IOT devices, farmers can collect and analyze data to make better decisions about irrigation, fertilization, and other factors that affect crop growth. Automation can also reduce labor costs and increase worker safety, while IOT technology can improve supply chain efficiency and reduce waste [8]. The adoption of IOT technology in agriculture is still in its early stages, but it has the potential to improve productivity and profitability for farmers, while also reducing the environmental impact of farming [9].

4 Cyborg Chickens

The concept of cyborg chickens in agriculture may seem like science fiction, but recent advancements in biotechnology and engineering have made it a reality. In this essay, I will discuss how cyborg chickens can be used in agriculture. A cyborg chicken is a chicken that has been implanted with electronic devices or sensors to monitor its health, behavior, and other characteristics [9]. For example, a cyborg chicken could be implanted with a sensor to monitor its body temperature, heart rate, or other vital signs. These sensors could then be used to detect the early signs of disease or stress, allowing farmers to intervene before the chicken becomes seriously ill or dies. Cyborg chickens could also be used to monitor environmental conditions in the poultry house, such as temperature, humidity, and air quality. By collecting data on these factors, farmers could make adjustments to improve the health and well-being of the chickens, which could lead to higher productivity and profitability. Another potential use for cyborg chickens is in the field of genetics [9]. By implanting electronic devices in the eggs of chickens, scientists could collect data on the genetic traits of the developing embryos [9]. This could lead to the identification of new genes and genetic variations that could be used to improve the health, productivity, and disease resistance of poultry [9]. Cyborg chickens could also be used to monitor and control the spread of infectious diseases in poultry. By implanting electronic devices in the chickens, farmers could track the spread of diseases and quarantine infected birds to prevent the spread of the disease to other flocks [9]. This could be especially useful in the case of highly infectious diseases such as avian flu or Newcastle disease, which can spread rapidly through poultry flocks [9].

In addition to the benefits to farmers, cyborg chickens could also have a positive impact on animal welfare. By monitoring the health and behavior of chickens, farmers could identify and address issues such as stress, disease, or poor living conditions that could impact the well-being of the birds. This could lead to improved animal welfare and a reduction in the use of antibiotics and other drugs to treat sick animals. Despite the potential benefits of cyborg chickens in agriculture, there are also concerns about the ethical implications of implanting electronic devices in animals. Some critics argue that this could lead to animal suffering, as the devices could cause pain or discomfort for the animals. Others worry that this could lead to a blurring of the line between humans and animals, and that it could lead to the commoditization of living creatures [9]. In conclusion, cyborg chickens have the potential to revolutionize the agriculture industry by providing farmers with real-time data on the health and behavior of their chickens. By monitoring environmental conditions, genetics, and the spread of diseases, cyborg chickens could lead to higher productivity, improved animal welfare, and a reduction in the use of antibiotics and other drugs to treat sick animals. However, there are also concerns about the ethical implications of implanting electronic devices in animals, and further research and discussion is needed to fully understand the potential risks and benefits of this technology.

5 Robotics in Agriculture

The use of robotics in agriculture has the potential to revolutionize the industry by increasing efficiency, reducing labor costs, and improving crop yields. Robots can be used for a wide range of tasks, including planting, harvesting, and monitoring crop health. In this paper, we will explore the various applications of robotics in agriculture, the benefits and limitations of these technologies, and the future of robotics in agriculture [9].

6 Applications of Robotics in Agriculture

6.1 Planting and Seeding [10]

Robots can be used to plant and seed crops with great precision and accuracy. Automated seeding machines can use GPS and other sensors to plant crops with consistent spacing and depth, which can improve yields and reduce the amount of seeds needed. Automated planting machines can also help to reduce labor costs and improve planting efficiency.

6.2 Harvesting [10]

Robots can be used for harvesting a wide range of crops, including fruits, vegetables, and grains. Automated harvesting machines can use sensors and computer vision to identify and pick ripe fruits and vegetables, reducing the need for manual labor. In addition, robots can work continuously, allowing for faster and more efficient harvesting.

6.3 Monitoring Crop Health [11]

Robots can be used to monitor the health of crops and identify issues such as nutrient deficiencies, disease, and pest infestations. Automated sensors can be used to collect data on soil moisture, temperature, and other environmental factors, which can help farmers to make informed decisions about fertilization and irrigation. Robots equipped with cameras and machine learning algorithms can also analyze images of crops to identify signs of disease or pest damage.

6.4 Weed Control [11]

Weeds can be a major problem for farmers, and traditional methods of weed control, such as herbicides and manual weeding, can be time-consuming and expensive. Robotics can be used for automated weed control, which can be more efficient and cost-effective. Robots equipped with machine learning algorithms can identify and target individual weeds, reducing the amount of herbicides needed and improving crop yields [11].

7 Benefits of Robotics in Agriculture

7.1 Increased Efficiency

Robots can work continuously and efficiently, which can improve farming efficiency and reduce labor costs. Automated machines can work faster than human labor, allowing for faster planting and harvesting [12].

7.2 Improved Crop Quality and Yield

Robots can be used to plant, monitor, and harvest crops with precision and accuracy, which can improve crop yields and reduce waste. Automated sensors and cameras can identify signs of disease or pest damage early, allowing farmers to take action to protect their crops [12].

7.3 Reduced Environmental Impact

Robots can be used to reduce the environmental impact of farming by reducing the use of pesticides and herbicides. Automated machines can target individual weeds and pests, reducing the amount of chemicals needed to control them.

7.4 Improved Safety

Robots can be used to perform dangerous or repetitive tasks, reducing the risk of injury for human workers. Automated machines can also work in hazardous environments, such as fields with steep inclines or other difficult terrain [12].

8 Limitations of Robotics in Agriculture [13]

High Initial Costs

The cost of acquiring and maintaining robotic systems can be high, which may make it difficult for small farmers to adopt these technologies [13].

- Limited Adaptability [13]

Robots may have difficulty adapting to changes in the environment, such as sudden changes in weather or variations in terrain [13].

Limited Autonomy

Robots may require human supervision and intervention, especially in complex or unexpected situations.

Technical Complexity

Robotic systems can be complex and require specialized technical knowledge for maintenance and repair [13].

9 Future of Robotics in Agriculture

The use of robotics in agriculture is likely to increase in the coming years as technology advances and costs decrease [13]. Advances in sensors, computer vision, and machine learning are likely to make robotic systems more adaptable and efficient. In addition, increased connectivity and data sharing can allow farmers to make more informed decisions about planting, irrigation, and harvesting. Some of the future applications of robotics in agriculture are: [13].

Autonomous Vehicles

Autonomous vehicles such as tractors, sprayers, and other equipment can help to reduce labor costs and improve efficiency. These vehicles can be equipped with sensors and GPS technology to navigate fields and perform tasks such as planting, fertilizing, and harvesting. Autonomous vehicles can also work around the clock, allowing for 24/7 farming operations [13].

Drones

Drones can be used for a variety of tasks in agriculture, including crop monitoring, mapping, and spraying. Drones equipped with sensors and cameras can collect data on crop health, identify areas of nutrient deficiency, and map fields for precision planting. Drones can also be used to apply fertilizers and pesticides with greater accuracy, reducing waste and improving crop yields [13].

Swarm Robotics

Swarm robotics involves the use of multiple robots working together to perform tasks such as planting, weeding, and harvesting. These robots can communicate with each other and share information, allowing for more efficient and precise operations. Swarm robotics can also reduce the need for human supervision, allowing for greater autonomy and flexibility.

Vertical Farming

Vertical farming involves growing crops in indoor environments using artificial light and controlled environments. Robots can be used to perform tasks such as planting, harvesting, and monitoring crop health in vertical farming systems. Vertical farming can improve crop yields and reduce the environmental impact of agriculture by reducing the need for pesticides and herbicides [14].

Big Data and AI

The use of big data and AI can help farmers to make more informed decisions about planting, irrigation, and harvesting. Automated sensors and cameras can collect data on crop health, weather patterns, and soil moisture levels, which can be analyzed by AI algorithms to provide recommendations for optimizing farming operations. These recommendations can be used to reduce waste, improve crop yields, and reduce costs [15, 16].

10 Challenges and Risks [16]

As with any technology, there are challenges and risks associated with the use of robotics in agriculture. Some of the challenges and risks include: [17].

Job Losses

The adoption of robotics in agriculture may lead to job losses for farm laborers. This can have negative impacts on local economies and communities, especially in areas where agriculture is a major employer [18].

Cyber security Risks

The use of robotics in agriculture involves the collection and storage of sensitive data, such as crop yields and soil moisture levels. These data are vulnerable to cyber attacks, which can have negative impacts on farmers and the agricultural industry [19].

Ethical Concerns [19]

The use of robotics in agriculture raises ethical concerns about the treatment of animals and the impact of technology on the environment [19]. There are also concerns about the concentration of power and control in the hands of large corporations that develop and own the technology [20].

Regulatory and Legal Issues

The use of robotics in agriculture is subject to regulatory and legal issues related to safety, liability, and privacy. There may also be issues related to intellectual property and ownership of the technology [20]. The use of robotics in agriculture has the potential to revolutionize the industry by increasing efficiency, reducing labor costs, and improving crop yields. Robots can be used for a wide range of tasks, including planting, harvesting, and monitoring crop health [20]. The benefits of robotics in agriculture include increased efficiency, improved crop quality and yield, reduced environmental impact, and improved safety [20]. However, there are also limitations and challenges associated with the use of robotics in agriculture, including high initial costs, limited adaptability, and technical complexity [20]. The future of robotics in agriculture is likely to involve the use of autonomous vehicles, drones, swarm robotics, vertical farming, and big data and AI [20]. As with any technology, there are also challenges and risks associated with the use of robotics in agriculture, including job losses, cyber security risks, ethical [20].

11 Conclusion

In conclusion, the integration of AI and IOT in agriculture, also known as high tech agriculture, has emerged as a promising solution to address the challenges facing the industry. The applications of these technologies in agriculture, such as precision farming, smart irrigation, automated pest detection, and livestock monitoring, have shown significant potential to optimize farming practices, reduce costs, and increase yields. Moreover, the implementation of these technologies can also improve the sustainability of agriculture by reducing the environmental impact, enhancing animal welfare, and ensuring food security. However, the implementation of high-tech agriculture requires significant investment and infrastructure development, and there are concerns regarding the potential social and ethical implications of relying on technology to make farming decisions. Therefore, it is crucial to prioritize the development and adoption of these technologies while ensuring that they do not lead to social and economic disparities, particularly in developing countries. Overall, the integration of AI and IoT in agriculture has enormous potential to transform the industry and address the challenges of food security, sustainability, and economic growth. Further research and development of these technologies are necessary to fully realize their potential, and it is essential to address the challenges of infrastructure and investment to enable their adoption on a global scale.

References

1. Anand, N., Patel, R.: A comprehensive review of IoT and AI-based smart agriculture. Int. J. Comput. Sci. Mob. Comput. **9**(5), 329–333 (2020)
2. Gautam, A., Kumar, V., Singh, R.: A review of IoT and AI-based smart agriculture. Int. J. Comput. Appl. **182**(45), 21–26 (2021)
3. Jadhav, V., Jadhav, A., Raut, R.: Role of artificial intelligence in high-tech agriculture: a review. J. Emerg. Technol. Innov. Res. **8**(1), 146–153 (2021)
4. Jain, N., Sharma, D.: A review on IoT and AI-based smart agriculture. J. Electr. Electron. Comput. Eng. **9**(1), 1–8 (2021)
5. Jena, S.K., Das, A.K.: A review on IoT and AI-based smart agriculture. Int. J. Comput. Sci. Eng. **8**(9), 8–14 (2020)

6. Jindal, V., Gupta, N.: A comprehensive review of IoT and AI-based smart agriculture. Int. J. Innov. Res. Comput. Commun. Eng. **8**(5), 723–728 (2020)
7. Kumar, V., Mishra, A., Joshi, A.: A review on advanced technologies and AI-enabled IoT applications in agriculture. J. Ambient. Intell. Humaniz. Comput. **12**(9), 11399–11410 (2021)
8. Mehta, S., Gupta, R., Singh, V.: IoT and AI-enabled smart agriculture: a review. Int. J. Adv. Res. Comput. Sci. **11**(1), 298–304 (2020)
9. Mohanraj, N., Sangeetha, S.: Advanced technologies and AI-enabled IoT applications in agriculture: a review. Int. J. Control Autom. **13**(2), 292–301 (2020)
10. Mondal, A., Das, S.K.: Role of IoT and AI in smart agriculture: a review. Int. J. Innov. Technol. Exploring Eng. **10**(6), 1207–1212 (2021)
11. Nayak, A., Singh, P.: Smart agriculture using IoT and AI: a comprehensive review. Int. J. Res. Eng. Sci. Manage. **3**(10), 491–496 (2020)
12. Rajput, S., Saini, S.: A comprehensive review on smart agriculture using IoT and AI. Int. J. Innov. Res. Sci. Eng. Technol. **10**(6), 3263–3271 (2021)
13. Ramakrishnan, S., Thangavel, K., Subramanian, S.: IoT and AI-enabled smart agriculture: a review. J. Netw. Comput. Appl. **162**, 102734 (2020)
14. Saravanakumar, G., Kumar, V.: A novel approach of IoT and AI enabled agriculture monitoring and control system. Int. J. Intell. Syst. Appl. **13**(6), 62–70 (2021)
15. Sharma, A., Jaiswal, R.: Review on smart farming: IoT and AI applications in agriculture. Int. J. Trend Sci. Res. Dev. **5**(4), 1004–1011 (2021)
16. Sharma, S., Sardana, S.: An integrated approach of IoT and AI in agriculture: a review. J. Ambient. Intell. Humaniz. Comput. **12**(5), 5567–5583 (2021)
17. Singh, S., Rathore, S., Jindal, R.: Smart agriculture using IoT and AI. Int. J. Adv. Sci. Technol. **30**(2), 8775–8781 (2021)
18. Tiwari, A., Singh, S.P.: Application of IoT and AI in smart agriculture: a comprehensive review. J. Intell. Fuzzy Syst. **39**(6), 9559–9569 (2020)
19. Vyas, S., Sharma, S., Singh, A.K.: A comprehensive review on the role of AI and IoT in smart agriculture. Int. J. Emerg. Technol. Eng. Res. **9**(1), 94–100 (2021)
20. Yadav, N., Yadav, A.: A review on IoT and AI based applications in agriculture. Int. J. Recent Technol. Eng. **9**(4), 3754–3759 (2020)

Futuristic Development in Health Care Using Digital Twin Technology and AIoT

Shivani Yadao and P. R. Anisha

Abstract In recent years, spurred by the COVID-19 pandemic, our world has seen a profound shift towards digitalization, with digital twins emerging as a key innovation. These virtual replicas of physical objects or processes, capable of collecting real-time data, hold promise across various sectors, notably in healthcare. With the growing demand for precise diagnosis and personalized treatment, digital twins offer innovative solutions. Despite current challenges in data management and simulation, they are poised to revolutionize healthcare, providing high-resolution patient models for tailored treatment, leveraging Big Data, IoT, and AI technologies.

Keywords Digitalization · Digital twins · Healthcare · Precise diagnosis · Personalized treatment · Innovation · Real-time data · Challenges · Revolutionize · High-resolution patient models · Big Data · IoT · AI technologies

1 Introduction

Artificial Intelligence (AI) and the Internet of Things (IoT) are revolutionary technologies, each excelling in various capabilities such as sensing, predicting, and assisting people. Overcoming unique challenges over time, they have merged to form a powerful force known as the artificial Internet of Things (AIoT), reshaping industries and providing new value to businesses and consumers. AIoT integrates AI into IoT systems, leveraging data-driven insights for enhanced efficiency and productivity, particularly through edge AI, which brings machine learning capabilities closer to data sources. This fusion aims to optimize IoT operations, improve

S. Yadao (✉) · P. R. Anisha
Faculty of Computer Science and Engineering, Stanley College of Engineering and Technology for Women, Hyderabad, India
e-mail: shivaniyadao@stanley.edu.in

P. R. Anisha
e-mail: pranisha@stanley.edu.in

B. K. Mishra and A. Rocha (eds.), *Future of AI, IoT, and Sustainability*, Information Systems Engineering and Management 23,
https://doi.org/10.1007/978-3-031-76286-4_2

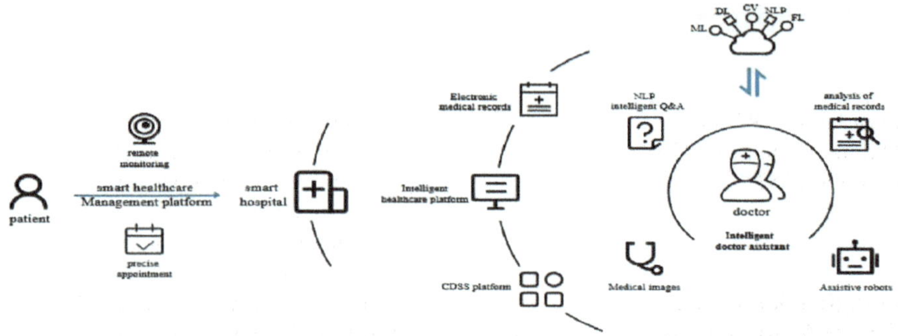

Fig. 1 Practice scenarios of the adoption of AIoT in healthcare

human machine interactions, and refine data management and analytics. By integrating AI into infrastructure elements and leveraging APIs for seamless communication, AIoT enables IoT devices to collect and analyze data, thereby revolutionizing decision-making processes and business operations. The concept of Digital Twins (DT) is also gaining traction, evolving from its initial definition to encompass digital replications of physical entities, including applications in health and wellbeing facilitated by advancements in communication, sensor, and big data technologies. Figure 1 depicts the way of AIoT in the field of healthcare.

The healthcare sector stands to gain significant advantages from AIoT integration. Through smart devices capable of real-time monitoring and data analysis, patient care will become more tailored, enabling medical professionals to make well-informed decisions. In medicine, where uncertainty is inherent and personalized treatment is paramount, conventional drug therapy's limitations highlight the importance of personalized medicine. Precision medicine, focusing on individual variability in genes, environment, and lifestyle, has gained increasing attention over the past decades. Leveraging digital technology, previous studies have constructed virtual physiological human bodies, aiming to bridge computational physiology with clinical practice. The emergence of digital twin (DT) applications has been set the stage by the developments in big data, cloud computing, virtual reality, and Internet of Things (IoT). This ultimately gives clinicians and researchers a finer scale to analyze disease occurrence and progression leading to more accurate diagnosis and treatment [1].

2 Structure and Characteristics of Digital Twin Technology

2.1 Structure of Digital Twin Technology

The DT's structural framework is formed through this interaction (Fig. 2). In the sense of execution of given tasks, gathering information with the help of sensors, this may often take the form of a product, a functional system, or an entire city. In this sense, many tools have been applied within different contexts for knowing about it through IoT technology; examples being two-dimensional codes among others. Nevertheless, it is crucial to note that there are four collection methods used here together with two codes two cards (one more certain than the other) plus sensors among others. We utilize above mentioned methods to perfect and control design and operation immediately through data collection and acting on processed information utilizing communication technology.

Virtual model is a digital representation of physical object that includes its variables and abilities thus showing its geometric properties together with behaviors in virtual space. VR/AR/MR three-dimensional computer generated image (3D CGI) tech supports seeing and joining the virtual and real things or situations. Links provide for connecting and contacting different parts in DT models. One way of meeting the demands for data relay for digital twins is via the usage of 5G communications tech which has high speed and almost no delay. When the virtual world and humans interact, the advancement of human-computer interaction technologies, as well as human-robot interaction and cooperation come into play. Considering this fact, DT data comes in different categories and involves their fusion, therefore enhancing significantly the models.

Big Data can use lots of information produced quickly to unravel and forecast daily tasks and decisions of the DT procedure. At the same time, block chain technology

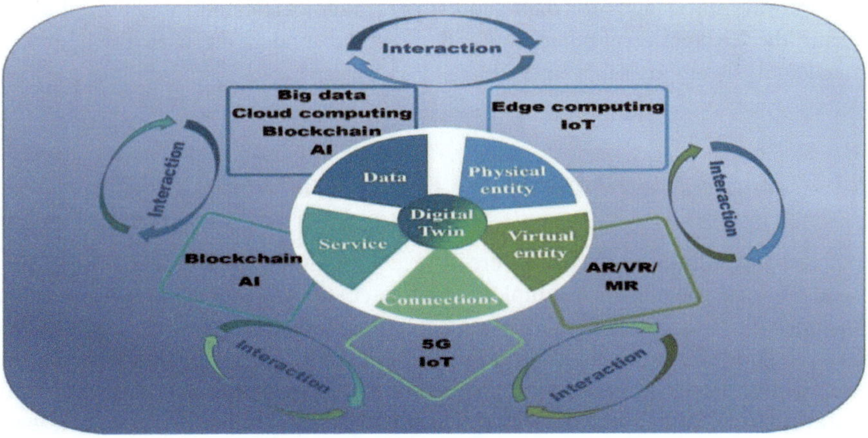

Fig. 2 Composition and technical imprints of the digital twin

can protect two-way data from being corrupted so that it is secure enough to function as a trusted security system when combined with DT.

The DT service model is composed of serviced for the physical entity and the virtual model that improves the accuracy of virtual model through adjusting the coefficients and maintenance of its operability and it is through this that it is able to perform data analysis, data fusion and deep learning of twin data via smart selection of best algorithms in such a way that responsiveness and accuracy of different services will be greatly increased in addition to the improvement in data value.

2.2 Simulation Traits of Digital Twins

The defining trait of simulation and Digital Twin is high fidelity, which plays a crucial role in maintaining exactitude of simulation and real-time data collection, which significantly influences the DT's quality. Data validation serves as a benchmark for the simulation, with continuous calibration in the field to enhance result accuracy. Sensor data and numerical simulation outcomes further contribute to the DT's accuracy. In addition to this, a well-constructed virtual modeled is dependent on high levels of standardization, robustness, modularity and lightweight design.

Standardizing communication encodings alongside communication protocols is beneficial when information is exchanged and integrated; non-modularity enables recombination or separation of each individual model, consequently offering adequacy [2]. Consequently, predicting patient data from a health care system onto a predictive framework will still lead us into combining inductive reasoning with deductive reasoning-oriented approaches through Digital Twins' key technical characteristics that combine inductive with deductive reasoning pathways so as to achieve precision medicine as the ultimate goal. Some of the first aspects of the DTs are already exerting influence over the clinical process; first among them is the division of workflows into stages for data acquisition, diagnostics, and treatment planning. During these stages, computational and statistical models are used for automated image analysis among other operations.

2.3 Technical Attributes of the Healthcare Digital Twin

A possible solution to this is to create a computation-based library of artificial blood pressure waveforms by employing a low-cost model and methods for teaching machines. For example, to obtain precise diagnoses, noninvasive catheters may help measure arterial flow through virtual FFR more accurately compared with physical ones, while surface ECGs may be mapped onto the heart's surface if desired [3]. A neural network is then made and trained. It is then used to predict unknown waveforms of blood pressure using waveforms available. Later on another neural network is trained to analyze this predicted waveform, which describes illness severity by the inverse model.

The foundation for other methods lies in a combination of online and offline techniques. During the offline step, these methods use a combination of training data and high-fidelity simulation models that generate simulation data to build an intelligent computational model (AI) and 3D-models of artificial organs. The online phase has four key subsystems: motion capture system (MCS), Inverse Kinematics (IK) technology, and artificial intelligence-based (AI) model and visualization software. The motion capture system combined with Inverse Kinematics (IK) technology is able to capture the real-time positions and body poses of human beings.

Integrating the Digital Twin (DT) of medical equipment and auxiliary medical devices creates a novel platform and experimental approach for personalized Services for health management and healthcare. Various sources like medical scans—CT, MRI, and ultrasound, wearable instruments helped create the accomplishment of the virtual patient by combining meticulously collected data. Further biochemical parameters and various modules such as messenger RNAs (mRNAs) are responsible for building the DT patient (Fig. 3). Moreover, Due to utilization of technology such as Digital Twin (DT) together with Big Data analytics this leads to creating simulations by means of very detailed models of patients and this helps identify where exactly one should be treated at, what kind of drugs one needs to take according to his/her disease pattern; there are even multiple variants as to methods of treatment itself—all making it feasible to obtain precision medical care, tailored individual attention, as well as smart healthcare services.

Moreover, governments and international regulatory authorities must put strict controls in place and agree on unified, efficient and workable standards. In order

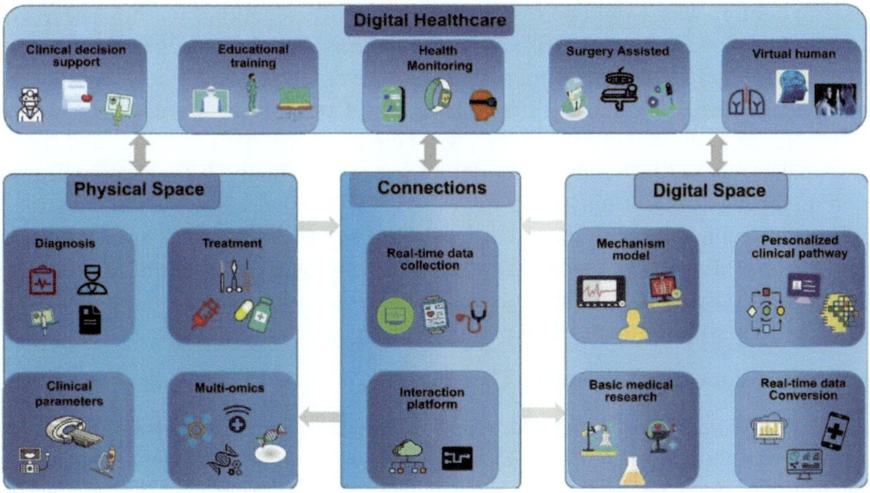

Fig. 3 Digital twin in the field of medicine and digital healthcare for personalized treatment

for the general public, health providers as well as scientific minders to embrace and endorse DT technology, awareness creation campaigns have to be activated.

2.4 Advantages and Disadvantages of the Digital Twin Technology in Health Care

Advantages in Health Care

A. **Individualized Medicine**

It is expected the Digital Twin (DT) will boost self-sufficiency in patients helping them attain equal treatment irrespective of their race or gender and personalized health care, which increases human well-being. It is possible to create a smart virtual partner that can always monitor and prevent the origin or development of illnesses. It is furthermore imagined that each individual will have their own DT someday. The DT offers complete integration through two avenues that complement each other: One rests on personalized mechanistic models from patient data crucial for fine-tuning the decision points in disease management; the other uses electronic records from large populations to create statistical models that track the patient's personal healthcare path across time from birth until death.

B. **Refinement and Risk Forecasting**

Digital Twins are taking medicine to the next level by making it possible for us to quantify and predict about health as well as sicknesses. This makes everything else different about treatments themselves on one hand but also raises concerns among doctors who would be lost without that stuff sometimes on those very same hands. Additionally, they can also be used in running hospitals and clinics effectively even as they help in maintaining patient care treatments. The DT makes it possible to forecast various scenarios in a virtual world prior to actual changes like bed schedules or treatment protocols; this will help cut down both expenses and risks. It also makes it possible to send the model information on methods for treating illnesses and medications that it may need for validation, thus enhancing treatment strategy optimization and allowing for an earlier discovery or prevention of diseases if it has not yet been diagnosed. Once DT is missing, employees working at hospitals will be left with only their experience as well as common sense despite which they don't seem like enough for them when planning things out.

C. **Surveillance and Accurate Therapy**

One solution that is powerful with regard to real time tracking, exact diagnosis and treatment is the Digital Twin (DT) technology. It is a platform which can monitor the health of the patients by use of devices worn on their body anywhere at any time, therefore allowing for early identification of deviations from norm for prescription purposes. Through computer algorithm-based process employing techniques and principles from bioinformatics, more effective treatment strategies custom-tailored to individual patient DT models could be identified; hence increasing survival rates and improving quality of life.

Disadvantages in Health Care

A. **Data Gathering and Integration**

The collection of data seems to be a major obstacle for using Digital Twins in medicine because it includes sensor measurements like sizes, sensor data, and functionality tests, among others. At present, patient data sources are distributed all over the place, and it is difficult to unify them. Analysis of textual data requires human power, as technology has not been developed for this. Not having real-time data on body stress while people are moving along with problems in combining different data types restricts the development of a fully grown DT model of a person's body.

B. **Precision in Simulation**

The accuracy of technology is responsible for the effectiveness of the simulation which greatly limits the medical sphere's possibility to adopt Digital Twins (DT), a significant limitation. In their creation, all models are simplified representations of actual entities and have restrictions. Depending somewhat on the fundamental assumptions are the resulting outcomes. It would be highly useful for developing computational models if more guidelines could be issued so enhanced regulators' ability to evaluate evidence. Moreover, future efforts are required to address obstacles such as standardized processes as well as information technology (IT) systems.

C. **Ethical and Social Risks**

Socioethical risks also accompany the implementation of Digital Twins (DT) in healthcare. Privacy concerns, notably the most significant, represent a primary drawback of DT adoption [4, 5]. In addition, the high price of DT healthcare services may lead to social imbalances and unfairness hence deepening the societal gap between the rich and the poor. It is believed that nascent technology is more flexible in part because basic science and clinical practices have not fully developed into the future [5]. For this reason, it would be necessary to enlighten people on how they can benefit from such systems and how they can be adopted. The truth is that good health makes individuals fear data privacy issues more than people facing cancer because they are dying.

3 Related Work

The literature indicates a growing interest in integrating IoT technology into healthcare, with studies such as [6] highlighting how devices such as wearable sensors, remote monitoring tools, and connected medical devices are transforming patient care by collecting vital data for continuous monitoring, thus reducing hospital readmissions and assisting in managing chronic diseases. However, research by [6] raises concerns about data security and privacy in IoT-based healthcare systems, emphasizing the importance of robust security measures.

The application of AI in healthcare is extensively discussed in the literature, showcasing its role in diagnostic procedures, treatment planning, and predicting patient outcomes. Pioneering studies, like those by [7], demonstrate AI's ability to analyze medical images and patient data to identify diseases more accurately than traditional methods. Additionally, AI's predictive analysis, as discussed in [8], plays a crucial role in preventive healthcare by identifying potential health risks before they escalate. Nonetheless, the literature also highlights challenges in AI implementation, such as the requirement for extensive datasets and concerns regarding the interpretability and decision-making process of AI systems.

Artificial Intelligence (AI) encompasses a range of subfields, including machine learning and deep learning. Machine learning utilizes algorithms that adapt autonomously to produce desired outputs based on provided inputs, operating independently without human intervention, as depicted in Fig. 4. Conversely, deep learning enables computers to self-learn through neural networks and unlabeled data. Deep learning employs layered algorithms to construct neural networks capable of learning, providing comprehensive insights into data. These multi-layered approaches operate via artificial neural networks, closely resembling the neural networks of the human brain. The complexity of learning increases with the addition of more neurons and hidden layers, mirroring the intricate connections in the brain. Coordinated learning models within deep learning can exclude specific features from labeled training datasets for multi-layered comprehension. Convolutional Neural Networks (CNNs), Recurrent Neural Networks (RNNs), and particularly large-scale CNNs play crucial roles in estimating coordinated learning processes (Table 1).

This article explores the applications of deep learning in healthcare, investigating its potential for individual disease predictions—forecasting an individual's illness based on their medical history—and community disease predictions, which involve estimating disease prevalence or epidemics within populations. A critical analysis of the literature reveals several gaps in current technology. While IoT devices excel in

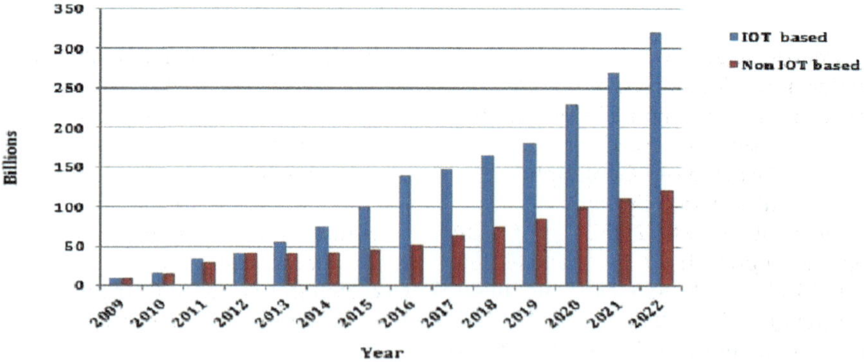

Fig. 4 Survey on AI and digital twin in health care

Table 1 Approaches for using AI and IoT and digital twin in healthcare

Year	Team	Applications of DT in the field of medicine	The attributes of DTs in medical field
2011	Niederer et al. [9]	Exploring how the effectiveness of CRT varies based on cellular-level mechanisms and organ function using mechanical models	It was employed to forecast a reduced response to CRT therapy in patients and to pinpoint innovative criteria for patient selection
2017	Cone Health team [10]	Heart Flow Analysis uses data from CT scans, machine learning, cloud-based performance calculations, and computational physiology	The procedure that was not invasive has helped doctors to find out how much heart blood flow is affected by any blockages
2018	Prakosa et al. [11]	Their use of DT technology was in regards to ventricular tachycardia connected with myocardial infarction	In a clinical procedure, the system improved guidance for electrode positioning by signaling moving optimal areas in patients
2018	The Philips [12]	The Philips HeartNavigator tool is a single image which combines CT scans to show a patient's heart	It allowed surgeons to see the location of the instruments in a 3D view that changed as they moved, which made it easier to plan surgeries ahead of time
2019	Chakshu et al. [13]	To create diagnostic tools, they constructed a DT model that incorporated head vibration and blood flow	Evaluating how serious carotid artery blockage is through comparing real-time detected in vivo vibrations with virtually obtained datasets from a human face video clip
2019	Mazumder et al. [14]	Using the pressure reflexes, a moving body was made using pressure regulators and dynamic equations	The possibility for large data sets to be generated regarding alterations that occur with respect to both blood pressure and blood flow patterns
2020	Suzuki et al. [15]	A multivariate logistic regression analysis was used to create a model that predicts the risk of rupture related to CT	Among these risk factors identified, the model included age, supercritical length and pressure loss coefficient
2020	Mussomeli et al. [2]	Takeda Pharmaceuticals started using DT technology in its manufacturing process in order to bring new treatment to people worldwide	DT models have excellent input and output biological response prediction capabilities and can eliminate or broaden drug production methods

(continued)

Table 1 (continued)

Year	Team	Applications of DT in the field of medicine	The attributes of DTs in medical field
2020	Erol et al. [16]	Atos and Siemens partnered with the pharmaceutical industry to develop DT models that have been advanced by IoT, AI, and many other cutting-edge technologies	Efficient physical DT models were utilized in enhancing the production process to cater for production and efficiency challenges
2020	Grosman et al. [3]	Creating a Decision Tree in the circumstance when making an assessment on blood sugar levels and computer usage policy for personalized customization	It was the first successful demonstration of DT's automation and personalization in insulin pump therapy
2020	Croatti et al. [17]	This is how they integrated agent technology, multi-agent systems as well as data technology into the treatment process of traumatized patients	This is the first application of agent-based decision tree in situations of severe trauma and as such, it allows for management according to a person
2020	Subramanian et al. [18]	A decision tree was designed by them which fused scientific data together with clinical source information	Drug discovery, biomarker identification, test development, screening, clinical trial optimization have been assisted by AI
2021	Golse et al. [3]	They formed an entire blood circulation system's model, with individual patient information automatically calibrated	The estimation of hepatic flow rate through the DT model helped in getting good results in forecasting post-operative portal hypertension
2021	He et al. [19]	Building a lumbar spine virtual clone with AI, motion tracking systems, reverse kinematics methods, among others	They created an integrated DT model that combines shape and performance to forecast the biomechanical characteristics of an actual lumbar spine
2021	Zohdi [18, 19]	Zohdi created a fusion of DT and machine learning frameworks that seamlessly integrates a genetic algorithm, associated with simplified equations which describe fluid-particles interactions	It was used to evaluate the positions and airflow volumes of various air exchange devices with the aim of improving air circulation systems

(continued)

Table 1 (continued)

Year	Team	Applications of DT in the field of medicine	The attributes of DTs in medical field
2021	Pilati et al. [20]	They designed a system for the tracking of diseases during vaccinations and had it tested in a clinical study	It established a mobile immunization center and allowed for realtime simulation, both of which increase the efficacy of vaccinations
2022	Aubert al. [13]	They used 3D X-ray scans of people to create a DT that would simulate how bones heal	To evaluate the hazard of recurrent fractures, you have to put in the greatest load while walking
2022	Hernigou et al. [21]	A dynamic theraband model (DT) was built with computerized tomography (CT) scans of the foot combined with artificial intelligence (AI) algorithms which significantly improved identifying foot orientation and evaluating compensation possibilities for subtalar joint rotations	Image-guided systems helped them reduce errors that arise from manually choosing anatomical markers
2022	Laybenbacher et al. [22]	They came up with a plan for making an AI for the immune system and this plan has four parts	This possibility has the power to alter the landscape of biomedical research and speed up the process of translating fundamental research into clinical practice
2022	Cydar [16]	The technology they employed was computer vision, machine learning kinetics wontonta loop with AI_GPU in the cloud" or "they used cutting edge cloud based computers also known as artificial intelligence GPU for would want	The powerful tool of artificial intelligence has improved the way doctors see inside the human bodies using surgery which has opened new pathways for quick recovery times after any operation
2022	FEops [10]	Data-driven insights were generated by combining AI-powered anatomical analyses with cardiac images to DT format	It bestowed doctors one-of-akind digital tools for employing the correct technology on correct patients at the correct moment
2022	Corify Care [13]	The mapping system brought together hardware and creative imaging and signal processing technologies	The application of this technology was useful in studying cardiac activity using a beat-to-beat approach that considers multi-chambers and provides 3D mapping

data collection, their integration with healthcare systems often faces interoperability challenges, as noted in [10]. Similarly, although AI has shown promise in data analysis, concerns regarding the ethical use of AI, biases in AI algorithms, and the need for transparency in AI decision-making are recurring themes in recent research. Furthermore, studies such as [9] highlight the lack of standardized regulations and guidelines for the combined use of AI and IoT in healthcare, which are crucial for ensuring patient safety and data privacy.

4 Implementation Methodologies

To establish how HDTs have advanced beyond traditional DTs and show how they differ, this section will give a brief introduction on what makes up the general architecture and technology base of standard DTs. In such a scenario, a standard DT will usually have around three levels.

The physical entity current state is perceived by a data collection layer with integrated sensors at first [22, 23]. The sensor technology type applied may depend on the application.

A second step involves an implementation of data storage and processing layer for aggregation and storage of the raw data that is collected from the weather sensors as well as for generating feedback based on prediction as well as simulation results [22, 23]. For efficient handling of large datasets databases such as MongoDB or MySQL are often used in contemporary data-intensive workflows. Hence models are employed to structure data stored within these databases so as to provide correct knowledge about the real entities and their characteristics.

Furthermore, this data is used by DTs to speculate on the next scenarios ocurring to the entity and disseminate relevant information. In other words, the utilization of AI facilitates precise and strong decisions [24]. Predictive deep learning models can, for instance, approximate certain states like products' remaining service life or human-related health indicators by means of specific data [25, 26]. Instead AI meets intelligent and robust data aggregation and feedback generation. Instead, Reinforcement Learning maps to the intelligent feedback of DT data directly through learning such optimal control schemes that employ reward systems on outcomes.

Another difference between digital twins and digitized models or digital shadows is that DTs have the potential of physically affecting real objects through their output and interaction layer [22, 23]. Furthermore, depending on application, this layer can provide its own feedback as well as decision-making services for relevant users. To enable direct interaction between physical entities, they create what are called data threads that span from their data processing units all way down to these entities. Optical, coaxial and twisted pair cables are some of the connection technologies that are commonly used whereas wireless communication systems use Zig-Bee, Bluetooth, Wi-Fi, ultra-wideband or NFC [27]. For interfacing with human users one can use web based applications e.g. [24] and virtual or augmented reality systems e.g. [4]. Web-based applications are compatible with a wider range of devices in contrast to VR and AR, which promote deep involvement and clearer data representation like displaying essential details on the targeted object (Boctor-Digoni, Roth, Chisalita, et al. 2019). Therefore, providing smart feedback becomes one of most important capabilities for digital twins (Buehler 2020), they make systems adapts dynamically or support human user interaction, if we think about desirability of deeper immersion. A typical Digital Twin framework is illustrated in Fig. 5.

In the realization of distributed testing applications, security plays a very important role up to the edges. DTs are volumes of sensitive data that are centralized substantial and highly specialized thus they are vulnerable to cyber-attacks [28]. There is however several strong data transmission and storage protection measures including data encryption through DES/AES algorithms, authentication protocols and modern block chain technologies [29]. Despite potentially affecting the overall performance of DTs their incorporation ensures security at every level.

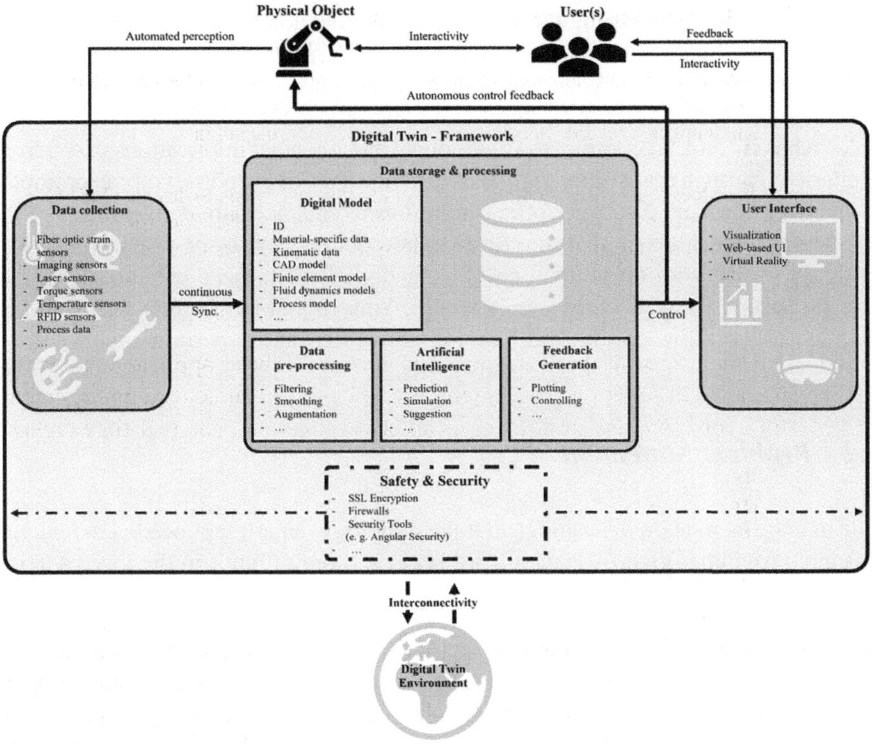

Fig. 5 Typical digital twin substructure in conventional application domain

5 Case Study

5.1 Case

Enhancing Patient-Centric Efficiency-Utilizing Digital Twin Technology for Process Optimization and Layout Design in Healthcare Settings.

The landscape of healthcare is undergoing significant transformation. Healthcare providers are facing challenges such as rapid advancements in science, escalating cost pressures, demographic shifts, and a transition towards value-based reimbursement models. Consequently, they seek enhanced tools to identify and leverage opportunities for improvement.

Workflow Simulation emerges as a potent solution for informed decision-making. It begins with the creation of a 3D computer model representing your clinical setting, encompassing areas like radiology, operating rooms, and emergency rooms. This model is then turned into a digital twin of your organization by incorporating financial and operational data to mimic workflows that occur in real life.

This digital replica of your operational environment enables the exploration of various scenarios and layout configurations. Workflow Simulation serves as a valuable "what-if" tool, forecasting the operational and financial implications of different scenarios. Through realistic animations and comprehensive reports, you can promptly evaluate options and discover optimal solutions to enhance care delivery.

Previously, achieving such improvements required months or years of trial and error. However, with the utilization of your digital twin, these enhancements can now be accomplished within days or weeks. Your digital twin offers a streamlined pathway to workflow optimization, cost-effectiveness, and superior patient care.

5.2 Problem Statement

The management team determined that the radiology department needed to undergo change. Providing effective patient care became more difficult in the face of aging infrastructure, growing clinical complexity, increased patient demand, and limited space.

The swift evolution of medical technology underscored the urgency for equipment modernization. Yet, perhaps most crucially, the mounting waiting times, interruptions, and delays were detrimentally affecting the patient experience.

5.3 Application Modules

A shift in paradigm away from Quality by Testing towards Quality by Design (QbD) is essential for crafting resilient and effective processes. QbD is a systematic, science-driven approach centered on understanding processes, facilitating continuous improvement through data-driven insights. Process Analytical Technology (PAT) plays a crucial role in implementing QbD, because it enables gathering data during service as well as its analysis on-the-fly to decide on what to do next. If PAT is applied, one can achieve bargain process knowledge, command, and decreased quality nonuniformity.

Apart from PAT, Modeling plays a crucial role in QbD implementation. It acts as a priceless instrument for predicting and fine tuning process parameters that will in turn enable development of robust processes capable of producing desired product qualities. It is possible to identify potential risks through simulating different process conditions and thus control them before implementation starts. Modeling is key to standardizing process parameters, reducing experimental workload, and removing the need for expensive and time-consuming hitand-miss tactics. This is because the full dynamic model needs an elaborate mathematical form based on which it can understand the dynamics of this process and they in turn lead to time-consuming solution methods, like sorption–sublimation ones. In order to improve efficiency of these models, simplified versions have been made based on pseudo stationary conditions in the first phase of drying at Primary Structural System while giving faster solutions with reduced complexity aspects. These models have enabled performance enhancement through process optimization. Figure 6 illustrates digital twin hierarchy with varied digital model.

Analysis and Evaluation of Data: To learn more about how radiology operations are now doing, data from the RIS (Radiology Information System) was analyzed. Siemens' design solution team also worked with the hospital's architects to evaluate current configurations and identify areas for improvement. This comprehensive assessment, spanning one week, involved on-site workshops, stakeholder interviews, and observation of processes is shown in Fig. 7.

Baseline digital twin model: Drawing from the findings of data analysis and evaluation, a digital replica of the department (baseline model) was generated.

Defining Scenarios: Two scenario-driven models were created to investigate variations in patient demand and complexity. The effects of these diverse scenarios were examined, analyzed, and deliberated upon by all stakeholders.

Introducing a Fresh Layout: Ultimately, a new layout and infrastructure were devised to ensure optimal workflows, heightened efficiency, and an enhanced experience for both staff and patients.

5.4 Outcome

The insights gained via Workflow Simulation enabled MPH to make well-informed decisions about workflows, procedures, and facility design. As an ideal catalyst for change and transformation, management might evaluate the operational consequences of several solutions before implementing them by finding the best course of action.

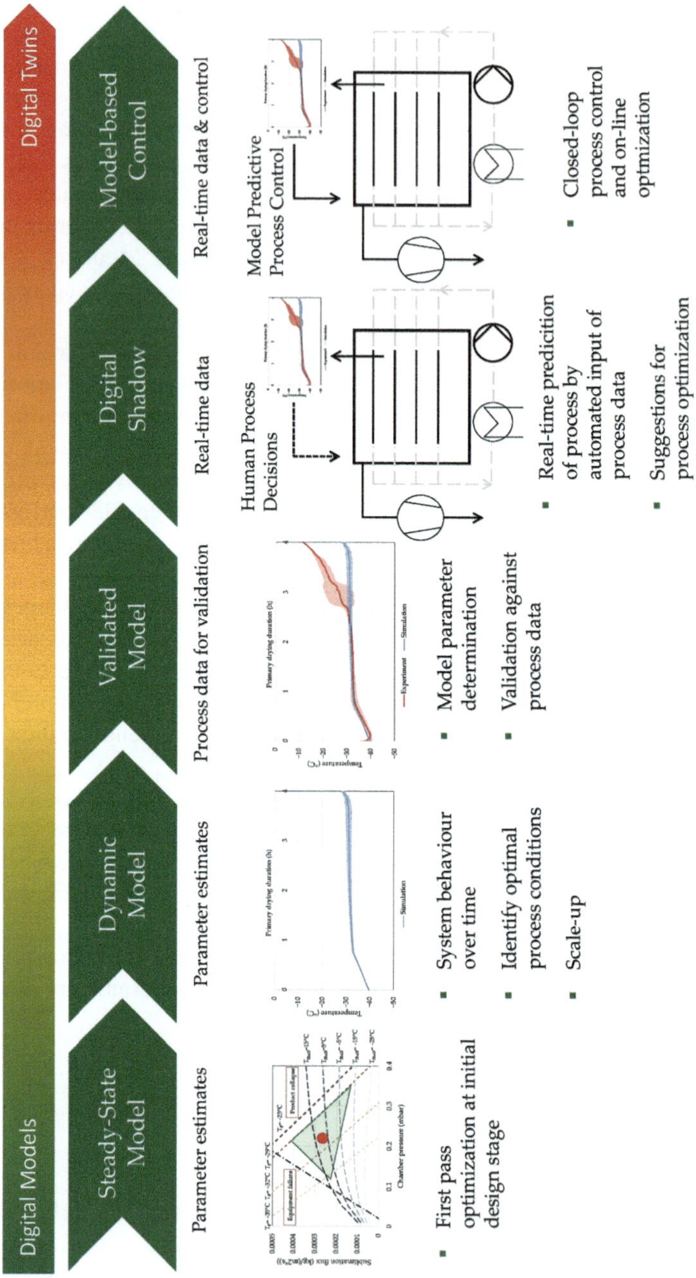

Fig. 6 Digital twin hierarchy with different digital model

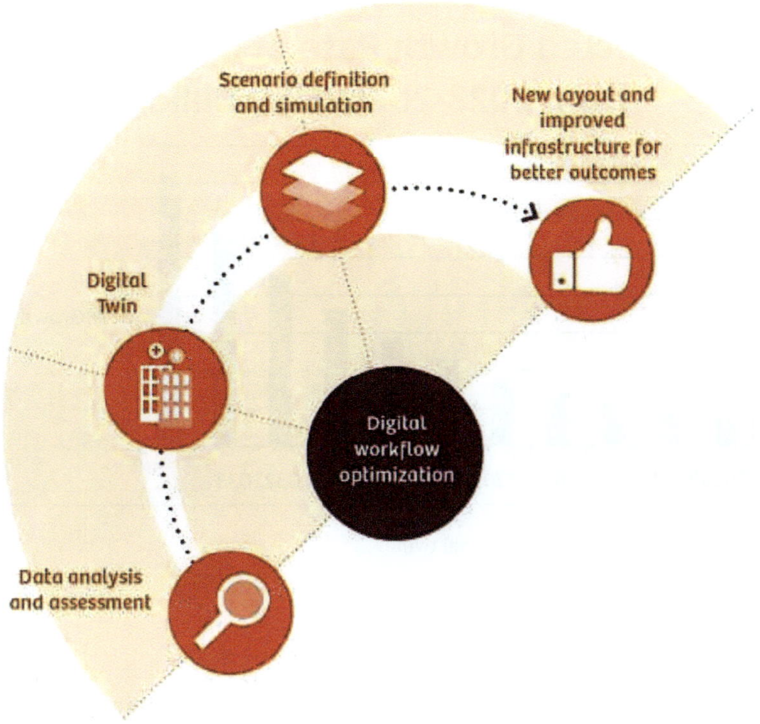

Fig. 7 Application modules for digital workflow optimization

Enhancing the patient experience was MPH's top priority. Patient changing facilities were redesigned to create a more comfortable atmosphere, which boosted throughput. Additionally, maneuvering areas were optimized to enhance healthcare workers' ability to attend to bedridden and wheelchair-bound patients.

These achievements unlock a plethora of benefits, including:

Enhanced patient experience
Improved process efficiency
Optimized resource allocation
Heightened clinical safety
Increased operational flexibility
Elevated staff satisfaction

Figure 8 depicts the future of healthcare is digital twin, with Workflow Simulation already proving its significance. Through the digital twin concept, providers can more easily achieve their objectives of value-based care, cost reduction, staff satisfaction, and an enhanced patient experience.

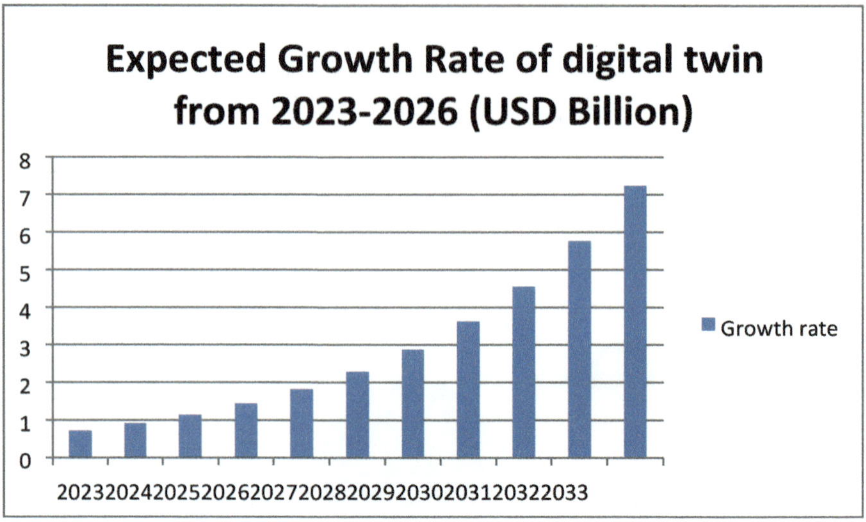

Fig. 8 Future of healthcare using digital twin

6 Conclusion and Future Work

A virtual version of a physical thing that can be tracked, examined, and enhanced is called a Digital Twin (DT). When something tangible is digitally represented, it becomes "live" and "dynamic," continuously updating as the actual object undergoes changes. It can also gather data from other machines and its environment. The expanding Internet of Things (IoT) and digitalization processes are enabling more individuals to access valuable information digitally. Without graphical restrictions, DT procedures in healthcare can improve access to medical education, information on self-care management, and other remote healthcare services.

But as digital technology is still in its infancy, a number of major obstacles and limits must be overcome before it can reach its full potential. These include financial constraints, the complexity and upkeep of data, a lack of standards and rules, and problems with communication and privacy. In light of these difficulties, future investigations ought to concentrate on: (1) simplifying computational complexity through simulation and modeling techniques, (2) leveraging 5G communications, (3) processing and analyzing IoT data through big data, AI, and machine learning, (4) ensuring interoperability, and (5) utilizing cloud computing.

In addition to its wide application in industry as a virtual replica of physical entities, DTs have found utility in the medical field. Being allowed by them, clinicians can anticipate different scenarios in a virtual environment before they put changes into effect thereby reducing risk and costs while originally applied to musculoskeletal system by our team, the dynamic interaction which characterizes dt technology between physical and digital spaces indicates some hope. Even though real-time biomechanical analysis offers advantages, data collection, fusion, and accurate simulation present challenges. As a result, previously proposed ligament constraints and soft-tissue material properties should not be included in human body biomechanical models. Thanks to developments in Big Data, IoT, and AI technologies, holistic digital twins of the human body can be developed for crisis alerts and real-time monitoring, particularly for senior citizens. Generally using (DT) technology may help medical professionals diagnose patients accurately, create pertinent treatment plans, and predict treatment outcomes, leading to individualized care in the future.

In organ transplantation, DT technology can also be used to estimate probability or resolve concerns about tissue adaption after transplantation. During the COVID-19 pandemic, the demand for hospital resources intensified, highlighting issues such as capacity and resource management. DTs of hospitals or departments can facilitate more accurate resource management and planning, optimizing processes and resource utilization through machine learning and AI techniques. The primary goals of digital twin research in the medical field have been to create organ replicas, comprehend cellular activity, and implement appropriate medications and therapies. It is anticipated that DT technology will advance with the development of sensor, IoT, and machine learning technologies, leading to the eventual production of a full-body DT system.

Additionally, DT solutions are anticipated to play a crucial role in solving management problems in hospital systems and medical resources, particularly in response to the pandemic of COVID-19. Additionally, it is anticipated that DT solutions will speed up the COVID-19 vaccine and medication development process.

This study concludes by reviewing significant DT research from a range of healthcare specialties and making predictions about how these technologies will advance in the future. Comprehensive research on DT technology from the viewpoints of aerospace, military, education, and security systems is possible in addition to healthcare. Furthermore, research can be focused on certain application areas that the healthcare industry has not previously investigated.

References

1. Silver Spring, M.: Paving the Way for Personalized Medicine: FDA's Role in a New Era of Medical Product Development (2013)
2. Mazumder, O., Roy, D., Bhattacharya, S., et al.: Synthetic PPG generation from haemodynamic model with baroreflex autoregulation: a digital twin of cardiovascular system. Annu. Int. Conf. IEEE Eng. Med. Biol. Soc. 5024–5029 (2019)
3. Pilati, F., Tronconi, R., Nollo, G., et al.: Digital twin of COVID-19 mass vaccination centers. Sustainability **13**, 7396 (2021)
4. Shamanna, P., Saboo, B., Damodharan, S., Mohammed, J., Mohamed, M., Poon, T., Kleinman, N., Thajudeen, M.: Reducing hba1c in type 2 diabetes using digital twin technology-enabled precision nutrition: a retrospective analysis. Diab. Therapy **11**, 2703–2714 (2020)
5. Hu, W., Zhang, T., Deng, X., Liu, Z., Tan, J.: Digital twin: a state-of-the-art review of its enabling technologies, applications and challenges. J. Intell. Manuf. Special Equipment (2021)

6. Boukhennoufa, A.A., Bensaali, F., Soheilian Esfahani, S.: A novel gateway-based solution for remote elderly monitoring. J. Biomed. Inform. **1**(09) (2020)
7. Duran-Vega, L.A., et al.: An IoT system for remote health monitoring in elderly adults through a wearable device and mobile application. Geriatrics **4**(2), 34 (2019)
8. Lopes, S.I., Pinho, P., Marques, P., Abreu, C., Carvalho, N.B., Ferreira, J.: Contactless smart screening in nursing homes: an IoT-enabled solution for the COVID-19 era. In: 2021 17th International Conference on Wireless and Mobile Computing Networking and Communications (WiMob), pp. 145–150 (2021)
9. Chow, E.J., Schwartz, N.G., Tobolowsky, F.A., Zacks, R.L., Huntington-Frazier, M., Reddy, S.C., et al.: Symptom screening at illness onset of health care personnel with SARS-Cov-2 infection in king county Washington. JAMA **323**(20), 20872089 (2020)
10. Tkachenko, R., Izonin, I., Chopyak, V., Kryvinska, N., Lotoshynska, N.: Piecewise-linear approach for medical insurance costs prediction using SGTM neural-like structure. In: Proc. IDDM, pp. 170–179 (2018)
11. FEops Institute: Altering the course of structural heart disease
12. Cone Health: HeartFlow Analysis patient case
13. Daubert, C., Behar, N., Martins, R.P., et al.: Avoiding non-responders to cardiac resynchronization therapy: a practical guide. Eur. Heart J. **38**, 1463–1472 (2017)
14. Corify Care Institute: The solution: Acorys®. [Internet] (2022)
15. Ahmed, H., Devoto, L.J.S.I.: The potential of a digital twin in surgery. Surg. Innov. (2020). https://doi.org/10.1177/1553350620975896
16. Raffort, J., Adam, C., Carrier, M., et al.: Fundamentals in artificial intelligence for vascular surgeons. Ann. Vasc. Surg. 65–260 (2020)
17. Cydar Medical: Surgical augmented intelligence
18. Golse, N., Joly, F., Combari, P., et al.: Predicting the risk of post-hepatectomy portal hypertension using a digital twin: a clinical proof of concept. J. Hepatol. **74**, 661–669 (2021)
19. Croatti, A., Gabellini, M., Montagna, S., et al.: On the integration of agents and digital twins in healthcare. J. Med. Syst. **44**, 161 (2020)
20. Devoto, L., Muscroft, S., Chand, M.: Highly accurate, patient-specific, 3-dimensional mixed-reality model creation for surgical training and decision-making. JAMA Surg. **154**, 968–969 (2019)
21. Mussomeli, A., Parrot, A., Umbenhauer, B., et al.: Digital Twins Bridging the Physical and Digital. Deloitte Insights (2020)
22. Erol, T., Mendi, A.F., Doğan, D. (eds): Digital transformation revolution with digital twin technology. In: 2020 4th International Symposium on Multidisciplinary Studies and Innovative Technologies (ISMSIT). IEEE (2020)
23. Aubert, K., Germaneau, A., Rochette, M., et al.: Development of digital twins to optimize trauma surgery and postoperative management. A case study focusing on tibial plateau fracture. Front. Bioeng. Biotechnol. **9**, 722275 (2021)
24. Fuller, A., Fan, Z., Day, C., Barlow, C.: Digital twin: enabling technologies, challenges and open research. IEEE Access **8**, 108952–108971 (2020)
25. Mohapatra, S., Bose, S.: An appraisal of literature for design and implementation of developing a framework for digital twin and validation through case studies. Heal. Technol. **10**, 1229–1237 (2020)
26. Han, Z., Shi, H., Liu, Y., Ba, L., Yuan, Q., Yang, M., Gao, X: Framework for remaining useful life prediction of casting roller in continuous casting machine based on digital twin. In: 2021 IEEE International Conference on Software Engineering and Artificial Intelligence (SEAI), pp. 17–22 (2021)
27. Mazhar Rathore, M., Shah, S.A., Shukla, D., Bentafat, E., Bakiras, S.: The role of AI, machine learning, and big data in digital twinning: a systematic literature review, challenges, and opportunities. IEEE Access **9**, 32030–32052 (2021)
28. Zhu, Z., Liu, C., Xun, X.: Visualisation of the digital twin data in manufacturing by using augmented reality. Procedia CIRP **81**, 898–903 (2019)

29. Bravo, A., Maier, A.M.: Immersive visualisations in design: Using augmented reality (ar) for information presentation. Proc. Design Soc. Design Conf. **1**, 1215–1224 (2020)
30. Jhawat, V., Gulia, M., Gupta, S., et al.: Integration of pharmacogenomics and theranostics with nanotechnology as quality by design (QbD) approach for formulation development of novel dosage forms for effective drug therapy. J. Control. Release **327**, 500–511 (2020)
31. Soni, A., Gowthamarajan, K., Radhakrishnan, A.: Personalized medicine and customized drug delivery systems: the new trend of drug delivery and disease management. Int. J. Pharm. Compd. **22**, 108–121 (2018)
32. Psaty, B.M., Dekkers, O.M., Cooper, R.S.: Comparison of 2 treatment models: precision medicine and preventive medicine. JAMA **320**, 751–752 (2018)
33. Viceconti, M., Hunter, P.: The virtual physiological human: ten years after. Annu. Rev. Biomed. Eng. **18**, 103–123 (2016)
34. Alexiadis, A., Simmons, M.J.H., Stamatopoulos, K., et al.: The virtual physiological human gets nerves! How to account for the action of the nervous system in multiphysics simulations of human organs. J. R. Soc. Interface. **18**, 20201024 (2021)
35. Li, L., Lei, B., Mao, C.: Digital twin in smart manufacturing. J. Ind. Inform. Integr. **26**, 100289 (2022)
36. Glaessgen, E., Stargel, D. (eds): The digital twin paradigm for future NASA and US Air Force vehicles. In: 53rd AIAA/ASME/ASCE/AHS/ASC Structures, Structural Dynamics and Materials Conference 20th AIAA/ASME/AHS Adaptive Structures Conference 14th AIAA (2012)
37. Khan, J.I., Khan, J., Ali, F., Ullah, F., Bacha, J., Lee, S.: Artificial intelligence and internet of things (AI-IoT) technologies in response to COVID-19 pandemic: a systematic review. IEEE Access **10**, 62613–62660 (2022). https://doi.org/10.1109/ACCESS.2022.3181605
38. Subramanian, K.: Digital twin for drug discovery and development—the virtual liver. J. Indian Inst. Sci. **100**, 653–662 (2020)
39. Chakshu, N.K., Carson, J., Sazonov, I., et al.: A semi-active human digital twin model for detecting severity of carotid stenoses from head vibration—a coupled computational mechanics and computer vision method. Int. J. Numer. Methods Biomed. Eng. **35**, e3180 (2019)
40. Chen, Y., Yang, O., Sampat, C., et al.: Digital twins in pharmaceutical and biopharmaceutical manufacturing: a literature review. Processes **8**, 1088 (2020)
41. Barricelli, B.R., Casiraghi, E., Fogli, D.: A survey on digital twin: Definitions, characteristics, applications, and design implications. IEEE Access **7**, 167653–167671 (2019)
42. Mashaly, M.: Connecting the twins: a review on digital twin technology & its networking requirements. Procedia Comput. Sci. **184**, 299–305 (2021)

Security, Privacy, and Interoperability in IoT Ecosystems

Rakesh Nayak, Umashankar Ghugar, and Satyabrata Das

Abstract The Internet of Things (IoT) has rapidly transformed the means we engage with innovation producing a large network of connected gadgets that add to the smooth performance of numerous elements of modern-day life. However, as the IoT ecosystem continues to broaden, problems relating security, privacy, and interoperability have become significantly prominent. This chapter explores the essential concerns of security, privacy, and interoperability within the IoT ecosystem, examining the challenges, potential risks, and innovative solutions that aim to address these concerns. Through an in-depth analysis of current trends and emerging technologies, this article supplies useful understandings right into the intricate and evolving nature of IoT communities, clarifying the approaches on the strategies and best practices that are crucial for guaranteeing the integrity, privacy, together with seamless integration of IoT devices and systems.

Keywords Security · Privacy · Interoperability · IoT ecosystems

R. Nayak (✉) · U. Ghugar
Department of CSE, OP Jindal University, Raigarh, India
e-mail: rakesh.nayak@opju.ac.in

U. Ghugar
e-mail: umashankar.e18640@cumail.in

S. Das
Department of CSE, University Institute of Engineering (UIE), Chandigarh University, Mohali 140413, Punjab, India
e-mail: sdash@gitam.edu

U. Ghugar
Department of CSE, GITAM (Deemed to Be University), Visakhapatnam, India

B. K. Mishra and A. Rocha (eds.), *Future of AI, IoT, and Sustainability*, Information Systems Engineering and Management 23,
https://doi.org/10.1007/978-3-031-76286-4_3

1 Introduction: Understanding IoT Ecosystems

In today's rapidly progressing technical landscape, the Internet of Things (IoT) has emerged as a transformative force, forming the method we communicate with the world around us. Understanding IoT ecosystems is necessary in understanding the elaborate internet of linked gadgets, data as well as processes that are changing our day-to-day day lives.

The Evolution of IoT: The concept of IoT has evolved from the very early days of internet-connected gadgets to an innovative network of smart devices, sensors, and actuators, all working in tandem to collect, exchange, and act upon data. This advancement has actually caused an abundant community that goes beyond standard borders together with has the prospective to redefine sectors as well as customer experiences.

A CAGR of 6% is predicted to see the overall number of internet users increase from 3.9 billion in 2018 to 5.3 billion by 2023, according to the Cisco Yearly Net Record (2018–2023) white paper [1]. In terms of population, this represents 51% of the global population in 2018 and 66% of worldwide populace infiltration by 2023 (Fig. 1).

By 2023 the customer share of the overall gadgets, consisting of both taken care of as well as mobile gadgets, will certainly be 74%, with service declaring the continuing to be 26%. Customer share will certainly expand at a somewhat slower price at a 9.1% CAGR loved one to business sector which will certainly expand at a 12.0% CAGR (Fig. 2).

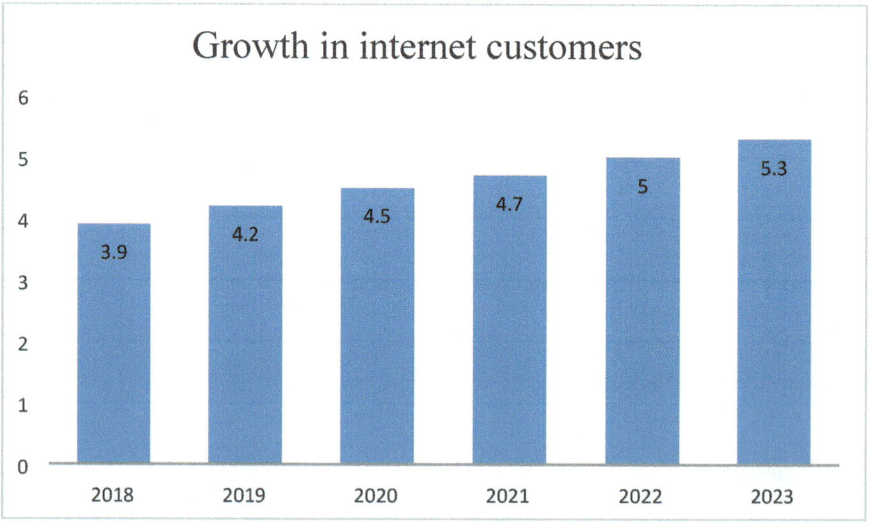

Fig. 1 Growth in internet customers (*Source* CISCO yearly net record)

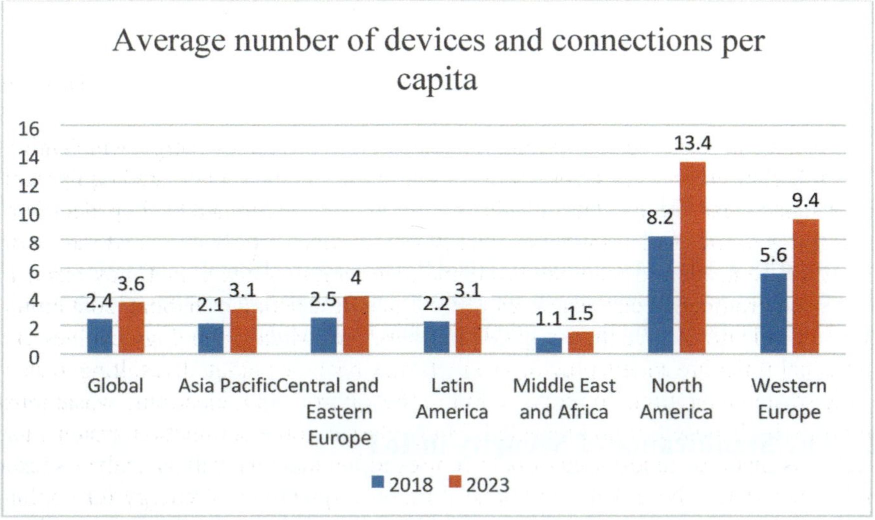

Fig. 2 Average number of devices and connections per capita (*Source* CISCO yearly net record)

At the core of IoT environments are the linked tools that interact with each various other to collect and also share information. These devices can range from everyday objects such as home appliances and wearables to intricate commercial and infrastructure. Additionally, IoT ecosystems incorporate cloud platforms, cloud computing, data analytics, and connectivity protocols that enable seamless communication and data processing.

The spreading of IoT environments has considerable implications across various industries. In manufacturing, IoT-enabled sensors and automation innovations are maximizing manufacturing processes and predictive maintenance. In healthcare, wearable devices and remote monitoring systems are improving health care and treatment outcomes. Moreover, in smart cities, IoT is helping with reliable source monitoring, traffic control, as well as public safety and security procedures.

While the potential of IoT ecosystems is vast, they also present challenges related to security, privacy, with interoperability. Ensuring the protection of sensitive data, addressing privacy concerns, and establishing standards for seamless device interaction are critical considerations in the development and deployment of IoT solutions. Nevertheless, these challenges additionally symbolize chances for advancement along with the production of durable, safe as well as easy to use IoT environments.

As IoT ecosystems remain to progress, a thorough understanding of their details is vital for businesses, policymakers, and individuals alike. Welcoming the potential of IoT while resolving its challenges will certainly lead the way for a future where interconnected devices and data-driven processes boost effectiveness, convenience, and sustainability across all aspects of our lives. Understanding IoT ecosystems not practically recognizing the technical aspects; it is about identifying the profound influence they have on our world and the opportunities they present for a smarter, more connected future.

The chapter is structured as follows: Sect. 2 provides the required background and related work. Section 3 describes the significance of security in IoT. Privacy Concerns in IoT Environments are discussed in Sect. 4. Section 5 describes Challenges of Interoperability in IoT Systems. Section 6 describes about Standards and Protocols for Interoperable IoT Devices. Detailed Data Protection and Compliance in IoT is described in Sect. 7. Risk Management in IoT Security and Role of Encryption in IoT Security are discussed in Sects. 8 and 9 respectively. Role of Encryption in IoT Security and User Authentication and Authorization in IoT us discussed in section 10 and 11 respectively. The last section is devoted to the conclusion.

2 The Significance of Security in IoT

In the age of interconnected devices and smart innovations, the Internet of Things (IoT) has actually reinvented the way we communicate with the world. However, with this interconnectivity comes the critical demand for robust security measures to safeguard sensitive information as well as guarantee the integrity of IoT environments [2].

Vulnerabilities in IoT Systems: The expansion of IoT devices has actually presented a myriad of vulnerabilities [3] that can be manipulated by malicious actors. From unauthorized access to accessibility data to the manipulation of connected devices, the possible security risks in IoT environment are diverse and prevalent. Without appropriate security measures, these susceptibilities pose a significant threat to the privacy, safety, and functionality of IoT devices also the data they collect and process.

Data Privacy and Protection: IoT devices typically accumulate and transmit sensitive data, ranging from personal information to critical operational information in commercial settings. The importance of security in IoT hinges on maintaining the privacy of this information along with safeguarding it from unauthorized access or breaches [4]. Without robust security procedures, the direct exposure of such information can result in effects consisting, including identity theft, financial loss, and reputational damage.

Implications for Industry and Infrastructure: In industrial IoT applications, the security of interconnected systems is critical, as any compromise can have major effects on production processes, supply chain operations, and even public safety. In addition, in smart cities, where IoT technologies are utilized for urban monitoring as well as infrastructure, the security of interconnected devices is important to avoid prospective disruptions and ensure the safety and well-being of residents.

Mitigating Security Risks: Dealing with the importance of security in IoT entails executing a diverse method approach that encompasses encryption, authentication mechanisms, secure firmware updates as well as proactive threat detection. In addition, the facility of industrywide criteria as well as best practices for IoT security is vital to ensure a consistent and robust security position throughout varied IoT communities.

The importance of security in IoT cannot be overemphasized. As the interconnectivity of devices becomes progressively prevalent, the requirement for extensive security measures to protect data, ensure privacy, and moderate risks is utmost. Dealing with these challenges will not just strengthen the reliability and trustworthiness of IoT ecosystems yet additionally facilitate the realization of their full potential in shaping a more connected and efficient future.

3 Privacy Concerns in IoT Environments

The Internet of Things (IoT) has actually changed the way we engage with modern technology, embedding connectivity right into everyday things along with developing an internet of interconnected devices. While this interconnectivity has actually produced many advantages, it also increases substantial privacy concerns [5] that necessitate focus and consideration. This section explores the complex privacy concerns in IoT environments as well as their effects on individuals, organizations, and society as a whole.

1. Data Collection and Usage: IoT devices accumulate vast quantities of information, including personal information, behavioural patterns, as well as environmental data. The comprehensive and often unnoticed collection of such information elevates concerns about user consent, transparency, and the possibility for abuse.
2. Personal Information Protection: The transmission and processing of personal data across IoT gadgets present a risk to individual privacy. Shielding personal information from unauthorized access, exploitation, and also profiling is a crucial concern in IoT environments.
3. Surveillance and Ethical Implications: The deployment of IoT devices, such as smart cameras together with environmental sensing units, in public and private spaces raises ethical questions about pervasive surveillance, the right to privacy, along with the potential misuse of surveillance data.
4. Trust and Transparency: Building trust in IoT environments requires transparent data practices, clear approval systems, and effective interaction concerning data usage. The lack of transparency can wear down trust and hinder the acceptance of IoT technologies.
5. Regulatory and Compliance Challenges: The rapid evolution of IoT has actually overtaken the advancement of extensive privacy regulations and standards. Navigating the intricate landscape of privacy regulations, data protection, and cross-border information flows offers substantial challenges for IoT stakeholders.

6. Information Security and Breach Risks: The security of interconnected devices is vital to prevent unauthorized gain of access and data breaches. The compromise of IoT devices can lead to the direct exposure of sensitive individual information, highlighting the intersection of privacy and security concerns.
7. Empowering Individuals: Providing individuals with devices for information monitoring educated authorization devices, and clear privacy policies outfits them to make informed decisions about their data within IoT environments.
8. Ethical Data Usage: The ethical use of information accumulated by IoT devices is a problem, as it entails appreciating individual privacy, avoiding discriminatory practices, and ensuring data is used for legitimate and transparent purposes.
9. Privacy by Design: Integrating privacy factors to consider right into the layout and also advancement of IoT options is important to proactively deal with personal privacy worries as well as alleviate dangers to specific personal privacy.
10. Collaborative Solutions: Addressing personal privacy worries in IoT settings needs collective initiatives amongst modern technology programmers, legislators, along with personal privacy supporters to guarantee that IoT modern technologies appreciate specific personal privacy civil liberties as well as associate honest concepts.

Privacy concerns in IoT environments are complicated and multifaceted, incorporating data collection, usage, security, along with ethical factors to consider. Addressing these problems requires a comprehensive approach that incorporates technological innovation, regulatory frameworks, and ethical guidelines to maintain specific personal privacy civil liberties as well as develop a reliable IoT landscape. By recognizing and resolving these concerns, IoT stakeholders can foster a privacy respectful atmosphere that leverages the capacity of IoT modern technologies while shielding private personal privacy.

4 Challenges of Interoperability in IoT Systems

Interoperability, or the capability of different gadgets and systems to interact, exchange information as well as run in a coordinated manner. It is a basic requirement for the success and scalability of IoT systems. Nonetheless, attaining smooth interoperability in the varied and complicated IoT landscape offers a number of several challenges. This section looks into the multifaceted challenges [10] of interoperability in IoT systems, clarifying the barriers that need to be dealt to recognize the complete capacity of IoT.

- Diverse Communication Protocols: IoT tools plus systems frequently make use of a variety of interaction methods varying from MQTT plus CoAP to HTTP plus AMQP, making it testing to develop smooth interoperability amongst diverse gadgets and also networks.

- Data Standardization: The diversity of data formats together with frameworks throughout IoT gadgets presents a considerable challenge for interoperability, as it requires for standardization as well as compatibility to make sure purposeful information exchange together with application.
- Device Fragmentation: The expansion of diverse IoT devices developed by different results in gadget fragmentation producing compatibility concerns as well as interoperability obstacles when trying to incorporate tools from different suppliers.
- Security and Privacy Concerns: Ensuring interoperability while preserving robust security measures and appreciating privacy requirements presents a complex challenge, as interoperable systems have to protect delicate information as well as reduce safety and security dangers throughout linked gadgets.
- Scalability and Performance: Achieving interoperability at scale demands durable efficiency along with scalability capacities to deal with the enhancing quantity of information, devices, and interactions within the IoT ecosystem, posing technical challenges in maintaining seamless interoperability.
- Legacy Systems Integration: Integrating legacy systems with modern IoT facilities presents interoperability challenges, as legacy systems might work on obsolete procedures and also lack the required user interfaces to perfectly interact with more recent IoT gadgets and also systems.
- Semantic Interoperability: Ensuring semantic interoperability, which includes constant analysis as well as understanding of information exchanged between IoT devices, calls for resolving difficulties pertaining to information semiotics, ontologies as well as context-aware information handling.
- Cross-Domain Interoperability: IoT systems frequently run throughout numerous domain names such as healthcare, smart cities, and industrial automation, each with its own special demands, standards, as well as regulations, presenting challenges in accomplishing interoperability throughout varied domain.
- Regulatory Compliance: Navigating regulatory requirements and compliance standards throughout various geographical areas together with sectors includes intricacy to accomplishing interoperability as IoT systems have to comply with differing guidelines without endangering interoperability.
- Ecosystem Collaboration: Encouraging collaboration and standardization initiatives amongst IoT industry stakeholders, consisting device manufacturers, software application programmers, as well as criteria companies is vital for dealing with interoperability obstacles plus supporting a integrated IoT ecological community.
- Power Management: IoT devices vary in power requirements and consumption, leading to challenges in integrating them into a unified system without compromising energy efficiency and performance.
- Device Discovery and Management: Identifying and managing the diverse array of IoT devices within a network presents challenges in maintaining visibility, control, and efficient utilization of resources.

- Standardization and Compliance: The lack of global criteria and also interoperability structures throughout IoT gadgets and also systems hinders smooth assimilation coupled with interaction, causing compatibility problems.
- Data Governance and Privacy: Managing the huge quantities of information created by interconnected IoT gadgets while upholding personal privacy laws and also information administration criteria includes intricacy to attaining interoperability.

The obstacles of interoperability in IoT systems are many as well as require detailed options that deal with technological, standard, safety, and collective aspects. Overcoming these challenges is vital in realizing the full potential of IoT, making it possible for smooth assimilation, information exchange as well as teamwork amongst varied IoT gadgets and also systems. By acknowledging as well as dealing with these obstacles the IoT sector can drive development, boost individual experiences along with open the transformative power of adjoined IoT systems.

5 Standards and Protocols for Interoperable IoT Devices

The Internet of Things (IoT) has actually become a transformative force, revolutionizing the means we communicate with technology and the world around us. As IoT ecosystems continue to expand, the demand for interoperability in between varied devices and systems ends up being significantly essential. Standards and protocols play a crucial duty in allowing seamless communication, data exchange, and interoperability within the complex web of IoT devices. This section will certainly look into the importance of requirements together with protocols for accomplishing interoperability in IoT gadgets discovering the vital structures as well as techniques that underpin this crucial element of IoT modern technology.

- Standardization: A Foundation for Interoperability Standardization creates the bedrock of interoperability in IoT devices, offering a common language coupled with structure for devices to communicate and collaborate effectively. Industry-wide standards guarantee that tools from various makers can act together perfectly, promoting a natural as well as related IoT ecosystem. Standards encompass a wide array of facets, consisting of interaction protocols, information layouts, security measures, and device management, among others.
- Communication Protocols: Enabling Seamless Data Exchange methods work as the linchpin of interoperability in IoT devices promoting the smooth exchange of information as well as commands between different devices and systems. Commonly embraced procedures such as MQTT (Message Queuing Telemetry Transport), CoAP (Constrained Application Protocol), and HTTP provide robust structures for effective information transmission making it possible for gadgets to connect throughout varied networks as well as systems.

- IoT Device Management Standards: Cohesive Operations Effective gadget administration is essential to accomplishing interoperability within IoT ecological communities. Gadget administration criteria and also procedures such as OMA DM (Open Mobile Alliance Device Management) and Lightweight M2M (LwM2M), enhance the provisioning, setup as well as tracking of IoT tools guaranteeing natural procedures as well as reliable source application throughout varied implementation.
- Data Formats and Interoperability Standards: Facilitating Meaningful Data Exchange Standardizing information layouts as well as frameworks is vital for making it possible for significant information exchange and also interoperability. Formats such as JSON (JavaScript Object Notation) and XML (eXtensible Markup Language) give an usual structure for standing for as well as transferring information throughout various gadgets and also systems, advertising interoperability and also smooth combination.
- Security Protocols: Safeguarding Interconnected Systems Security is a vital problem in IoT communities along with durable safety and security procedures are crucial for guaranteeing safe and also compatible gadget interaction. Protocols such as TLS (Transport Layer Security) and DTLS (Datagram Transport Layer Security) give security and also verification systems to guard information exchange minimizing safety dangers as well as strengthening the interoperability of IoT devices.
- Industry Confederations and Collaborative Standards: Development Industry consortia as well as collective initiatives play an important duty in driving the growth as well as fostering of interoperability requirements plus procedures. Organizations such as the Open Connectivity Foundation (OCF), the Industrial Internet Consortium (IIC), and the Thread Group spearhead initiatives to establish interoperable frameworks, fostering collaboration and convergence across diverse IoT domains.
- Challenges and Future Directions: While standards and protocols develop the foundation of indivisible IoT devices, challenges continue in accomplishing smooth combination throughout varied ecological communities. Addressing the intricacy of legacy systems, navigating the landscape of diverse standards, as well as ensuring backward compatibility are among the challenges that necessitate concerted efforts and innovation. Additionally, modern technologies such as 5G and edge computing are poised to influence the evolution of IoT interoperability, shaping the future landscape of standards and protocols.

Standards and protocols are indispensable enablers of interoperability in IoT devices, providing a cohesive framework for devices to communicate, collaborate, and exchange data seamlessly. As IoT ecosystems continue to evolve, the development and adoption of robust standards and protocols will be instrumental in unlocking the full potential of interconnected IoT devices, driving innovation, and fostering a truly interconnected and interoperable IoT landscape.

6 Data Protection and Compliance in IoT

In today's interconnected world, the Internet of Things (IoT) has actually come to be an important component of our daily lives. From smart homes and wearable devices to industrial applications, IoT innovation has actually changed the way we interact with our environment. Nonetheless, with the increasing quantity of information being produced and shared by IoT devices, concerns about data protection and compliance have actually pertained to the leading edge. This section explores the essential aspects of data protection and compliance in IoT, focusing on the challenges and best practices for ensuring the security and privacy of IoT data.

1. The Growing Importance of Data Protection in IoT: As the variety of connected devices remains to rise, the quantity of information being produced by IoT devices is expanding significantly. This data includes a wide range of sensitive information, including personal details, location data, and even health-related information. For that reason, making sure the protection of this information has become a critical concern for both consumers and organizations.
2. Regulatory Landscape and Compliance Requirements: The governing landscape for IoT data protection is complex and constantly evolving. Numerous regulations such as the General Data Protection Regulation (GDPR) in the European Union and the California Consumer Privacy Act (CCPA) in the United States enforce strict requirements for the collection, storage, and processing of data. IoT stakeholders have to browse these regulations to guarantee compliance and avoid potential legal consequences.
3. Security Challenges in IoT Environments: IoT gadgets are usually susceptible to security violations due to factors such as inadequate authentication mechanisms, weak encryption, and obsolete firmware. Malicious actors can make use of these susceptibilities to get unapproved accessibility to sensitive data, posing significant risks to both individuals' entities and organizations.
4. Data Encryption and Confidentiality: Implementing robust encryption mechanisms is important for safeguarding IoT data against unauthorized accessibility and interception. By securing information using encryption data both at rest and in transit, IoT stakeholders can alleviate the threat of data violations and ensure the confidentiality of sensitive information.
5. Authentication and Access Control: Effective authentication and access control mechanisms are necessary for stopping unauthorized individuals or devices from accessing IoT information. Executing strong authentication protocols as well as user access monitoring practices can substantially decrease the threat of information compromise.
6. Privacy by Design Principles: Adhering to privacy by design concepts entails integrating data factors to consider right into every stage of the IoT system advancement lifecycle. By focusing privacy together with security from the outset, IoT developers can alleviate possible potential risks and build consumer trust.

7. Data Minimization and Purpose Limitation: Adhering to the concepts of information minimization as well as purpose limitation is vital for making certain that IoT data is accumulated and made use of just for particular legit objectives. By restricting the range of information collection together with handling companies. By restricting the range of information collection and processing, organizations can decrease the possible influence of information violations plus personal privacy violations.
8. Secure Data Storage and Transmission: IoT stakeholders should utilize secure data storage and transmission techniques to shield sensitive information from unauthorized gain of access. This consists of using utilizing secure cloud storage solutions, implementing secure communication protocols, and frequently upgrading tool firmware to attend to recognized susceptibilities.
9. Compliance Audits and Assessments: Regular compliance audits and evaluations are important to verify that IoT systems abide by pertinent data protection regulations and industry standards. By carrying out detailed assessments of their data protection practices, organizations can determine and address potential compliance gaps proactively.
10. Vendor and Supply Chain Management: Many IoT implementations involve a complex ecosystem of vendors and partners, each adding to the overall system. Making sure the information protection practices of these third-party entities is critical to maintaining a secure IoT environment. Organizations ought to execute robust vendor management processes and legal arrangements to apply information defense needs throughout the supply chain.
11. User Awareness and Consent: Educating users regarding the data collection methods of IoT devices and acquiring notified consent for data processing are essential components of data protection and compliance. Transparent communication with users regarding the purpose and scope of data collection fosters trust and empowers people to make educated choices concerning their information.
12. Incident Response and Data Breach: Management Despite preventive measures, data violations can still happen in IoT environments. Developing robust incident response strategies and data violation administration treatments allows companies to reduce the influence of protection events as well as satisfy their responsibilities under information defense guidelines.
13. International Data Transfers and Cross-Border Compliance: For IoT implementations that include the transfer of data across international borders, compliance with data protection laws in various territories is a complex challenge. Should browse the details of cross-border information transfers while making sure that information stays protected in accordance to appropriate guidelines.
14. Ethical Considerations in IoT Data Usage: Beyond legal and regulatory requirements, ethical considerations play a crucial role in IoT data protection. Responsible and ethical data practices are instrumental in building consumer trust and fostering a positive relationship between IoT providers and their users.
15. Industry Collaboration and Best Practices: Promoting collaboration among industry stakeholders and sharing best practices for data protection in IoT can drive constant enhancement in security and privacy standards. Industry

consortia, standards bodies, and collaborative initiatives play a pivotal role in advancing the state of IoT data protection.
16. Emerging Technologies and Data Protection: As IoT remains to progress arising modern technologies such as blockchain plus homomorphic security deal cutting-edge strategies to improving information security

7 Risk Management in IoT Security

The Internet of Things (IoT) has swiftly changed the means we connect with modern technology, embedding connectivity into everyday objects, from smart home devices to industrial machinery. While this connection brings extraordinary comfort together with performance it additionally presents a host of safety and security threats. With the expansion of interconnected devices, the need for durable threat administration in IoT security has actually never ever been a lot more vital.

Risk management strategies for IoT security

1. **Asset Inventory and Vulnerability Assessment**: Organizations should maintain a comprehensive inventory of IoT devices and conduct regular vulnerability assessments to identify potential entry points for security breaches. Understanding the assets within the IoT ecosystem is critical for creating an effective risk management plan.
2. **Access Control and Authentication**: Implementing strong access controls and authentication mechanisms is essential for preventing unauthorized access to IoT devices and networks. This includes the use of robust password policies, multi-factor authentication, and role-based access control.
3. **Encryption and Data Protection**: All communication between IoT devices and networks should be encrypted to protect sensitive data from interception or tampering. Additionally, data at rest should be secured through encryption to prevent unauthorized access in the event of a breach [9].
4. **Continuous Monitoring and Incident Response**: Real-time monitoring of IoT devices and networks can help organizations detect anomalies and potential security incidents. A well-defined incident response plan should be in place to address security breaches promptly and minimize their impact.
5. **Regulatory Compliance and Standards**: Adhering to industry-specific regulations and security standards is crucial for IoT risk management. Compliance with regulations such as the General Data Protection Regulation (GDPR) and adherence to standards like ISO/IEC 27,001 demonstrate a commitment to robust security practices.

The proliferation of IoT devices presents both opportunities and challenges for organizations seeking to manage security risks effectively. By implementing robust risk management strategies, including asset inventory, access control, encryption, continuous monitoring, and compliance with regulations, organizations can mitigate the inherent security risks associated with IoT. As the IoT ecosystem evolves, proactive risk management will be instrumental in safeguarding the integrity and security of interconnected devices, ensuring a resilient and secure IoT landscape.

8 Role of Encryption in IoT Security

Security plays an essential function in guaranteeing the protection and also personal privacy of information in the IoT. As IoT tools end up being significantly incorporated right into different elements of our lives, the requirement for durable file encryption systems comes to be much more vital than ever before. This section discovers the role of encryption in IoT security is paramount for ensuring the protection of sensitive data and the integrity of communications within IoT ecosystems. Here are some key insights drawn from various sources:

- **Data Privacy and Integrity**: Encryption serves to safeguard the privacy of data transmitted between IoT devices and systems, preventing unauthorized access and ensuring that sensitive information remains confidential [6]. Additionally, encryption contributes to the integrity of IoT communications by verifying the authenticity of transmitted data, thereby preventing unauthorized alteration or manipulation [6].
- **Challenges and Considerations**: The management of encryption keys in IoT environments with a high volume of devices generating encrypted data presents a significant challenge [6]. It's important for organizations to ensure that their cryptography choices offer sufficient protection for the specific use cases they are securing [6]. Additionally, the implementation of encryption must be carefully executed to prevent compromise, especially in devices with limited processing power.
- **Types of Encryption Algorithms**: Various encryption algorithms are utilized in IoT security, including symmetric and asymmetric encryption methods. The Advanced Encryption Standard (AES) is a widely used encryption algorithm in IoT, known for its robust security features [7].
- **Future Considerations**: As the IoT landscape continues to evolve, the importance of encryption in protecting data privacy and security is expected to increase. The implementation of secure and efficient encryption systems will play a vital role in safeguarding sensitive information from cyber threats as the number of connected devices continues to grow [8].

Encryption plays a critical role in ensuring the security and privacy of data within IoT ecosystems, and its effective implementation is essential for the continued advancement and protection of IoT technologies.

9 User Authentication and Authorization in IoT

The expansion of Internet of Things (IoT) devices has significantly broadened the extent of connection and also communication in between individuals and smart devices. However, as the IoT ecosystem continues to grow, the requirement for robust user authentication and authorization mechanisms becomes increasingly critical. This section explores the world of user authentication and authorization in IoT, highlighting the significance, challenges, and best practices associated with safeguarding individual interactions within IoT environments.

- Importance of User Authentication and Authorization in IoT: User authentication and authorization play crucial function in guaranteeing the security and privacy of IoT systems. It entails confirming the identification of users and determining their level of access to IoT devices and data. By applying strong authentication and authorization mechanisms, IoT ecosystems can alleviate the danger of unauthorized access, data breaches, and malicious activities.
- Authentication in IoT: Authentication is the procedure of confirming the identification of individuals or devices attempting to access IoT resources. In IoT environments, individual verification can take different types, consisting of password-based verification, biometric verification, two-factor authentication along with device-based verification. Each method intends to verify the legitimacy of the user or device seeking access to IoT systems.
- Challenges in User Authentication for IoT: IoT environments existing one-of-a-kind challenges for user authentication, such as source restrictions in IoT tools, varied interface and also the demand for smooth individual experiences. Additionally, the large range of IoT implementations presents intricacies in handling customer qualifications, guaranteeing protected interaction networks as well as preserving verification honesty throughout intertwined tools.
- Authorization in IoT: Authorization involves granting specific privileges and access rights to authenticated users within the IoT ecosystem. It specifies what activities users are allowed to carry out and what sources they can access. Reliable authorization systems make certain that individuals have ideal authorizations to connect with IoT tools and also information, while protecting against unapproved tasks.
- Role-Based Access Control (RBAC) in IoT: Role-Based Access Control (RBAC) is a preferred authorization model in IoT, where accessibility legal rights are appointed based upon the functions of customers within the system. RBAC eases accessibility monitoring by classifying individuals' right into functions as well as connecting them with predefined consents therefore streamlining the permission procedure in intricate IoT environments.
- Best Practices for User Authentication and Authorization in IoT: Implementing solid cryptographic protocols, multi-factor authentication, safe and secure interaction networks, and regular credential updates. Leveraging biometric authentication and behavioral improve individual confirmation and also scams discovery. Leveraging blockchain modern technology to develop tamper-proof individual

identifications plus accessibility control in IoT atmospheres. Executing continuous tracking as well as violation discovery to recognize unapproved accessibility efforts as well as protection violations in real-time.

Future Trends in User Authentication and Authorization in IoT: The combination of expert system and machine learning for versatile individual verification coupled with anomaly discovery. Innovations in biometric authentication technologies, such as face acknowledgment as well as behavior biometrics, to improve individual confirmation in IoT. The introduction of decentralized identification options leveraging blockchain as well as dispersed journal modern technology to develop protected as well as interoperable customer identifications throughout IoT communities.

Individual verification as well as consent are necessary elements of IoT protection making sure that just genuine individuals accessibility IoT gadgets plus information while preserving the stability coupled with discretion of info. By attending to the difficulties as well as welcoming sophisticated verification as well as consent strategies IoT ecological communities can develop a protected and also relied on atmosphere for individual communications leading the way for the proceeded development as well as advancement in the world of linked tools.

10 Conclusion

In conclusion, user authentication and authorization important parts of IoT protection, making certain that just genuine individuals gain access to IoT tools together with information while preserving the stability plus discretion of information. By addressing the challenges and embracing advanced authentication and authorization techniques, IoT ecosystems can establish a secure and trusted environment for user interactions, paving the way for the continued growth and innovation in the realm of connected devices. As the IoT landscape remains to progress it is important for companies and also designers to focus on the execution of durable verification plus permission systems to protect individual identifications coupled with protect the honesty of IoT ecological communities. With the fostering of sophisticated modern technologies coupled with finest methods the future holds encouraging potential customers for improved customer verification as well as permission within the IoT domain name eventually adding to a much safer and also much more resistant IoT atmosphere for individuals as well as companies alike.

References

1. https://www.cisco.com/c/en/us/solutions/collateral/executive-perspectives/annual-internetreport/white-paper-c11-741490.html
2. Eustis, A.G.: The Mirai botnet and the importance of IoT device security. In: Latifi, S. (eds) 16th International Conference on Information Technology-New Generations (ITNG 2019).

Advances in Intelligent Systems and Computing, Vol. 800. Springer, Cham (2019). https://doi.org/10.1007/978-3-030-14070-0_13
3. Ramadan, R.: Internet of things (IoT) security vulnerabilities: a review. PLOMS AI **2**(1) (2022)
4. Ahmed, G.: Improving IoT privacy, data protection and security concerns. Int. J. Technol. Innov. Manage. (IJTIM) **1**(1) (2021)
5. Aqeel, M., Ali, F., Iqbal, M.W., Rana, T.A., Arif, M., Auwul, M.R.: A review of security and privacy concerns in the internet of things (IoT). J. Sens. (2022)
6. https://www.techtarget.com/iotagenda/tip/Learn-the-basics-of-cryptography-in-IoT
7. https://www.nabto.com/iot-data-encryption-algorithm-guide/
8. https://sigmaos.com/tips/startups/internet-of-things-iot-terms-explained-data-encryption
9. Nanda, A.K., Nayak, R., Awasthi, L.K.: NTRU with Gaussian integer matrix. Int. J. Adv. Res. Comput. Sci. Softw. Eng. (IJARCSSE) **5**(2), 359–365 (2015)
10. Bhoi, S.K., Ghugar, U., Dash, S., Nayak, R., Bagal, D.K.: Exploring the security landscape: a comprehensive analysis of vulnerabilities, challenges, and findings in internet of things (Iot) application layer protocols. Migr. Lett. **21**(S6), 1326–1342 (2024)

Advancements in Smart Farming: Leveraging Artificial Intelligence and Internet of Things for Sustainable Agricultural Management

Haytham M. Salem, Linda R. Schott, and Ali M. Ali

Abstract In the realm of modern agriculture, farmers across the globe are increasingly adopting smart farming practices, harnessing a wide array of cutting-edge technologies. Notably, the integration of Artificial Intelligence (AI) and the Internet of Things (IoT) has permeated various sectors, including agriculture, catalyzing a wave of revolutionary advancements. These digital innovations have brought about transformative changes in agriculture, providing intelligent systems that are capable of monitoring, controlling, and visualizing diverse farm operations in real-time, with a level of intelligence that rivals that of human experts. This chapter delves into the numerous potential applications of IoT and AI in the evolution of smart farm machinery, fertilizer application, irrigation systems, pest and weed control, and more. The primary objective is to comprehensively explore recent advancements in intelligent agricultural management, highlighting the benefits and potential of AI and IoT technologies in promoting sustainability. By emphasizing IoTbased agricultural management techniques for sustainable farming, this review aims to shed light on the latest technologies and innovations that are poised to enhance the productivity and efficiency of food producers. By adopting smart farming technologies, stakeholders in the agriculture sector can make well-informed decisions regarding technology investments, thereby bolstering food production while upholding environmental sustainability.

Keywords Smart Farming · Artificial Intelligence · Sustainable Agriculture · Agricultural Machinery · Smart Irrigation · Precision Fertilization

---

H. M. Salem (✉)
State Key Laboratory of Ecological Safety and Sustainable Development in Arid Lands, Xinjiang Institute of Ecology and Geography, Chinese Academy of Sciences, Urumqi 830011, China
e-mail: eng_haytham1982@yahoo.com; haytham@ms.xjb.ac.cn

L. R. Schott
Department of Soil and Water Systems, Twin Falls Research and Extension Center, University of Idaho, Twin Falls, ID, USA

A. M. Ali
Department of Soil Fertility and Microbiology, Desert Research Center, Cairo, Egypt

B. K. Mishra and A. Rocha (eds.), *Future of AI, IoT, and Sustainability*, Information Systems Engineering and Management 23,
https://doi.org/10.1007/978-3-031-76286-4_4

1 Introduction

The United Nations (UN) is a strong proponent of sustainable development goals (SDGs) that are specifically designed to promote a revolutionary agricultural sector. Cutting-edge technologies such as Artificial Intelligence (AI) and the Internet of Things (IoT) are playing a pivotal role in transforming various aspects of agriculture, ranging from the cultivation of crops to the processing of food products. As the global population is anticipated to grow by an additional 2 billion individuals by the year 2050, the Food and Agriculture Organization (FAO) emphasizes the critical importance of advancing agricultural techniques to meet the escalating demands for food [12]. The integration of AI-powered IoT systems and sophisticated machine learning algorithms is proving to be instrumental in boosting agricultural output while simultaneously reducing wastage [20]. Moreover, the implementation of smart farming technologies harnesses digital solutions to streamline agricultural operations, thereby enhancing efficiency, productivity, and sustainability (Fig. 1). The widespread adoption of smart farming initiatives not only contributes to the expansion of the digital economy but also opens up new market opportunities while addressing pressing issues like climate change through enhanced transparency. By incorporating AI and IoT technologies into conventional farming practices, smart farming facilitates data-driven decision-making processes that significantly improve agricultural efficiency and productivity. Positioned as a fundamental component of Agriculture 4.0, the fourth industrial revolution in agriculture, smart farming advocates for intelligent cultivation methods and advanced crop management techniques [19, 39, 64].

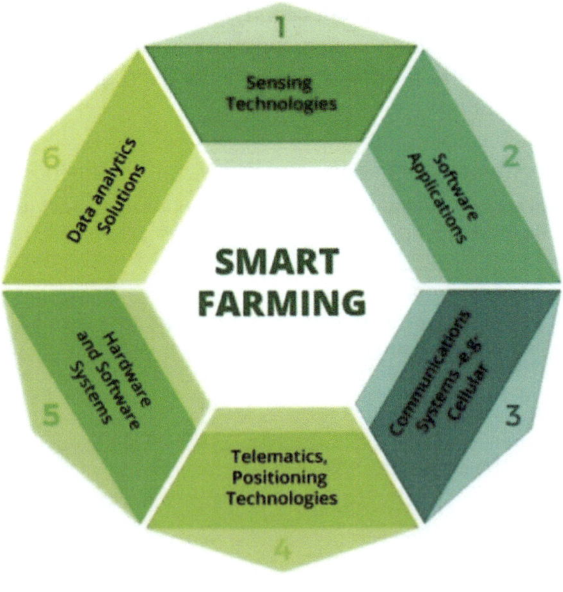

Fig. 1 The context of smart farming

The eradication of hunger, as outlined in the second UN SDG2, can be significantly advanced by incorporating AI and IoT technologies into the agricultural sector. These innovative solutions not only boost agricultural productivity and efficiency but also equip farmers with crucial insights and decision-making tools based on real-time data analysis [14, 22]. Furthermore, IoT-enabled monitoring devices play a vital role in the early detection of diseases and pests, enabling farmers to take swift action to prevent crop damage. Moreover, the incorporation of AI and IoT technologies is crucial in achieving SDG-2 by promoting sustainable agricultural practices that address hunger [63, 65]. AI systems guide farmers in the judicious use of resources like water, electricity, and fertilizers, thereby minimizing their environmental impact and conserving natural resources. Additionally, IoT sensors provide valuable data on the effects of climate change on crops and monitor soil conditions, empowering farmers to adapt their strategies and enhance crop resilience in the face of changing environmental conditions.

The combination of AI and IoT in the field of smart farming has the ability to greatly enhance the stability and diversity of ecological functions that are vital for supporting food production and agricultural ecosystems [8, 34]. By utilizing these advanced technologies, farmers are able to make better decisions and implement more efficient farming techniques, resulting in practices that are more sustainable and make better use of resources. As a result, the integration of AI and IoT in farming plays a crucial role in establishing agricultural ecosystems that are healthier and more resilient, which are essential for ensuring long-term food production and protecting the natural environment.

The agricultural sector is currently witnessing substantial changes in IoT technologies, with a particular emphasis on wireless network infrastructure, device advancements, and security concerns. The integration of AI into farming practices is becoming more prevalent, as various institutions and countries are embracing AI and IoT strategies to improve analysis, predictions, and overall efficiency. Although there have been significant research endeavors in AI-driven IoT smart farming, there is still a need for further investigation in this field, as highlighted by recent studies conducted by Yazdinejad et al. [72] and Lee et al. [29].

Smart farming utilizes technology to oversee and control the various aspects of crop production. The IoT plays a crucial role in connecting remote devices like robots and sensors, enabling the monitoring of soil characteristics, moisture levels, climate conditions, and other pertinent factors. This integration of IoT allows for automated decision-making processes based on data analysis. As a result, smart farming can now be extended to urban environments, including rooftops and indoor settings. However, effective data management is vital for the successful implementation of smart farming practices in both urban and rural areas.

Agriculture heavily relies on irrigation, especially in areas with low annual rainfall like those below 500 mm. Irrigation has been a fundamental practice in regions spanning Africa, Asia, and the Americas to sustain agricultural activities. Recent research studies by Kumar and Sharma [26], Singh et al. [62], and Maraveas et al. [35] have introduced various innovative systems for managing smart agriculture farming. These systems are designed to monitor essential factors such as humidity, temperature, and soil moisture to improve the overall efficiency and productivity of farming operations. By gathering and analyzing data on these essential parameters, farmers can make well-informed decisions regarding their agricultural practices and techniques. The implementation of these proposed systems has the potential to transform traditional farming methods, optimize the use of resources, and ultimately increase crop yields while promoting positive environmental outcomes. The continuous development and refinement of smart agriculture technologies are essential to address the challenges faced by modern farmers and to promote sustainable agricultural practices for the future.

The primary objective of this chapter is to conduct a comprehensive analysis of the recent developments in intelligent agricultural management and underscore the advantages and potential of AI and IoT technologies in promoting sustainability. The chapter's examination is centered on the application of IoT-based agricultural management methods for sustainable farming techniques. The aim is to offer valuable insights into the most recent technologies and advancements that can aid food producers in enhancing their crop yield and overall effectiveness. By embracing smart farming technologies, stakeholders in the agricultural industry can make well-informed decisions regarding technology investments, thereby improving food production while also ensuring environmental sustainability. The incorporation of AI and IoT in agriculture demonstrates significant potential in enhancing production monitoring and sensing capabilities.

2 Typical Applications of IoT in Smart Farming

2.1 Intelligent Agricultural Machinery

Intelligent agricultural machinery has brought about a significant transformation in farming practices by providing efficient, standardized, and interactive operations for a wide range of agricultural tasks, such as cultivation, sowing, fertilization, and harvesting. These machines are capable of autonomously executing these tasks while simultaneously gathering crucial data on soil conditions, water quality, and crop health. This data collection process supports precision agriculture and promotes sustainable farming practices [32]. The integration of IoT technologies further enhances the functionality of agricultural machinery by enabling fast and accurate data collection, processing, and transmission. By utilizing sensing, positioning, and wireless networks, intelligent terminals mounted on vehicles and remote service platforms are created to deliver comprehensive agricultural machinery service information [56]. Lin et al. [31] developed an agricultural machinery monitoring system that utilizes Bluetooth technology and embedded systems to facilitate data collection, processing, and wireless transmission. In the context of fruit picking applications, Onishi et al. [44] devised an apple-picking robot equipped with a single-lens multi-box detection method and stereo camera technology to detect and locate fruits in three-dimensional space. Similarly, Wang et al. [69] proposed an agricultural machinery operation parameter collector that leverages IoT technology to combine data collection devices with IoT infrastructure. This innovation not only streamlines agricultural operations by reducing manual labor and minimizing subjective errors but also enhances the reliability of data collection, thereby fostering scientific management practices in agriculture.

2.2 Smart Irrigation Systems

The advancement of water management practices has the potential to significantly impact global society by promoting sustainable water resources and agriculture, as highlighted by Kumar et al. [27]. This shift towards more efficient water management can pave the way for a sustainable future, with far-reaching benefits for the environment, economy, and society as a whole. Eco-friendly agriculture and profitable farming heavily rely on effective water resource management, necessitating a comprehensive approach that encompasses all stages from planning to monitoring, as emphasized by El-Beltagi et al. [15]. By implementing robust water resource management systems, the agricultural sector can undergo a transformative process that ensures long-term sustainability for both communities and businesses, as noted by Xiang et al. [71]. The integration of smart irrigation systems, which utilize sensors to collect crucial data and analytics on irrigation processes, plays a pivotal role in empowering farmers and agronomists to make well-informed decisions and optimize agricultural practices, as illustrated in Fig. 2. These sensor-based systems consist of interconnected components such as data players and sensors, as discussed by Almalki et al. [6] and Pham et al. [49].

Integrated analyses of meteorological and sensor data are easily accessible to users through user-friendly web applications, offering valuable recommendations for enhancing fieldwork efficiency. The utilization of IoT-based smart irrigation systems holds the potential to transform traditional farming methods by precisely monitoring soil moisture levels, weather patterns, and water consumption [1, 42]. By ensuring that crops receive the precise amount of water they need, smart irrigation systems contribute to increased crop yields, improved efficiency, reduced wastage, and the conservation of water resources. Furthermore, the implementation of smart irrigation systems not only benefits farmers by enhancing productivity but also has positive implications for the environment and consumers, as it results in the production of fresher, higher-quality agricultural produce. Embracing smart irrigation technology is crucial for the advancement of agricultural practices, as it replaces labor-intensive manual irrigation methods like flooding and watering cans with automated systems and IoT technology. These advanced systems enable real-time monitoring and control through the use of moisture sensors and wireless communication, offering a more efficient and sustainable approach to irrigation management [58].

A novel and cost-effective smart irrigation system that leverages IoT is proposed by Nawandar and Satpute [40]. The system is composed of three key modules: the

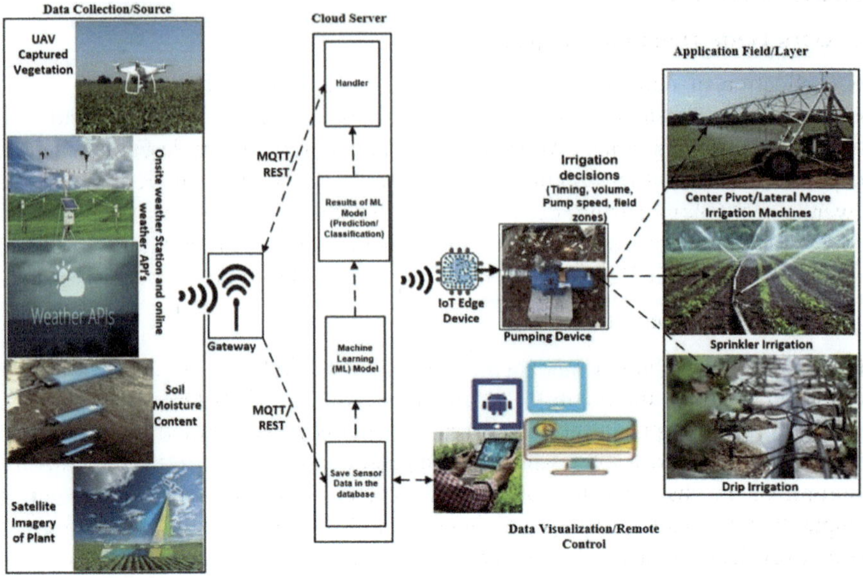

Fig. 2 The smart irrigation systems, showcasing their functionality and design

sensor module, irrigation module, and sensor communication module. The sensor module collects crucial data pertaining to crops, plantations, and soil conditions. This data is then processed using a sophisticated neural network model, enabling the system to make intelligent decisions. These decisions are subsequently transmitted to both the irrigation module and sensor communication module. Based on the received data, these modules determine whether to activate or deactivate the water mechanism at specific locations. The system also maintains a record of incoming data, with the sensor communication module logging it through the MQTT broker. In the current setup, an AtMega 328 microcontroller and Zigbee communications are utilized for monitoring and controlling sensor data. Additionally, each field's central node stores the sensor data in a cloud database. To provide users with real-time updates on automatic irrigation, reports are transmitted through an Android application [60].

In a study conducted by Alomar and Alazzam [7], an advanced irrigation system that incorporates IoT technology and fuzzy logic was successfully implemented. This system involves the monitoring of various environmental factors such as air climate, humidity, soil wetness, and light intensity through the use of sensors. The core component of this system is a Mamdani fuzzy controller, which is responsible for regulating the flow of water based on the real-time data collected by the sensors. By integrating IoT, Dynamic Programming-Based Long Short-Term Memory Recurrent Neural Networks (DPLRN), and fuzzy computation, a self-sustaining irrigation ecosystem is created, ensuring optimal water usage and crop health. The operational simulation of this system relies on input variables such as soil wetness, atmospheric conditions, and crop status. Data collected by sensors is transmitted to the cloud via a Raspberry Pi IoT gateway, while DPLRN algorithms analyze crop health and determine the most suitable timing for irrigation using fuzzy logic controllers, as highlighted by Patel et al. [46]. Additionally, the implementation of an efficient farm tracking framework is achieved by combining RFID technology and sensors with the LoRaWAN protocol, as demonstrated by Saha et al. [55]. This framework addresses key agricultural challenges including fungal attacks, cattle monitoring, proper watering, soil assessment, crop inspection, and productivity enhancement. By utilizing temperature and humidity sensors, the system ensures precise irrigation practices while also promoting water conservation. Overall, the integration of IoT, fuzzy logic, DPLRN, and advanced sensor technologies in agriculture represents a significant advancement towards sustainable and efficient farming practices.

Mohammed et al. [37] have introduced a novel method that employs DPLRN for the purpose of regulating irrigation systems based on soil moisture and structure. This innovative approach integrates soil images and moisture sensors to gather relevant data, which is then processed in realtime using video processing techniques to determine the irrigation requirements and moisture levels of the crops. In their study, the researchers utilize a VGG-19 model to classify both the type of crop and the images, which assists in accurately predicting the irrigation needs of the crops. By leveraging TensorFlow Lite, the experiments are conducted on cost-effective computational hardware, allowing for efficient evaluation of the proposed model using a proprietary dataset. This comprehensive methodology showcases the potential of utilizing advanced technologies in optimizing irrigation practices for improved crop yield and resource management.

2.3 Fertilizer Application

Improper use of fertilizers has a significant negative impact on agricultural productivity, leading to decreased yields. To address this issue, it is crucial to conduct soil analysis before applying fertilizers in order to determine the specific nutrient requirements of crops [57]. However, many farmers tend to skip this step due to the complexity of laboratory procedures. In recent times, there has been a push for farmers to embrace digitalization and adopt technology-driven farming practices. The IoT technology provides smart solutions for fertilizer application. A nitrogen (N), phosphorus (P), and potassium (K) sensor, which utilizes light-emitting diodes (LED), light-dependent resistors, and resistors, can accurately measure the levels of N, P, and K in the soil [28, 38]. This sensor operates based on Colorimetric and photoconductivity principles, and the data collected is transmitted directly to a processing unit or system-on-chip for analysis. The subsequent analysis can be performed through fog computing or edge computing. Fog computing involves processing the data directly at the sensor, while edge computing utilizes an edge server connected via the internet. Fuzzy logic is employed to analyze the data and determine the optimal quantities of fertilizers required. Cloud services like Google

Cloud Platform offer scalable and uninterrupted service, enabling the delivery of recommended fertilizer quantities to farmers' mobile phones via SMS.

Ali [4] and Ali et al. [5] conducted a study where they utilized active canopy sensors to evaluate crop agronomic information in order to enhance N fertilizer management in wheat. They developed and validated algorithms that relied on data from the GreenSeeker canopy reflectance sensor and chlorophyll meter. The developed algorithms aimed to optimize fieldspecific N fertilizer management by taking into account the spatial and temporal variations of N throughout the growth season. By implementing the sensor-guided N dosage estimated by the algorithms, they were able to achieve yields comparable to general recommendations, but with a reduced amount of fertilizer N. This approach resulted in improved N recovery and agronomic efficiencies, thus highlighting the effectiveness of sensor-based algorithms in optimizing N fertilizer management in wheat.

Bachuwar et al. [10] put forward a methodical approach that aims to gather real-time data from agricultural sites. This approach involves collecting various parameters such as light intensity, soil moisture, temperature, and humidity. The main objective of this approach is to provide localized access to crucial information regarding soil temperature, humidity, wetness, and light intensity. To achieve this, the researchers utilized an ESP8266-12E module, which enables the transmission of sensor data over a Wi-Fi network using the IEEE 802.11 protocol. On a similar note, Rahman et al. [52] introduced a different technique that focuses on generating comprehensive data sets. These data sets include temperature, humidity, and nutrient content (N, P, K) based on readings from soil sensors. The ultimate goal of this method is to contribute to the establishment of a soil fertility data repository by leveraging IoT technologies.

Lavanya et al. [28] presented an innovative framework centered on the Internet of Things (IoT), which integrates Light Dependent Resistor (LDR) and Light Emitting Diodes (LED) to develop a unique NPK sensor. This sensor enables the detection of deficiencies in N, P, and K in soil by employing the Mamdani inference process. Through the utilization of this framework, farmers receive timely notifications regarding the precise quantities of fertilizers required, as determined by the sensor data. Postolache et al. [50] introduced a method to assess changes in soil nutrients for enhanced drainage and fertilizer management. Their approach involves a device capable of measuring NPK levels in soil, as well as soil temperature, conductivity, pH, and moisture. This comprehensive system aims to improve agricultural practices by providing valuable insights into soil conditions.

Ahmad et al. [3] devised a solar fertigation control mechanism that merges weather-based predictions with real-time sensor data to efficiently control watering and fertilization procedures. Through the integration of a hybrid model and a crop repository, this system supervises plant water conditions, soil moisture levels, and water quality, thereby enhancing agricultural productivity. Zhang et al. [74] implemented a continuous monitoring system for evaluating soil water content and nutrient status in citrus orchards. This system enables fruit farmers to make informed decisions regarding fertilization and irrigation strategies, leading to improved crop yields and resource utilization. Pratama et al. [51] introduced an IoT-enabled device designed to simplify citrus seed planting by utilizing NPK indicators to assess soil nutrients for citrus seedlings. The results are transmitted to the Thingspeak cloud platform, assisting farmers in optimizing citrus cultivation practices and enhancing overall agricultural efficiency.

2.4 Pest and Weed Control

The application of herbicides for weed control can have negative effects on both human health and the environment, making it a challenging task to find a precise and effective method. Traditional weeding methods often involve applying herbicides uniformly across the entire field, regardless of the presence of weeds. This approach not only increases costs but also contributes to greenhouse gas emissions. To overcome these challenges, a promising solution is to develop a site-specific system that integrates IoT, robotics, and advanced image analysis techniques [9, 13, 70]. In order to detect weeds in fields, both Infrared and RGB imaging sensors can be utilized. The images captured from the fields undergo a series of preprocessing steps, including resizing, color space transformation, and normalization, to prepare them for segmentation. Segmentation is the process of grouping associated pixels together to form connected objects with similar properties. This allows for the separation of plants from soil and other backgrounds, isolating the vegetation for further analysis. Once the segmentation is complete, essential features are extracted from the segments. These features encompass various aspects such as biological morphology, spectral features, visual textures, and spatial contexts.

However, the extraction of these features may result in increased dimensionality, which requires the selection of essential feature combinations. This can be achieved using algorithms like Stepwise Linear Discriminant Analysis (SWLDA), Principal Component Analysis (PCA), or Linear Discriminant Analysis (LDA). In order to classify the weeds and crops, machine learning-based or deep learning-based algorithms can be employed.

Weed management systems that are based on IoT make use of processing components or system-on-chip devices such as Raspberry Pi. These devices are combined with sensors, cameras, and sprayers to effectively apply herbicides. One of the key components of these systems is the integration of AI models that are specifically designed for weed classification. These models often utilize convolutional neural networks (CNNs) to accurately identify and differentiate between weeds and crops. Once the AI model classifies the weeds, the sprayer is activated accordingly. This IoT-driven approach to weed control has shown great potential for automation, which can lead to significant reductions in herbicide usage and address health-related concerns associated with excessive herbicide application [13].

Robotic weed control systems have been developed to effectively identify and manage weeds in various agricultural fields. In the case of onion fields, Arakeri et al. [9] demonstrated the integration of machine learning-based convolutional recurrent neural networks (MCLRN) and IoT technology to detect weeds. Once the weeds are detected, these systems utilize appropriate herbicides for targeted spraying, ensuring efficient weed control. Similarly, in chili fields, Islam et al. [17] employed a combination of random decision forest (RADF), k-nearest neighbors (KNN), and support vector machine (SVM) algorithms, along with MCLRN and image processing techniques, to develop a weed detection system. MATLAB software was utilized for the implementation of these systems. For weed identification in soybean crops, Razfar et al. [53] introduced a vision-based system that utilizes DPLRN. This system incorporates three distinct convolutional neural network (CNN) models and MobileNetV2 architecture within a five-layer DPLRN framework, enabling accurate weed identification and management in soybean crops.

Traditional methods of pest control heavily depend on chemical pesticides, which have sparked concerns among consumers due to their potential negative effects on both soil quality and human health. By harnessing the power of IoT, a wide array of innovative strategies can be implemented to manage pests more effectively [41, 61, 68]. Weather conditions play a critical role in influencing pest populations, and IoT technology enables precise remote monitoring of weather patterns at a relatively low cost. Furthermore, the use of AI systems and image processing for monitoring crop health can aid in the early identification of potential pest threats. Recent advancements in pest control have also seen the development of early warning models driven by big data analytics, which gather real-time field data on soil quality, weather conditions, and environmental factors to predict pest outbreaks more accurately.

The utilization of an AI-enabled IoT system, in conjunction with DPLRN technology proposed by Chen et al. [11], presents a promising method for examining crop growth trends and forecasting pest occurrences in agricultural settings. This innovative system integrates a neural network algorithm called "You Only Look Once (YOLO)" to precisely detect pests within images. In research conducted by Vijayalakshmi et al. [67], a pest detection and management system integrated into an IoT framework is introduced. This system employs a Raspberry Pi microcontroller to capture images of plants when specific environmental parameters, such as temperature and humidity, surpass predetermined thresholds. Subsequently, the captured images are cross-referenced with a database, and if pests are recognized, farmers are promptly notified to implement suitable pest control strategies. This proactive approach not only aids in early pest detection but also facilitates timely intervention to mitigate potential crop damage.

A novel approach to address the issue involves the utilization of a Multi-Agent System (MAS) within an Edge AI-IoT architecture, orchestrated by Kubernetes, as proposed by Debauche et al. (2020). This system employs advanced AI algorithms to effectively identify plant diseases and pests. Once detected, it triggers the activation of an irrigation pivot to administer appropriate treatment to the affected crops. In a similar vein, Materne and Inoue (2018) employed machine learning algorithms, including K-nearest neighbors (KNN) and linear regression, in a plantation environment to develop a prediction model. The primary objective of this model is to forecast disease and insect outbreaks, harnessing the strengths of algorithms like MCLRN and the linear regression framework to ensure accurate predictions.

Cutting-edge spraying devices, like air-assisted sprayers, ultralow volume (ULV) sprayers, canopy sprayers, ultrasonic sensor-based sprayers, and electrostatic sprayers, provide enhanced spray effectiveness as highlighted in recent studies [2, 33]. These advanced technologies not only decrease application rates but also lessen the frequency of sprays needed in agricultural and horticultural settings, ultimately leading to a reduction in pesticide consumption, lowering the chances of over-application, and minimizing the environmental impact on soil and groundwater contamination.

3 Advancements in Smart Farming Technologies

The emergence of smart farming technologies has brought about a remarkable transformation in the agricultural landscape, revolutionizing the way farmers operate and enhancing sustainability and efficiency. These cuttingedge technologies encompass a diverse range of tools, such as AI-driven analytics and precision agriculture techniques, which empower farmers to make well-informed decisions based on data, ultimately maximizing crop yields [21]. Additionally, the integration of IoT smart devices and sensors has significantly improved the monitoring and management of agricultural operations, paving the way for the adoption of more sustainable and efficient farming practices.

3.1 Software Applications for Smart Farming

Smart farming applications have been specifically designed to optimize and improve the efficiency and productivity of agricultural practices. These innovative applications effectively transform farmers' mobile devices into comprehensive tools for managing and cultivating crops. By providing crucial information and resources, they enable farmers to make informed decisions and effectively address crop-related issues. These applications offer diagnostic capabilities that help identify and resolve common agricultural challenges, offering practical solutions tailored to the specific needs of farmers. Additionally, these applications prioritize accessibility by providing content in multiple languages, ensuring that farmers from diverse backgrounds and geographical locations can benefit from their features and functionalities.

With a vast repository of information on various crops, smart farming apps empower farmers to make well-informed decisions aimed at maximizing crop productivity. By harnessing the capabilities of these applications, the agricultural sector can achieve greater effectiveness, higher production output, and increased profitability, ultimately enhancing the overall quality of life for farmers. These apps are easily accessible and can be conveniently downloaded from smartphone app stores, making them readily available to farmers. To provide a comprehensive overview of the software applications utilized in smart farming, Table 1 presents a detailed

Table 1 A compilation of software applications developed for the domain of smart farming

Software	Key features	Purpose	References
Farm manager	Observes, oversees, and modifies all field data encompassing crop details, planting conditions, and spraying activities	Assists farmers in managing and monitoring field data related to planting and spraying activities	[45]
Soil analysis	Observes mineral composition, soil texture, and depth levels	Enables farmers to observe soil attributes to enhance crop planning	[47]
Agrippa	Generates digital field maps, monitors crop growth, and identifies objects within the farm area	Assists agronomists in mapping fields and monitoring crop development	[54]
Semios	Provides instantaneous event alerts and network connectivity for monitoring orchards in real time	Offers real-time monitoring of frost, orchard pests, insects, and irrigation systems	[66]
ThingSpeak	Enables the development of personalized applications and facilitates data transfer to MATLAB R2019a for analysis	Empowers farmers to design personalized applications for monitoring fields and crops	[48]

summary of their specific purposes and key features. These applications encompass a wide range of functionalities, including soil analysis, field data management, pest control, crop mapping, custom application development, and real-time orchard monitoring.

3.2 Drones or UAV

Unmanned aerial vehicles, commonly known as drones, have seen a surge in their utilization within the realm of precision agriculture. These drones can be operated either autonomously or by remote pilots, carrying out a variety of tasks such as crop mapping, monitoring, and spraying, as highlighted by Kleinschroth et al. [23]. Equipped with an array of sensors including thermal, multi-spectral, or RGB cameras, drones provide farmers with a bird's eye view of their fields, offering a unique perspective that aids in decision-making processes. The high-resolution thermal sensors on drones are particularly noteworthy as they enable real-time crop data collection, even in low-light conditions such as night time, allowing farmers to closely monitor their crops and livestock, ultimately leading to informed decision-making and operational efficiency improvements. As emphasized by Hafeez et al. [16], drones have now become indispensable tools in precision agriculture, granting farmers access to crucial information that would otherwise be difficult to obtain.

This technological advancement has revolutionized the agricultural sector, empowering farmers with the ability to optimize their operations and maximize their yields through data-driven insights provided by drones.

The utilization of agricultural drones with cameras has revolutionized sustainable farming practices by providing immediate insights, improving operational efficiency, reducing costs, and promoting sustainability within the industry. By automating various tasks through drones, farmers are able to manage their time more effectively, ultimately increasing crop yields while minimizing negative impacts on the environment. Cutting-edge drones such as the VoloDrone offer real-time updates on the progress of harvesting activities, while spraying drones present farmers with cost-efficient and effective solutions for applying pesticides and fertilizers [30]. These technological advancements are crucial for the progress of sustainable agriculture, as they not only help protect the environment but also contribute to the development of a healthier food system overall. By integrating drones equipped with cameras into farming practices, farmers can make informed decisions based on the data collected, leading to more sustainable and environmentally friendly agricultural operations.

Drones have become a significant innovation in the agricultural sector, particularly in the realm of spot spraying. This cutting-edge technology empowers farmers to precisely target specific areas within their fields for spraying, eliminating the necessity of applying chemicals across the entire field, a practice commonly associated with traditional crop dusters. Although drones do not completely supplant conventional crop dusters, they bring additional value to agricultural operations, especially in efficiently accessing confined areas. Nevertheless, the integration of drone spraying into farming practices necessitates a thorough examination of legal frameworks, and farmers must secure the requisite approvals before implementing such technology [59]. Figure 3 showcases the myriad applications of drones in smart farming, encompassing activities such as harvesting, spraying, mapping, and sensing. This visual representation underscores the versatility and potential of drones in revolutionizing modern agricultural practices, paving the way for enhanced efficiency and sustainability in the industry.

Agricultural mapping or land surveying drones are sophisticated, computer-controlled aerial devices equipped with high-resolution cameras, enabling farmers to gather detailed data about their crops. The use of drones for field mapping has revolutionized the agricultural industry, offering a more efficient and safer alternative to traditional methods. Drone-based surveys of agricultural land can collect data within minutes, whereas conventional approaches often take several hours and come at a higher cost [25]. For farmers embracing smart farming practices, agricultural robot-drones automate laborious and repetitive tasks, freeing up valuable time to focus on maximizing overall yields [24]. These robotdrones are commonly used for tasks such as harvesting and weed control [43]. With advanced technology, centralized control centers can dispatch technicians with a fleet of pickers as needed. Once deployed, robot-drones efficiently select, pick, and package only ripe fruits, preparing them for market. By operating continuously and providing real-time updates on harvesting progress, duration, quantity harvested, and market prices, robot-drones are driving a significant transformation in the agricultural sector.

Fig. 3 Application of drones in smart farming

Their ability to streamline operations and improve efficiency makes them an invaluable tool for modern farmers looking to optimize their productivity and profitability.

3.3 Cloud-Based Data Analytics for Decision-Making

The importance of acquiring accurate and comprehensive data in modern agriculture cannot be overstated when it comes to maximizing productivity. Thankfully, the introduction of AI-powered smart farming platforms has provided farmers with an effective tool to manage and analyze agricultural data efficiently [36]. These platforms utilize sophisticated machine learning algorithms and user-friendly interfaces to enable farmers to extract valuable insights and make well-informed decisions based on data. By combining data from various sources such as Sentinel and Landsat satellites, agronomic parameters, soil moisture measurements, historical yield datasets, weather information, and soil composition data, these platforms leverage AI and machine learning models to transform inputs into actionable analytics that are stored in cloud-based repositories. Cloud computing plays a crucial role in precision agriculture by supporting activities such as yield forecasting, weather monitoring, real-time data collection, remote surveillance, and financial calculations. This, in turn, enhances production efficiency, water management practices, and enables precise assessment of farms and fields. The integration of AI-driven technologies in agriculture is revolutionizing the way farmers approach data management and decision-making processes, ultimately leading to more sustainable and productive farming practices.

The emergence of cloud platforms has revolutionized the landscape of data storage, processing, and management across various industries, particularly those reliant on data generated by IoT devices. Cloud platforms play a crucial role in numerous IoT systems due to their scalability, adaptability, and cost-effectiveness in handling large volumes of data [18, 73].

Outlined below are some of the most prominent IoT cloud platforms, each accompanied by a brief overview:

1. Ubidots Platform: A user-friendly cloud-based platform specifically designed for collecting, processing, and visualizing IoT data.
2. AWS IoT: A secure and scalable cloud platform tailored to effectively manage, secure, and analyze IoT data.
3. Google Cloud IoT: A cloud-based platform that provides IoT data management and analysis tools compatible with a wide range of devices.
4. Microsoft Lab IoT Suite: A cloud-based IoT data management and analysis platform equipped with robust analytics capabilities.
5. IBM Watson IoT: An IoT data management and analysis platform that leverages cloud infrastructure and supports various devices and protocols.

These IoT cloud platforms offer a diverse range of features and functionalities to meet the specific needs of businesses and industries utilizing IoT technology.

4 Challenges and Opportunities

Despite the transformative potential of technology in agriculture, a major obstacle arises from the lack of technical expertise among farmers when it comes to utilizing technology-based equipment. Overcoming this hurdle requires a focus on user-centered design in the creation of agricultural systems, especially in terms of user interface design for digital tools and the availability of resources in languages that farmers understand. Moreover, issues related to the cost and quality of devices and sensors present challenges for small-scale farmers who are considering the adoption of advanced technologies. Ensuring that these technologies are both affordable and reliable is crucial for their widespread acceptance and use in agricultural practices. Additionally, the reliability of IoT solutions is critical, as any malfunctions or breakdowns in the system can have a significant impact on agricultural operations. Therefore, it is essential to address these concerns and develop robust IoT solutions that can withstand the demands of agricultural environments.

The effective management and security of data in IoT applications play a critical role, especially considering the widespread deployment of small devices that have limited processing and storage capabilities. In order to optimize network performance, it is imperative to implement efficient data management strategies. Achieving interoperability among different IoT devices is crucial for seamless operation, although synchronizing devices from various manufacturers can pose challenges. Furthermore, the growing amount of data generated by IoT devices calls for scalable approaches to data processing and storage. Cloud-based architectures present a viable solution by providing significant computing power and storage capacity to effectively handle this exponential growth.

The effectiveness of AI algorithms in decision-making is contingent upon the quality and quantity of data that is accessible for training the models. This underscores the significance of acquiring substantial amounts of highquality data. Moreover, the prompt delivery of messages and information is crucial for real-time IoT solutions, as it enables timely action to be taken. Despite the challenges associated with these factors, the integration of IoT and AI systems exhibits immense potential for automating and optimizing agricultural practices, thereby creating and capturing more value. Looking towards the future, the advancement of 5G technology is poised to significantly augment the capabilities of IoT systems. This will be achieved through the provision of faster internet speeds and the reduction of communication delays. Furthermore, the promise of IoT in agriculture will be further strengthened by the availability of cheaper and more widely accessible sensors and embedded technologies. Additionally, the continuous growth of artificial intelligence and advanced algorithms will continue to enhance the decision-making abilities of intelligent applications, thereby driving further innovation within the agricultural sector.

5 Conclusion

Traditional agricultural methods are struggling to keep up with the demands of modern food production and quality standards, especially with the rapid growth of the global population. In response to this challenge, researchers have turned to intelligent technologies, with a particular focus on IoT-based farming practices. These technologies offer predictability, automation, efficiency, and sustainability as solutions to enhance agricultural practices.

This chapter offers a comprehensive overview of digitization and automation in agriculture, with a specific emphasis on AI and IoT. In the realm of digitization and automation in agriculture, IoT and AI play a pivotal role, influencing every aspect of farming activity. From monitoring crops to deploying autonomous harvesting robots, these technologies are gradually transforming agriculture, simplifying tasks for farmers and speeding up processes.

Key conclusions from this chapter include:

- AI and IoT are significantly improving modern agriculture by automating and controlling farming operations.
- The possible applications of digitization and automation in agriculture include smart farm machinery, intelligent irrigation systems, pest and weed control mechanisms, and optimized fertilizer application.
- IoT and AI-enabled solutions are becoming more cost-effective and reliable, indicating an inevitable adoption across the agricultural sector.

- Continued investment in research and development of AI and IoT technologies is essential to fully unlock the potential of smart farming.
- Establishing a supportive ecosystem for smart farming will lead to a more sustainable future for agriculture and global food security.

References

1. Abioye, E.A., Abidin, M.S.Z., Mahmud, M.S.A., Buyamin, S., Ishak, M.H.I., Abd Rahman, M.K.I., Ramli, M.S.A.: A review on monitoring and advanced control strategies for precision irrigation. Comput. Electron. Agric. **173**, 105441 (2020)
2. Ahmad, F., Khaliq, A., Qiu, B., Sultan, M., Ma, J.: Advancements of spraying technology in agriculture. Technol. Agric. **33** (2021)
3. Ahmad, U., Alvino, A., Marino, S.: Solar fertigation: a sustainable and smart IoTbased irrigation and fertilization system for efficient water and nutrient management. Agronomy **12**(5), 1012 (2022)
4. Ali, A.M.: Development of an algorithm for optimizing nitrogen fertilization in wheat using GreenSeeker proximal optical sensor. Exp. Agric. **56**(5), 688–698 (2020)
5. Ali, A.M., Ibrahim, S.M., Hassany, W.M., El-Sadek, A.N.: Fixed time corrective dose fertilizer nitrogen management in wheat using atLeaf meter and leaf colour chart. Exp. Agric. **57**(4), 232–243 (2021)
6. Almalki, F.A., Soufiene, B.O., Alsamhi, S.H., Sakli, H.: A low-cost platform for environmental smart farming monitoring system based on IoT and UAVs. Sustainability **13**(11), 5908 (2021)
7. Alomar, B., Alazzam, A.: A smart irrigation system using IoT and fuzzy logic controller. In: 2018 Fifth HCT Information Technology Trends (ITT) (pp. 175–179). IEEE (2018)
8. AlZubi, A.A., Galyna, K.: Artificial intelligence and internet of things for sustainable farming and smart agriculture. IEEE Access (2023)
9. Arakeri, M.P., Kumar, B.V., Barsaiya, S., Sairam, H.V.: Computer vision based robotic weed control system for precision agriculture. In: 2017 International Conference on Advances in Computing, Communications and Informatics (ICACCI), pp. 1201–1205. IEEE (2017)
10. Bachuwar, V.D., Shligram, A.D., Deshmukh, L.P.: Monitoring the soil parameters using IoT and Android based application for smart agriculture. In: AIP Conference Proceedings, vol. 1989, no. 1. AIP Publishing (2018)
11. Chen, C.J., Huang, Y.Y., Li, Y.S., Chang, C.Y., Huang, Y.M.: An AIoT based smart agricultural system for pests detection. IEEE Access **8**, 180750–180761 (2020)
12. Colizzi, L., Caivano, D., Ardito, C., Desolda, G., Castrignanò, A., Matera, M., Shi, H.: Introduction to agricultural IoT. In: Agricultural Internet of Things and Decision Support for Precision Smart Farming, pp. 1–33. Academic Press (2020)
13. Dankhara, F., Patel, K., Doshi, N.: Analysis of robust weed detection techniques based on the Internet of Things (IoT). Procedia Comput. Sci. **160**, 696–701 (2019)
14. De Abreu, C.L., van Deventer, J.P.: The application of artificial intelligence (AI) and internet of things (IoT) in agriculture: a systematic literature review. In: Southern African Conference for Artificial Intelligence Research, pp. 32–46. Springer, Cham (2022)
15. El-Beltagi, H.S., Basit, A., Mohamed, H.I., Ali, I., Ullah, S., Kamel, E.A., Ghazzawy, H.S.: Mulching as a sustainable water and soil saving practice in agriculture: a review. Agronomy **12**(8), 1881 (2022)
16. Hafeez, A., Husain, M.A., Singh, S.P., Chauhan, A., Khan, M.T., Kumar, N., Soni, S.K.: Implementation of drone technology for farm monitoring & pesticide spraying: a review. Inform. Process. Agric. (2022)
17. Islam, N., Rashid, M.M., Wibowo, S., Xu, C.Y., Morshed, A., Wasimi, S.A., Rahman, S.M.: Early weed detection using image processing and machine learning techniques in an Australian chilli farm. Agriculture **11**(5), 387 (2021)

18. Jagatheesan, M., Janaki, G.: Weather monitoring system using IoT for smart farming. ECS Trans. **107**(1), 17439 (2022)
19. Javaid, M., Haleem, A., Singh, R.P., Suman, R.: Enhancing smart farming through the applications of agriculture 4.0 technologies. Int. J. Intell. Netw. **3**, 150–164 (2022)
20. Kakamoukas, G., Sarigiannidis, P., Maropoulos, A., Lagkas, T., Zaralis, K., Karaiskou, C.: Towards climate smart farming—a reference architecture for integrated farming systems. In: Telecom, vol. 2, no. 1, pp. 52–74. MDPI (2021)
21. Karunathilake, E.M.B.M., Le, A.T., Heo, S., Chung, Y.S., Mansoor, S.: The path to smart farming: innovations and opportunities in precision agriculture. Agriculture **13**(8), 1593 (2023)
22. Khan, R., Dhingra, N., Bhati, N.: Role of artificial intelligence in agriculture: a comparative study. In: Transforming management with AI, big-data, and IoT, pp. 73–83. Springer International Publishing, Cham (2022)
23. Kleinschroth, F., Banda, K., Zimba, H., Dondeyne, S., Nyambe, I., Spratley, S., Winton, R.S.: Drone imagery to create a common understanding of landscapes. Landsc. Urban Plan. **228**, 104571 (2022)
24. Krishnan, A., Swarna, S.: Robotics, IoT, and AI in the automation of agricultural industry: a review. In: 2020 IEEE Bangalore Humanitarian Technology Conference (B-HTC), pp. 1–6. IEEE (2020)
25. Kulbacki, M., Segen, J., Knieć, W., Klempous, R., Kluwak, K., Nikodem, J., Serester, A.: Survey of drones for agriculture automation from planting to harvest. In: 2018 IEEE 22nd International Conference on Intelligent Engineering Systems (INES), pp. 000353000358. IEEE (2018)
26. Kumar, N., Sharma, B.: Opportunities and challenges with WSN's in smart technologies: a smart agriculture perspective. In: Handbook of wireless Sensor Networks: Issues and Challenges in Current Scenario's, pp. 441–463 (2020)
27. Kumar, R., Kumar, A., Gupta, M.K., Yadav, J., Jain, A.: Solar tree-based water pumping for assured irrigation in sustainable Indian agriculture environment. Sustain. Prod. Consumption **33**, 15–27 (2022)
28. Lavanya, G., Rani, C., GaneshKumar, P.: An automated low cost IoT based fertilizer intimation system for smart agriculture. Sustain Comput.: Inform. Syst. **28**, 100300 (2020)
29. Lee, C.L., Strong, R., Dooley, K.E.: Analyzing precision agriculture adoption across the globe: a systematic review of scholarship from 1999–2020. Sustainability **13**(18), 10295 (2021)
30. Li, H., Quan, L., Guo, Y., Pi, P., Shi, Y., Lou, Z., Chen, Q.: Improving agricultural robot patch-spraying accuracy and precision through combined error adjustment. Comput. Electron. Agric. **207**, 107755 (2023)
31. Lin, F.T., Kuo, Y.C., Hsieh, J.C., Tsai, H.Y., Liao, Y.T., Lee, H.C.: A selfpowering wireless environment monitoring system using soil energy. IEEE Sens. J. **15**(7), 3751–3758 (2015)
32. Ma, Y.Q., Sun, X.: Intelligent agricultural machinery equipment and technology. Agric. Equip. Technol. **46**(01), 4–6 (2020)
33. Mahmud, M.S., Zahid, A., He, L., Martin, P.: Opportunities and possibilities of developing an advanced precision spraying system for tree fruits. Sensors **21**(9), 3262 (2021)
34. Mandapuram, M., Mahadasa, R., Surarapu, P.: Evolution of smart farming: integrating IoT and AI in agricultural engineering. Glob. Discl. Econ. Bus. **8**(2), 165–178 (2019)
35. Maraveas, C., Piromalis, D., Arvanitis, K.G., Bartzanas, T., Loukatos, D.: Applications of IoT for optimized greenhouse environment and resources management. Comput. Electron. Agric. **198**, 106993 (2022)
36. Mendes, J.A.J., Carvalho, N.G.P., Mourarias, M.N., Careta, C.B., Zuin, V.G., Gerolamo, M.C.: Dimensions of digital transformation in the context of modern agriculture. Sustain. Prod. Consumption **34**, 613–637 (2022)
37. Mohammed, S.W., Soora, N.R., Polala, N., Saman, S.: Smart water resource management by analyzing the soil structure and moisture using deep learning. In: IOT with Smart Systems: Proceedings of ICTIS 2022, vol. 2, pp. 709–719. Springer Nature Singapore, Singapore (2022)
38. Monteiro-Silva, F., Jorge, P.A., Martins, R.C.: Optical sensing of nitrogen, phosphorus and potassium: a spectrophotometrical approach toward smart nutrient deployment. Chemosensors **7**(4), 51 (2019)

39. Monteleone, S., Moraes, E.A.D., Tondato de Faria, B., Aquino Junior, P.T., Maia, R.F., Neto, A.T., Toscano, A.: Exploring the adoption of precision agriculture for irrigation in the context of agriculture 4.0: the key role of internet of things. Sensors **20**(24), 7091 (2020)
40. Nawandar, N.K., Satpute, V.R.: IoT based low cost and intelligent module for smart irrigation system. Comput. Electron. Agric. **162**, 979–990 (2019)
41. Nayagam, M.G., Vijayalakshmi, B., Somasundaram, K., Mukunthan, M.A., Yogaraja, C.A., Partheeban, P.: Control of pests and diseases in plants using IoT technology. Measure. Sens. **26**, 100713 (2023)
42. Obaideen, K., Yousef, B.A., AlMallahi, M.N., Tan, Y.C., Mahmoud, M., Jaber, H., Ramadan, M.: An overview of smart irrigation systems using IoT. Energy Nexus **7**, 100124 (2022)
43. Oliveira, L.F., Moreira, A.P., Silva, M.F.: Advances in agriculture robotics: a state-of-the-art review and challenges ahead. Robotics **10**(2), 52 (2021)
44. Onishi, Y., Yoshida, T., Kurita, H., Fukao, T., Arihara, H., Iwai, A.: An automated fruit harvesting robot by using deep learning. Robomech J. **6**(1), 1–8 (2019)
45. Park, J.K., Park, E.Y.: Monitoring method of movement of grazing cows using cloud-based system. ECTI Trans. Comput. Inform. Technol. **15**(1), 24–33 (2021)
46. Patel, P., Patel, Y., Patel, U., Patel, V., Patel, N., Oza, P., Patel, U.: Towards automating irrigation: a fuzzy logic-based water irrigation system using IoT and deep learning. Model. Earth Syst. Environ. **8**(4), 5235–5250 (2022)
47. Patil, N., Khairnar, V.D.: Smart farming system using IoT and cloud. In: Computer Networks and Inventive Communication Technologies: Proceedings of Fourth ICCNCT 2021, pp. 215–232. Springer Singapore (2022)
48. Penchalaiah, N., Nelson Emmanuel, J., Suraj Kamal, S., Lakshmi Narayana, C.V.: IoT based smart farming using thingspeak and MATLAB. In: ICCCE 2020: Proceedings of the 3rd International Conference on Communications and Cyber Physical Engineering, pp. 1273–1295. Springer Nature Singapore, Singapore (2020)
49. Pham, H.G., Chuah, S.H., Feeny, S.: Factors affecting the adoption of sustainable agricultural practices: findings from panel data for Vietnam. Ecol. Econ. **184**, 107000 (2021)
50. Postolache, S., Sebastião, P., Viegas, V., Postolache, O., Cercas, F.: IoT-based systems for soil nutrients assessment in horticulture. Sensors **23**(1), 403 (2022)
51. Pratama, H., Yunan, A., Candra, R.A.: Design and build a soil nutrient measurement tool for citrus plants using NPK soil sensors based on the internet of things. Brilliance: Res. Artif. Intell. **1**(2), 67–74 (2021)
52. Rahman, A., Budianta, D.: Data warehouse design for soil nutrients with IoT based data sources. In: 2019 International Conference on Informatics, Multimedia, Cyber and Information System (ICIMCIS), pp. 181–186. IEEE (2019)
53. Razfar, N., True, J., Bassiouny, R., Venkatesh, V., Kashef, R.: Weed detection in soybean crops using custom lightweight deep learning models. J. Agric. Food Res. **8**, 100308 (2022)
54. Saba, T., Rehman, A., AlGhamdi, J.S.: Weather forecasting based on hybrid neural model. Appl. Water Sci. **7**, 3869–3874 (2017)
55. Saha, H.N., Chakraborty, S., Roy, R.: Integration of RFID and sensors in agriculture using IOT. In: AI, Edge and IoT-Based Smart Agriculture, pp. 361–372. Academic Press (2022)
56. Salem, H.M., Muñoz-García, M.Á., Rodríguez, M.G.: Soil physical properties and soil water tension monitoring by wireless sensor network after reservoir and minimum tillage practices. Soil Res. **59**(3), 309–317 (2020)
57. Salem, H.M., Schott, L.R., Piaskowski, J., Chapagain, A., Yost, J.L., Brooks, E., Johnson-Maynard, J.: Evaluating intra-field spatial variability for nutrient management zone delineation through geospatial techniques and multivariate analysis. Sustainability **16**(2), 645 (2024)
58. Salem, H.M., Valero, C., Muñoz, M.Á., Rodríguez, M.G., Silva, L.L.: Shortterm effects of four tillage practices on soil physical properties, soil water potential, and maize yield. Geoderma **237**, 60–70 (2015)
59. Santini, A., Bartolini, E., Schneider, M., de Lemos, V.G.: The crop growth planning problem in vertical farming. Eur. J. Oper. Res. **294**(1), 377–390 (2021)

60. Saraf, S.B., Gawali, D.H.: IoT based smart irrigation monitoring and controlling system. In: 2017 2nd IEEE International Conference on Recent Trends in Electronics, Information & Communication Technology (RTEICT), pp. 815–819. IEEE (2017)
61. Saranya, K., Dharini, P.U., Darshni, P.U., Monisha, S.: IoT based pest controlling system for smart agriculture. In: 2019 International Conference on Communication and Electronics Systems (ICCES), pp. 1548–1552. IEEE (2019)
62. Singh, D.K., Sobti, R., Jain, A., Malik, P.K., Le, D.N.: LoRa based intelligent soil and weather condition monitoring with internet of things for precision agriculture in smart cities. IET Commun. **16**(5), 604–618 (2022)
63. Teh, D., Rana, T.: The use of internet of things, big data analytics and artificial intelligence for attaining UN's SDGs. In: Handbook of Big Data and Analytics in Accounting and Auditing, pp. 235–253. Springer Nature Singapore, Singapore (2023)
64. Torky, M., Hassanein, A.E.: Integrating blockchain and the internet of things in precision agriculture: analysis, opportunities, and challenges. Comput. Electron. Agric. **178**, 105476 (2020)
65. Varriale, V., Cammarano, A., Michelino, F., Caputo, M.: The role of digital technologies in production systems for achieving sustainable development goals. Sustain. Prod. Consump. (2024)
66. Verma, K., Chandnani, N., Bhatt, G., Sinha, A.: Internet of Things and Smart Farming. Internet of Things and Its Applications, pp. 283–303 (2022)
67. Vijayalakshmi, B, Ramkumar, C, Niveda, S, Pandian, S.C.: Smart pest control system in agriculture. In: 2019 IEEE International Conference on Intelligent Techniques in Control, Optimization and Signal Processing (INCOS), pp. 1–4. IEEE (2019)
68. Wang, X., Jannesari, V.: Towards a crop pest control system based on the Internet of Things and fuzzy logic. Telecommun. Syst. 1–13 (2024)
69. Wang, Z.Q., Yun, Y.L., Qin, Z.Z.: Design and test of data collector for agricultural machine operation parameters based on internet of things. Agric. Mech. **42**(01), 75–79 (2020)
70. Wu, X., Aravecchia, S., Lottes, P., Stachniss, C., Pradalier, C.: Robotic weed control using automated weed and crop classification. J. Field Robot. **37**(2), 322340 (2020)
71. Xiang, X., Li, Q., Khan, S., Khalaf, O.I.: Urban water resource management for sustainable environment planning using artificial intelligence techniques. Environ. Impact Assess. Rev. **86**, 106515 (2021)
72. Yazdinejad, A, Zolfaghari, B, Azmoodeh, A, Dehghantanha, A, Karimipour, H, Fraser, E., Duncan, E.: A review on security of smart farming and precision agriculture: security aspects, attacks, threats and countermeasures. Appl. Sci. **11**(16), 7518 (2021)
73. Zerihun, B.M., Olwal, T.O., Hassen, M.R.: Design and analysis of IOT based modern agriculture monitoring system for real-time data collection. In: Computer Vision and Machine Learning in Agriculture, vol. 2, pp. 73–82. Singapore: Springer Singapore (2022)
74. Zhang, X., Zhang, J., Li, L., Zhang, Y., Yang, G.: Monitoring citrus soil moisture and nutrients using an IoT based system. Sensors **17**(3), 447 (2017)

Towards Sustainable Living: Eco-Friendly Design and Environmental Impacts of IoT Home Automation Systems

Subrata Paul, Anirban Mitra, Piyal Roy, Amitava Podder, and Saptarshi Kumar Sarkar

Abstract In an era marked by increasing environmental concerns and the rapid advancement of technology, the quest for sustainable living has become more critical than ever. At the forefront of this movement are IoT (Internet of Things) home automation systems, which promise not only convenience but also significant opportunities for environmental stewardship. These systems leverage interconnected devices, sensors, and data analytics to optimize resource usage, reduce energy consumption, and minimize environmental impact in residential spaces. This chapter explores the integration of sustainability principles and environmental considerations into the design and implementation of IoT-based home automation systems. Beginning with an overview of IoT technology and its applications in home automation, the chapter emphasizes the importance of addressing environmental impacts in the development of IoT solutions. Understanding the environmental implications of traditional home automation systems sets the stage for discussing sustainable design principles, including energy efficiency, resource conservation, and lifecycle analysis. Key components of eco-friendly IoT home automation systems are outlined, with a focus on selecting energy-efficient devices and incorporating renewable energy sources. Energy management and optimization strategies, such as smart scheduling and adaptive control, are examined for their potential to reduce energy consumption and enhance sustainability. The chapter also delves into data analytics techniques for environmental monitoring, demonstrating how real-time data analysis can inform resource usage and promote sustainability. Through case studies and practical applications, the environmental benefits and energy savings associated with sustainable IoT home automation systems are highlighted. Lastly, the chapter addresses challenges and future directions, calling for continued innovation and collaboration to advance sustainability in IoT technology.

S. Paul (✉) · P. Roy · A. Podder · S. K. Sarkar
Department of CSE-AI, Brainware University, Barasat, West Bengal, India
e-mail: sp.cs@brainwareuniversity.ac.in

A. Mitra
Department of CSE, Amity University, Newtown, Kolkata, West Bengal, India

B. K. Mishra and A. Rocha (eds.), *Future of AI, IoT, and Sustainability*, Information Systems Engineering and Management 23,
https://doi.org/10.1007/978-3-031-76286-4_5

Keywords Internet of Things · Home automation systems · Sustainibility · Data Analytics · Energy management

1 Introduction

IoT, or the Internet of Things, is an interconnected system of real-world items or "things" incorporated with sensors, software, and various additional technologies that communicate and share data with different systems and devices via the internet. This technology allows elements to be managed and controlled from home, resulting in improved efficiency, simplicity, and digitization in a variety of fields, involving home automation.

In home automation, IoT technology enables different household gadgets and systems to interact with one another, collect data, and be managed remotely via a centralised hub or smartphone application. Some prevalent IoT applications in home automation involve IoT-enabled light bulbs and lighting systems that can be regulated remotely via smartphone apps or voice commands. Customers may modify brightness, colour, and schedules, resulting in energy savings and increased ambiance. IoT-connected thermostats acquire customer preferences and modify temperature settings consequently to provide the greatest convenience and save money on energy. They can be regulated online and provide energy consumption data. IoTbased safety devices include smart cameras, motion sensors, door/window sensors, and smart locks. Users may track their residence in real time, obtain alerts, and gain remote access to it. IoT technology allows everyday appliances like refrigerators, ovens, washing machines, and dishwashers to connect to the internet [1]. Users may remotely track appliance status, obtain notifications, and control settings, increasing simplicity and energy efficiency. IoT devices improve household entertainment encounters by allowing for effortless integration as well as management of audio/video systems, media streaming services, and smart TVs via a centralised structure or using voice commands. IoT devices are capable of tracking health parameters such as heart rate, blood pressure, and sleep patterns, offering helpful data to medical professionals and allowing for remote monitoring of patients for elderly individuals and those with persistent illnesses. IoT-based energy management systems track consumption of energy in real time, identifying ineffectiveness and optimising consumption habits to minimise utility expenses and impact on the environment. IoT technology transforms home automation by increasing comfort, ease, environmental sustainability, and security, as well as establishing the way to subsequent innovative services and applications [2].

The concept of sustainability refers to satisfying current needs without jeopardising the capacity of future generations to satisfy their own. It includes environmental, social, and economic dimensions, with the goal of striking an equilibrium among human endeavours and the natural environment in order to ensure everyone's longterm health and happiness. Sustainability is important in the setting of IoT-based solutions because it has the ability to address a wide range of environmental and societal issues while encouraging effectiveness as well as responsible utilisation of resources. Sustainability connects with IoT in the following ways:

- Resource Efficiency: IoT technologies can help make better use of natural assets like energy, water, and materials. Smart energy management systems, for instance, can optimise constructing consumption of electricity while reducing waste and carbon emissions. In a similar way smart irrigation system in agriculture are capable of tracking soil moisture levels and weather forecasts and provide water exactly whenever and wherever it is required, thereby preserving water resources.
- Environmental Monitoring: IoT sensors and devices can gather real-time information about environmental variables such as air quality, water quality, and biodiversity. This information might be used for monitoring the environment and management, such as identifying origins of pollution, mitigating hazards to the environment, and protecting ecosystems.
- Sustainable Transportation: IoT-based solutions may encourage environmentally friendly travel by optimising traffic flow, improving transportation management, and enabling the use of electric automobiles. For example, smart traffic management systems might decrease congestion and emission levels by overseeing traffic signals and supplying drivers with real-time traffic information.
- Circular Economy: IoT technologies can help with the shift to a business model that is circular by improving resource tracking, surveillance, and management across their entire lifecycle. For instance, IoT-enabled asset tracking systems may enhance product and material accountability, rendering them simpler to recycle, refurbish, or repurpose when they reach the conclusion of their lifespan.
- Social Impact: IoT-based solutions can benefit society by tackling issues that include healthcare access, education, and responding to disasters. For example, remote patient monitoring systems may enhance the quality of healthcare and convenience, particularly in areas where services are scarce. Similarly, IoT-enabled emergency management platforms can improve warning systems and collaboration during emergencies, lowering risks and improving resilience.

Sustainability is a crucial consideration in the development and deployment of IoT-based solutions, as they have the potential to contribute significantly to environmental protection, resource conservation, and social equity. By integrating sustainability principles into IoT design, implementation, and operation, we can create more resilient, inclusive, and environmentally friendly systems for the benefit of current and future generations [3, 4].

2 Environmental Impacts of Traditional Home Automation

Traditional home automation systems, although offering consumers with comfort and ease, may additionally have a substantial impact on energy usage and the environment. Numerous classic home automation systems depend upon networked gadgets and sensors that are continually in communication with one another as well as centralised hubs or servers. This continual interaction consumes energy, which might lead to higher electricity consumption. Furthermore, certain automated equipment, for instance smart thermostats or lighting systems, might demand backup electricity even if not in use, which increases energy usage. Metals, polymers, and rare earth elements are needed to manufacture home automation systems, that may result in serious environmental consequences that include destruction of habitat, contamination, and release of greenhouse gases. Furthermore, when technology evolves and new capabilities are added, previous gadgets may become outdated, resulting in electronic waste generation. To reduce harm to the environment, electronic waste must be properly disposed of and recycled. Many classic home automation systems use platforms that are cloud-based or data centres to maintain as well as analyse sensor and device data. These data centres need massive quantities of energy for cooling and power servers, which contributes to carbon emissions and destruction of the environment. Furthermore, the transfer of information among devices and the cloud consumes energy, which raises the system's overall effect on the environment.

Traditional home automation systems frequently capture confidential data about owners' behaviour, choices, and everyday activities. While this information can be utilised to optimise energy use and enhance the user experience, it also poses issues regarding confidentiality. Unauthorised utilisation of private data not merely violates the right to privacy, but it may additionally result in security violations and the possibility of exploitation of information. In some circumstances, typical home automation systems can become incompatible, forcing homeowners to operate various gadgets using several specialised platforms or protocols. The necessity for various communication protocols and duplicated equipment can result in inconsistencies, greater complexity, and greater electrical usage.

In order to reduce the energy expenditure and environmental repercussions, manufacturers, governments, and consumers need to think about principles of sustainable design, resource-efficient technology, and ethical end-of-life management techniques. This involves developing devices with minimal standby energy consumption, optimising protocols for communications for lowering data transmission overhead, fostering interoperability and standardisation, and implementing strong data privacy and security safeguards. Furthermore, raising public consciousness regarding the negative environmental effects of home automation systems can assist consumers in making better educated decisions and implementing sustainable behaviours [5].

2.1 Addressing Environmental Concerns: The Urgent Call for Sustainable Alternatives

The environmental consequences of standard home automation systems highlight the urgent requirement for sustainable solutions that prioritise resource effectiveness, conservation of energy, and responsibility for the environment. There are multiple explanations when environmentally friendly options are critical to mitigating these impacts.

i. Reduced Energy Consumption: Sustainable solutions prioritise electricity efficient design, inexpensively elements, and clever control algorithms. Sustainable solutions reduce the electricity demands of home automation systems, thereby lowering emissions of carbon and easing demand on energy infrastructure.
ii. Minimising Resource Use: Sustainable options seek to reduce resource consumption across the home automation product lifetime, from manufacture to disposal. This involves employing reused supplies, designing for long-term durability, and establishing ethical end-of-life management practices including recycle and refurbishing. By limiting resource extraction and waste output, alternative methods help to conserve natural resources and reduce environmental deterioration.
iii. Encouraging Renewable Energy Incorporation: Sustainable alternatives frequently include elements which allow interaction with energy from renewable sources which include solar power. Sustainable home automation systems might decrease carbon emissions and reliance on fossil fuels by allowing households to produce their own electricity using clean, renewable resources.
iv. Improving Interoperability and Compatibility: Sustainable solutions prioritise openness and compatibility among diverse devices and systems, decreasing the requirement for exclusive platforms and minimising redundancy. Sustainable alternatives improve interactions and the transfer of information by fostering interoperability, resulting in increased efficiency and lower energy use.
v. Raising Awareness among Customers and Participation: Sustainable solutions allow consumers to arrive at more educated decisions about the harmful effects of home automation systems. Sustainable alternatives encourage consumers to prioritise environmentally conscious options and activities by raising knowledge about the significance of sustainability and giving accessible details regarding product lifetime consequences. vi. Promoting Creative thinking and Partnerships: Sustainable alternatives encourage industry players to innovate and collaborate, resulting in the creation of novel technologies, operating models, and regulations that prioritise sustainability. They also help to ensure permanent environmental durability and viability by encouraging continual development and developing an environmental responsibility culture.

Sustainable alternatives to typical home automation systems serve an important role in reducing the impact on the environment, increasing energy conservation, and furthering sustainable development. By embracing sustainable design principles, prioritizing resource efficiency, and fostering consumer engagement, sustainable alternatives can help create a more environmentally friendly and socially responsible future for home automation [6].

3 Design Principles

Sustainable design concepts refer to an integrated strategy to developing goods, systems, and environments that reduce the impact on the environment, save resources, as well as encourage resilience over time. These concepts incorporate factors such as energy efficiency, resource conservation, and lifespan analysis to guarantee that architecture judgements prioritise sustainability from beginning to end. The most important concepts will be addressed throughout this section.

Energy efficiency is a fundamental concept which concentrates on reducing energy usage across the lifecycle of an item or system. These approaches in environmentally friendly design entail maximising energy consumption during the manufacture, operation, and disposing stages. This consists of developing goods with low-power components, applying effective monitoring algorithms, and making the best possible utilisation of renewable energy resources. Energy-efficient design decreases emissions of carbon and environmental effect while also lowering operational expenses and improving performance as a whole. Furthermore, resource conservation strives to reduce the consumption of renewable resources that include water, materials, and minerals while improving productivity and reducing waste. Sustainable design concepts support for solutions such as employing recyclable or renewable materials, planning for long-term durability, and incorporating closed loop or circular economy models. Sustainable design lowers environmental deterioration, enhances ecosystem health, and encourages responsible resource management by preserving resources and reducing waste output. Lifecycle analysis is additional organised approach for assessing the environmental effects of a product, process, or system over its full the lifespan, spanning raw material extraction to manufacture, distribution, usage, and disposal. To assess the entire environmental footprint and find areas for enhancement, LCA takes into account elements that include energy consumption, utilisation of resources, emission levels, and waste generation. Sustainable design concepts use lifecycle analysis in order to guide design decisions, prioritise environmental hotspots, and develop solutions for lowering environmental effect throughout the product's existence.

By incorporating these sustainable design concepts into manufacturing and process of decision-making, architects, technologists, and policymakers can produce solutions that reduce environmental impact, preserve assets, and encourage long term sustainability. Finally, sustainable design principles are critical for dealing with global issues that involve climate change, resource shortages, and deterioration of the environment, as well as promoting a more sustainable and fairer environment for all [7].

3.1 Applying Sustainable Design Principles to IoT Home Automation

Using principles of sustainable design in the construction of IoT-based home automation systems may end up resulting in systems that happen to be not only economical and practical, but also ecologically as well as socially responsible. These design ideas can be incorporated into the manufacturing process through a specific regulated technique.

- Energy efficiency: IoT devices are designed with low-power microcontrollers, sensors, and connectivity modules to reduce energy usage while in service. Wireless IoT networks employ optimised communication protocols that minimise information transfer complexity and energy consumption. Sleep modes and battery management strategies are used to reduce dormant electrical consumption while devices are not in use. Harvesting electricity approaches that involve solar power or kinetic energy are being investigated for sustaining IoT devices, minimising dependency on batteries and extending operational lifetimes.
- Resource Conservation: IoT devices are made with recycled or renewable resources in order to reduce extracting resources and environmental effect. IoT devices are built with flexible parts that allow for easy repair, upgrades, and reuse, extending their lifespans and minimising e-waste generation. Closed-loop or circular economy methodologies are used to recycle and repurposed components from terminal illness IoT devices, thus minimising waste and increasing the effectiveness of resources [8].
- Lifecycle Analysis: The ecological impact of IoT-based home automation systems is evaluated through lifecycle evaluations, which consider elements which include consumption of energy, consumption of resources, emission levels, and trash generation. LCA observations are utilised to guide choices regarding design, prioritise environmental areas of greatest need, and find ways to reduce environmental effect across the life cycle of the item. Feedback through lifecycle evaluations is used to promote continual enhancement of the development of goods, manufacturing procedures, and terminal illness management techniques, assuring sustainability over time performance.
- Interoperability and Compatibility: Industry norms and protocols improve openness and compatible among IoT devices and sites, eliminating complexity and the demand for exclusive solutions. Open-source innovation and cooperation should be encouraged to promote inventiveness, openness, and compatibility in the IoT ecosystem, allowing for a broader acceptance of environmentally friendly solutions.

Incorporating these environmentally friendly design concepts throughout the design of IoT-based home automation systems as seen in Fig. 1 allows programmers to develop technologies that additionally offer ease of use and effectiveness, but also

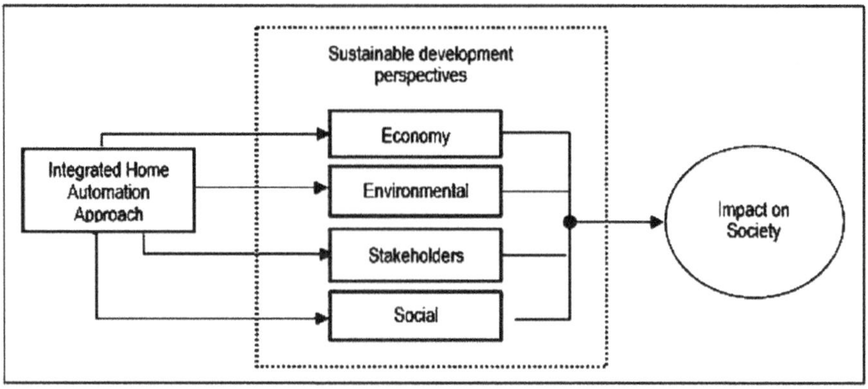

Fig. 1 Impact of integrated home automation to society [9]

reduce the ecological footprint, preserve assets, and encourage sustainable development over the long term. These activities help to establish a more resilient, fair, and ecologically sustainable future for smart homes and communities.

4 Components of an Eco-Friendly IoT Home Automation System

An Internet of Things-based home automation system is made up of several networked elements that operate together to allow for remote surveillance, oversight, and automating of household gadgets and services. This section describes the key elements necessary to operate such an arrangement.

Sensors are gadgets that sense and transform modifications to the environment onto electrical impulses. Sensors in a home automation system detect characteristics including temperature, humidity, motion, light, and air quality. Temperature, motion, humidity, and ambient light sensors are examples of frequently used sensor types. Actuators are devices which translate electrical impulses into physical actions. In home automation, actuators are employed for controlling and manipulating components such as lights, switches, motors, and valves. Examples of actuators include relays, servo motors, solenoid valves, and smart switches.

Microcontrollers are compact, low-power processing devices that function as the computational centre of an IoT system. They collect information gathered by sensors, process it, and then transmit instructions to actuators in accordance with established rules or feedback from humans. Microcontrollers may additionally link to the internet for interaction with different gadgets and cloud-based applications. Arduino, Raspberry Pi, and ESP8266/ESP32 are among the most popular microcontroller systems for home automation. Communication protocols specify the guidelines and norms for data transmission between IoT devices and systems. Wi-Fi, Bluetooth, Zigbee,

Z-Wave, and LoRaWAN are prominent wireless communication protocols employed in home automation for connecting sensors, actuators, and control hubs. Wired connections can also be made using Ethernet or Powerline Communication (PLC). The gateway or hub is a centralised point of control that manages and coordinates interactions among IoT devices in the home automation system. It could comprise physical elements such as microcontrollers, wireless modules, and networking interfaces, in addition to software components like firmware and application software. The gateway/hub may also offer protection, storage of data, and remote management via apps for smartphones or web interfaces [10, 11].

Cloud services provide remote administration, monitoring, and administration of IoT gadgets and information via the internet. Cloud platforms can store, analyse, and analyse enormous amounts of information from sensors, apply powerful automation algorithms, and integrate with a third-party application and service providers. Examples of cloud platforms for IoT include AWS IoT, Google Cloud IoT, Microsoft Azure IoT, and IoT platforms offered by device manufacturers. The user interface enables consumers to engage with and operate their IoT-based home automation system. This may involve smartphone apps, online portals, touchscreens, voice assistants, and physical control panels. The user interface includes functionality that includes device status monitoring, scheduling, rule-based automation, and manual device and setting control.

By combining these critical components, an IoT-based home automation system may enhance consumers' convenience, comfort, conservation of energy, and protection while also allowing for remote control and oversight of household equipment and systems.

4.1 Prioritizing Energy Efficiency: Key Considerations for Minimizing Environmental Impact

The use of devices that are energy-efficient, sensors, and communication protocols is critical for reducing the negative environmental effects of IoT-based home automation systems. Every component can improve the effectiveness of energy in several different methods.

4.1.1 Devices and Appliances

Choose Energy-Star-certified devices. Choose Energy Star-certified home appliances and gadgets, which meet the Environmental Protection Agency's (EPA) high environmentally friendly standards. Appliances with higher efficiency use less energy to operate, lowering electric consumption as well as decreasing their ecological impact.

Choose Low-Power Devices: Make the home automation system's gadgets and appliances as energy-efficient as possible. Seek for gadgets with environmentally friendly elements, power reserve modes, and resource-saving capabilities to reduce electricity usage, especially while inactive or not in use.

Consider Smart Energy Management: Use smart plugs, smart power strips, and energy tracking devices to monitor energy consumption and optimise associated device performance. Incorporate automated schedules, occupancy sensors, and energy-saving modes to minimise wasteful energy usage and boost general effectiveness.

4.1.2 Sensors

Optimise sensor placement: Sensors should be positioned intelligently to save energy while increasing effectiveness. Install motion sensors in areas with significant traffic and turn on illumination only when required, decreasing energy consumption.

Select low-power sensors: Employ sensors having power-efficient qualities to save energy while collecting precise information. Look for sensors with customisable sampling rates, sleep modes, and efficient communication protocols to save battery power while lowering the ecological impact.

Implement Sensor Fusion: Leverage sensor fusion approaches to aggregate data from many sensors, improving precision while consuming less energy. Incorporating complementing sensors and algorithms allows users to achieve greater effectiveness with fewer components while conserving energy and resources.

4.1.3 Communication Protocols

Choose low-power wireless technologies. Choose wireless communication protocols intended for low-power operation, which include Zigbee, Z-Wave, or Bluetooth Low Energy. The aforementioned protocols provide reliable connectivity while using low energy, which makes them perfect for battery-operated IoT devices.

Implement Sleep Modes and Power Management: Use sleep modes, duty cycling, and power management strategies to reduce the energy consumed while wireless transmission. Gadgets can adopt low-power modes while idle and wake up regularly for sending or acquire data, decreasing total energy consumption while maintaining connectivity.

Consider Range and Coverage: Select protocols for communication that provide adequate reach and protection for the home automation requirements while using the least amount of energy. Optimise the positioning of repeaters, routers, and gateways to guarantee consistent communication with a low power consumption [12].

By using environmentally friendly gadgets, sensors, and communication protocols, consumers may reduce the negative environmental effects associated with their IoT-based home automation systems while also benefiting from increased functionality, convenience, and productivity. Furthermore, making investments in energy-efficient gadgets might result in savings on expenses over time and help to create an environmentally friendly future.

5 Proposed System

As seen in Fig. 2, the Node MCU unit functions as the central microcontroller and controlling unit of the system, orchestrating the operation of various appliances. Users configure appliance behavior through a dedicated mobile application, which interprets commands either from voice input or switch configurations. Through a Wi-Fi-enabled wireless network, the mobile application transmits signals to the Node MCU unit, which hosts a built-in Wi-Fi module, facilitating communication with devices and receiving commands wirelessly. Upon receiving signals, the Node MCU utilizes a relay to activate or deactivate the connected appliances, thereby executing the user's commands. Physical connections link the Node MCU, relay, and the appliances, forming a cohesive system. A power supply unit ensures adequate power distribution to the microcontroller, relay, and final appliances, ensuring uninterrupted functionality. Additionally, a display unit provides users with visual feedback regarding the application's status.

Leveraging the Android operating system's open-source capabilities, the proposed system seamlessly integrates with mobile devices. The system capitalizes on the accessibility of built-in sensors within Android devices, enhancing functionality and user experience. Control implementation is versatile, accommodating user commands through two distinct modes within the application. In switch mode, users utilize radio buttons to control household appliances, with the application providing real-time status updates. Alternatively, voice mode enables users to issue commands vocally, leveraging the smartphone's microphone to capture audio inputs. These

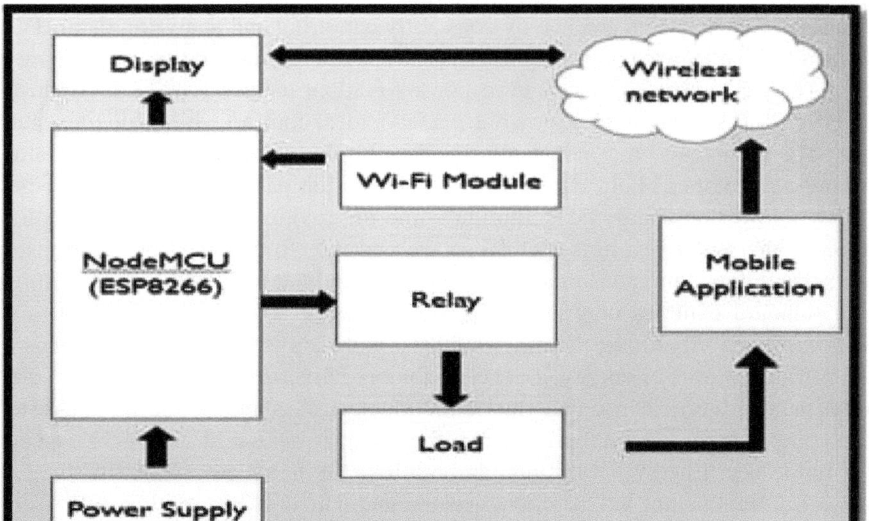

Fig. 2 Block diagram of the proposed system

inputs are then transmitted to Google servers, which process the data and return actionable responses. The system further analyzes the received string data, facilitating seamless execution of user commands and enhancing overall usability and convenience.

6 Energy Management and Optimization Strategies

Energy management approaches are critical in increasing the energy effectiveness as well as sustainability of IoT-based home automation systems. The present section discusses major energy management approaches such as smart scheduling, adaptive control, and energy harvesting. Smart scheduling entails programming gadgets and appliances to run according to established periods or customised regulations. Smart scheduling can assist minimise energy usage and utility expenses by timing device usage to coincide with instances of lower demand for energy or lower power rates. Smart thermostats, for example, may modify temperature settings based on occupancy and time of day, thereby increasing the system's cooling and heating efficiency. In a comparable manner automated lighting systems are capable of turning lights on and off based on an arrangement or the surrounding light levels, reducing wasteful energy consumption. Climatic forecasts, occupant trends, and customer preferences may all be considered by complex scheduling algorithms in order to constantly alter schedules and optimise consumption of energy in real time.

Adaptive control techniques entail continuously modifying device functioning depending on real-time sensor and environmental data. It can maximise energy effectiveness as well as convenience by constantly capturing and analysing data. HVAC systems with adaptive control algorithms, such example, can modify temperature set points as well as speeds of fans in response to indoor temperature, humidity levels, and occupant's status. This allows one preserve ideal comfort while reducing energy use. These systems may constantly modify luminance and colour temperatures according to natural light circumstances, occupation patterns, and customer preferences, lowering energy consumption while improving atmosphere and aesthetic appeal. Energy harvesting systems capture and transform renewable energy from the surroundings into consumable electrical energy that may supplement or substitute standard batteries or grid power sources. Solar power, kinetic energy (such as motion and vibration), thermal gradients, and radio frequency (RF) signals are all prevalent forms of energy collection. Energy harvesting devices, which include solar panels, piezoelectric generators, thermoelectric modules, and RF harvesters, can collect energies from these various sources and preserve it in cells or capacitor for future use. Energy harvesting can be utilised in home automation systems for powering lightweight IoT devices, sensors, and actuators, decreasing the need for throwaway batteries and increasing operating lifetime. As an instance, solar panels can power wireless sensors put outdoors, whereas motion sensors can create energies from user activity or environmental vibrations. Consumers may optimise the utilisation of energy in their IoT-based home automation systems by utilising smart scheduling, adaptive control, and energy harvesting approaches, resulting in lower energy usage, increased efficiency, and sustainability. These strategies allow adaptive and proactive administration of equipment and systems, resulting in increased comfort, convenience, and ecological responsibility [13].

6.1 Contribution to Energy Savings and Environmental Sustainability

Smart scheduling, adaptive control, and energy harvesting are three energy management technologies that greatly improve reducing energy consumption and environmental responsibility in IoT-based home automation systems.

Smart scheduling: Smart scheduling reduces the waste of energy throughout idle intervals by allowing equipment to work whenever necessary. For example, scheduling lights to switch off immediately while the daytime or while rooms are unattended saves electricity. Smart scheduling optimises consumption of energy by coordinating the device's activity with instances of reduced demand for energy or lower power rates. By arranging highly energy-intensive operations that include laundry, dishwashing, or HVAC operation during off-peak hours, households can save money on power and other utility bills. Smart scheduling decreases the load on the electricity network and the requirement for power generation based on fossil fuels by lowering energy use during high-demand intervals. This, in turn, reduces the release of carbon and helps to mitigate the effects of climate change and pollutants in the air.

Adaptive Control: These algorithms continuously track conditions in the environment, occupation patterns, and customer preferences and alter the device's settings in instantaneous fashion. This dynamic optimisation maximises energy economy while ensuring ideal comfort. For example, adaptive HVAC control can change temperature set points and fan speeds in accordance with the interior environment and occupant status, lowering energy consumption while maintaining satisfaction. It provides accurate oversight over the operation of the device, guaranteeing that energy is consumed only when and wherever it is required. Adaptive control reduces wasteful consumption of energy and improves global system effectiveness by adjusting device settings in response to real-time sensor feedback. These solutions provide more versatility and adaptability than standard preset scheduling or settings. This adaptability enables consumers to respond to varying circumstances outside, occupancy structures, and energy prices, thereby optimising their energy consumption in dynamic home contexts.

Energy Harvesting: With these methods, homeowners may produce renewable energy from natural sources that include sunlight, motion, and temperature differences. By utilising renewable energy, energy harvesting decreases dependency on fossil fuels and electrical power from the grid, resulting in decreased emission of carbon and environmental effect. It removes the necessity for rechargeable batteries or connection to the grid in low-power applications Internet of Things devices, sensors, and actuators. This decreases waste from batteries and the environmental effect of battery manufacturing, shipping, and removal. Energy harvesting increases the functional lifespan of IoT devices by offering a consistent source of renewable energy, minimising the need for battery exchanges and waste generation. This fosters long-term conservation of resources and a sustainable environment [14, 15].

These energy management solutions improve energy efficiency, decrease expenses, and ecological responsibility in IoT-based home automation systems by optimising energy consumption, eliminating waste, and encouraging the use of energy sources that are renewable. Smart scheduling, adaptive control, and energy harvesting strategies can help householders build more effective, at ease, and environmentally conscious living environments.

7 Integration of Renewable Energy Sources

Integrating renewable energy sources like solar panels and wind turbines with IoT home automation systems has various advantages, notably reduced dependency on fossil fuels, cheaper energy costs, and improved sustainability for the environment. Renewable energy sources can be connected with IoT home automation systems in a number of different ways:

- Solar panels produce electricity from sunshine and can be connected to the electrical system via on-grid devices. Solar panels in an IoT home automation system can create renewable energy throughout the day, which is subsequently utilised for lighting household gadgets, charge batteries, or send surplus electricity back towards the electrical grid for credit. Solar panels installed off-grid store additional electricity in batteries to utilise when solar energy production is low or there is a power disruption. Solar energy systems that operate off-grid in an IoT home automation system can supply additional power for important equipment and allow for selfsufficient operation even in remote or off-grid environments. IoT devices and sensors can be employed to evaluate solar panel performance, track energy generation, and optimise energy consumption in accordance with solar availability. Smart inverters, energy metres, and surveillance systems can offer immediate information on solar generation, battery storage phases, and residential energy use, allowing consumers to reap all of the advantages of solar energy.
- Wind Turbines: Small-scale windmills can supplement solar panels by producing electricity from wind energy, particularly in areas with abundant wind resources. In an IoT home automation system, turbines powered by wind can augment solar power and improve overall energy resiliency. Wind turbines, like solar panels, are capable of being integrated with IoT sensors for monitoring systems for monitoring wind speed, turbine performance, and energy generation. This information

can be utilised to optimise turbine functioning, change blade pitch angles, and maximise energy output dependent on the conditions of the wind.

- Integration into Energy Management Systems: Renewable sources of energy, which include solar panels and wind turbines, can be effortlessly linked into IoT-based energy management systems to improve energy usage, storage, and distribution. Smart control devices, energy administration software, and IoT platforms may coordinate the functioning of alternative power sources, energy storage systems, and residential gadgets, reducing grid dependency and increasing renewable energy personal consumption. Advanced algorithms for energy management can prioritize the utilization of renewable energy resources when they are available, store any surplus energy in batteries for future use, and dynamically adjust energy consumption based on real-time energy availability and user preferences. This integration not only enhances energy efficiency but also leads to reduced utility costs and contributes to environmental sustainability by maximizing the utilization of renewable energy.

Additionally, remote monitoring and maintenance systems based on IoT technology allow for the remote monitoring and management of renewable energy assets such as solar panels and wind turbines. Through real-time monitoring of performance metrics, detection of faults, and implementation of predictive maintenance algorithms, optimal operation can be ensured, potential issues can be identified early on, and downtime can be minimized, thereby maximizing the reliability and longevity of renewable energy systems [16].

By incorporating renewable energy sources like solar panels and wind turbines into IoT-enabled home automation systems, homeowners can effectively decrease their carbon footprint, decrease energy expenses, and enhance energy resilience, all while enjoying the advantages of a smart and sustainable home.

7.1 The Role of Renewable Energy Integration in Advancing Sustainability and Reducing Dependence on Conventional Power

The incorporation of renewable energy sources like solar panels and wind turbines into a typical household setting serves to bolster sustainability efforts while decreasing dependence on conventional power outlets.

- Decreased Carbon Emissions: Adoption of solar panels and wind turbines in household energy systems diminishes the need for electricity derived from fossil fuels like coal, natural gas, and oil. Solar panels utilize sunlight, a plentiful and clean renewable energy source, emitting no greenhouse gases or air pollutants. Meanwhile, wind turbines tap into wind energy, further diminishing reliance on fossil fuel-based electricity. With a greater reliance on renewable energy sources, households markedly lessen their carbon footprint, aiding in the fight against climate change and air pollution.

- Energy Autonomy and Resilience: By producing electricity onsite from renewable sources, households become less reliant on centralized power grids and external energy providers. Solar panels and wind turbines offer decentralized energy solutions, empowering households to locally and independently generate their own electricity. This self-sufficiency improves resilience against power outages, grid malfunctions, and energy supply disruptions, ensuring uninterrupted access to vital services and comfort during emergencies. Moreover, surplus energy produced from renewables can be stored in batteries or returned to the grid for compensation, offering households flexibility and economic advantages.
- Financial Savings and Long-Term Gains: While there are initial expenses associated with installing solar panels and wind turbines, the long-term benefits outweigh the upfront investment. Once installed, these renewable energy systems have minimal operational costs and require little maintenance compared to traditional power facilities. Over time, households can realize substantial savings on electricity bills, particularly as energy prices escalate and renewable technologies become more cost effective. Additionally, government incentives, tax breaks, and net metering initiatives may further incentivize the adoption of renewable energy, offering financial assistance and hastening the return on investment for homeowners.
- Environmental Responsibility: Integrating renewable energy sources reflects a commitment to environmental stewardship and sustainable lifestyles. By opting for clean, renewable energy over fossil fuels, households contribute to the preservation of natural resources, the protection of ecosystems, and the reduction of environmental harm. Renewable energy generation imposes minimal environmental impact compared to traditional power generation methods, which often entail resource extraction, pollution, habitat destruction, and ecosystem disturbance. As more households transition to renewable energy, they collectively make a substantial positive impact on global environmental sustainability and aid in the construction of a more resilient and equitable future for all.

The integration of renewable energy sources like solar panels and wind turbines elevates sustainability efforts, diminishes reliance on conventional power sources, and empowers households to shift toward cleaner, more resilient, and environmentally conscious energy systems.

7.2 Data Analytics for Environmental Monitoring

Data analytics techniques play a vital role in monitoring environmental conditions like indoor air quality, temperature, and energy usage across different environments, including residential, commercial, and industrial spaces. These techniques involve gathering, processing, analyzing, and interpreting data from sensors and devices to gain insights, detect patterns, trends, and anomalies, and make informed decisions for enhancing energy efficiency, comfort, and sustainability. This section discusses several common data analytics techniques used for environmental parameter monitoring.

Descriptive analytics involves summarizing and visualizing historical data to comprehend past trends and patterns in environmental conditions. This technique helps stakeholders understand baseline conditions, recognize seasonal fluctuations, and evaluate long-term trends. For instance, time-series charts, histograms, and heatmaps can visually represent trends in indoor temperature, air quality, and energy usage over time, assisting stakeholders in identifying patterns and irregularities.

Diagnostic analytics focuses on identifying the underlying causes of environmental issues or anomalies by analyzing historical data and relationships between variables. This technique aids stakeholders in understanding why certain environmental parameters deviate from expected norms. For example, correlation analysis can reveal relationships between indoor air quality and factors like ventilation rates, temperature, or occupancy levels, facilitating the diagnosis of potential sources of indoor air pollution or discomfort.

Predictive analytics employs statistical and machine learning models to forecast future trends and behaviors based on historical data. This enables stakeholders to anticipate changes in environmental conditions and take proactive measures to address them. Time-series forecasting models, for instance, can predict future indoor temperature, air quality, or energy usage based on past data and weather forecasts, helping stakeholders optimize HVAC settings and energy consumption.

Prescriptive analytics goes beyond predicting future outcomes to suggest actions or interventions for optimizing environmental conditions. This technique offers actionable insights and recommendations for enhancing energy efficiency, comfort,
18 and sustainability. For instance, optimization algorithms can recommend optimal HVAC settings, lighting schedules, or ventilation strategies based on real-time data and user preferences, aiding stakeholders in minimizing energy usage while maintaining indoor comfort levels.

Anomaly detection techniques identify abnormal or unexpected patterns in environmental data that may indicate equipment malfunctions, system failures, or unusual events. This helps stakeholders detect and diagnose issues in real-time, enabling timely interventions and corrective actions. Statistical methods, machine learning algorithms, and rule-based systems can be employed to detect anomalies in indoor temperature, air quality, or energy consumption data, alerting stakeholders to potential problems and preventing costly downtime or energy wastage.

By employing these data analytics techniques to monitor environmental conditions such as indoor air quality, temperature, and energy usage, stakeholders can acquire valuable insights, optimize system performance, and promote energy efficiency, comfort, and sustainability across various environments. These techniques facilitate informed decision-making, proactive management, and continual improvement, fostering healthier, more efficient, and sustainable living and working environments for occupants [17].

7.3 Harnessing Real-Time Data Analytics for Sustainable Resource Optimization and Continuous Improvement

Real-time data analytics is instrumental in enhancing resource utilization and uncovering avenues for advancing sustainability across diverse sectors, encompassing energy management, water conservation, waste reduction, and environmental monitoring. This is achieved through a structured approach:

- Energy Management: Real-time data analytics facilitates continuous monitoring of energy consumption patterns, equipment efficiency, and environmental factors like temperature, humidity, and occupancy. Analysis of this data enables stakeholders to pinpoint energy-intensive processes, equipment inefficiencies, and opportunities for savings. Automated responses, triggered by real-time energy analytics, such as adjusting HVAC settings or optimizing lighting schedules, curtail energy wastage and cut utility expenses. Predictive analytics models forecast future energy demands and suggest preemptive actions to optimize usage and mitigate peak demand.
- Water Conservation: Real-time data analytics allows for real-time monitoring of water consumption, leaks, and water quality metrics such as pH and turbidity. Analysis of this data aids in detecting leaks, inefficiencies, and avenues for water conservation. Automated responses driven by real-time water analytics, like adjusting irrigation schedules or optimizing water treatment processes, minimize water waste and preserve water resources. Predictive analytics models predict future water demands and recommend proactive measures to optimize usage and prevent shortages or contamination.
- Waste Reduction: Real-time data analytics facilitates the tracking of waste generation rates, recycling efforts, and waste composition in real-time. Analysis of this data identifies areas of high waste generation, recycling inefficiencies, and opportunities for waste reduction. Automated responses, guided by real-time waste analytics, such as optimizing waste collection routes or enhancing recycling programs, reduce waste generation and boost recycling rates. Predictive analytics models forecast future waste generation trends and propose proactive measures to improve waste management practices and reduce environmental impact.
- Environmental Monitoring: Real-time data analytics enables continuous monitoring of environmental factors such as air quality, temperature, humidity, and pollution levels. Analysis of this data helps identify sources of pollution, environmental risks, and avenues for mitigation. Automated responses, triggered by real-time environmental analytics, such as activating air purification systems or adjusting ventilation rates, safeguard public health and safety during environmental emergencies. Predictive analytics models forecast future environmental trends and recommend proactive measures to mitigate risks and enhance overall sustainability [18].

Real-time data analytics furnishes valuable insights, actionable recommendations, and automated responses to optimize resource utilization and foster sustainability enhancements. Leveraging real-time data analytics empowers stakeholders to achieve heightened efficiency, resilience, and sustainability across diverse applications and industries, thereby contributing to a more sustainable future for all.

8 Case Studies and Practical Applications

Within the present section two case studies will be discussed highlighting successful implementations of sustainable IoT home automation systems:

8.1 Smart Home with Renewable Energy Integration

This case study showcases a smart home equipped with renewable energy sources, energy-efficient devices, and IoT-based automation systems to achieve sustainability and energy independence.

The Features of this system include the following:

- Solar Panels: The smart home is equipped with rooftop solar panels to generate renewable electricity from sunlight.
- Energy Storage: Battery storage systems are installed to store excess solar energy for use during periods of low solar generation or grid outages.
- Energy-Efficient Devices: Energy-efficient appliances, smart thermostats, and LED lighting are used throughout the home to minimize energy consumption.
- IoT Automation: IoT sensors, controllers, and actuators are deployed to monitor energy usage, optimize HVAC settings, and automate lighting and appliance control based on occupancy and environmental conditions.
- Integration with Cloud Platform: Data from IoT devices are transmitted to a cloud-based platform for real-time monitoring, analysis, and optimization of energy usage.
- User Interface: Homeowners can access the smart home system through a smartphone app or web portal to monitor energy generation, consumption, and savings, as well as control devices remotely.

Benefits provided by these systems include the following:

- Reduced Energy Bills: By generating renewable energy onsite and optimizing energy usage through IoT automation, the smart home significantly reduces electricity bills.
- Energy Independence: The integration of renewable energy sources and energy storage systems enhances energy resilience and reduces reliance on grid electricity.

- Environmental Sustainability: The smart home minimizes carbon emissions, fossil fuel consumption, and environmental impact by utilizing clean, renewable energy sources and energy-efficient technologies.
- Enhanced Comfort and Convenience: IoT automation enhances comfort and convenience by adjusting temperature, lighting, and appliance settings based on user preferences and environmental conditions [19–21].

8.2 Smart Apartment Building with Energy Management System

This case study highlights a smart apartment building equipped with an energy management system (EMS) and IoT-based automation to optimize energy usage and promote sustainability among residents.

The features of this system include the following:

- Energy Monitoring: Each apartment unit is equipped with smart meters and sensors to monitor electricity, water, and gas usage in realtime.
- Centralized EMS: An EMS centrally monitors and controls energy usage across all apartment units, common areas, and building systems, such as HVAC, lighting, and elevators.
- Demand Response: The EMS implements demand response strategies to reduce peak energy demand, minimize utility costs, and alleviate strain on the electrical grid.
- Tenant Engagement: Residents are provided with access to a web portal or mobile app to track their energy usage, set energy-saving goals, and receive personalized recommendations for reducing consumption.
- Energy-Efficient Upgrades: The building undergoes energy-efficient upgrades such as LED lighting retrofits, HVAC system optimization, and insulation improvements to further reduce energy consumption.
- Renewable Energy Integration: Solar panels or wind turbines may be installed on the building's rooftop to generate renewable energy for onsite consumption.

Benefits provided by these system include the following:

- Lower Operating Costs: The EMS optimizes energy usage and reduces utility costs for building owners and tenants through demand response, energy efficiency measures, and renewable energy integration.
- Improved Tenant Satisfaction: Residents benefit from lower utility bills, increased comfort, and access to energy-saving tips and tools, enhancing overall satisfaction and quality of life.
- Environmental Impact: The smart apartment building reduces carbon emissions, energy consumption, and environmental footprint, contributing to sustainability goals and regulatory compliance.

- Scalability and Replicability: The success of the smart apartment building model can be replicated in other multifamily residential buildings, promoting widespread adoption of sustainable practices in urban living environments [22–24].

These case studies demonstrate the potential of sustainable IoT home automation systems to optimize resource usage, reduce environmental impact, and enhance quality of life for residents. By integrating renewable energy sources, energy-efficient technologies, and IoT-based automation, these systems offer a blueprint for creating smarter, more sustainable homes and communities.

8.3 Examining the Environmental, Energy, and User Impacts of Sustainable IoT Home Automation Systems

Let's analyze the environmental benefits, energy savings, and user experiences associated with the two deployments outlined in the case studies:

8.3.1 Smart Home with Renewable Energy Integration

Environmental Benefits: By generating renewable energy onsite from solar panels, the smart home significantly reduces its carbon footprint compared to traditional grid-powered homes. The use of renewable energy sources minimizes reliance on finite fossil fuels, conserving natural resources and reducing environmental degradation associated with resource extraction and combustion. The smart home mitigates air and water pollution by reducing emissions from fossil fuel-based power generation and minimizing waste associated with energy production and consumption.

Energy Savings: The integration of renewable energy sources and energy-efficient devices leads to significant reductions in electricity bills for homeowners, resulting in long-term cost savings and return on investment. IoT automation optimizes energy usage by adjusting HVAC settings, lighting, and appliance operation based on occupancy and environmental conditions, further reducing energy waste and enhancing efficiency.

User Experience: Residents experience greater comfort and control over their home environment through IoT automation, allowing them to customize settings and schedules to meet their preferences and lifestyle. Homeowners benefit from reduced energy bills, increased property value, and potential incentives or rebates for renewable energy installations, enhancing the financial appeal of sustainable home automation systems. Users feel empowered and environmentally conscious by living in a home that prioritizes sustainability, renewable energy, and energy efficiency, fostering a sense of pride and responsibility towards environmental stewardship.

8.3.2 Smart Apartment Building with Energy Management System

Environmental Benefits: The smart apartment building reduces environmental impact at scale by optimizing energy usage across multiple residential units, common areas, and building systems, amplifying the environmental benefits compared to individual homes. The deployment fosters a sense of community and collective responsibility towards sustainability among residents, encouraging collaboration and participation in energy-saving initiatives.

Energy Savings: The implementation of demand response strategies reduces peak energy demand and utility costs for building owners, resulting in significant energy savings and operational efficiencies. Energy-efficient upgrades and optimization measures lead to further reductions in energy consumption and operating expenses, improving the financial viability of the building and enhancing long-term sustainability.

User Experience: Residents enjoy lower utility bills, increased comfort, and improved living conditions in a smart and sustainable living environment, leading to higher tenant satisfaction, retention, and referrals. Tenants are empowered to monitor their energy usage, set conservation goals, and participate in energy-saving initiatives through user-friendly interfaces and personalized recommendations, fostering a sense of ownership and involvement in sustainability efforts. The smart apartment building enhances community resilience, livability, and attractiveness, attracting environmentally conscious residents and contributing to a positive social and environmental impact in the neighborhood.

The deployments of sustainable IoT home automation systems offer a range of environmental benefits, energy savings, and positive user experiences, demonstrating the potential for smart and sustainable living solutions to create value for homeowners, tenants, and communities alike.

9 Challenges and Future Directions

While sustainable IoT home automation systems offer significant advantages, several obstacles hinder their widespread adoption. One key challenge is the substantial initial costs involved in purchasing and installing renewable energy systems, energy-efficient devices, and IoT infrastructure. Many homeowners and building owners may find these upfront expenses prohibitive, despite recognizing the longterm savings and environmental benefits. Moreover, implementing these systems often demands technical expertise in renewable energy, building automation, and IoT technologies, which many consumers lack. The absence of standardized protocols and interoperability between various IoT devices and platforms further complicates matters, leading to compatibility issues and potential conflicts during integration.

Concerns over data privacy and security also present significant hurdles, as IoT devices collect and transmit sensitive information about occupants and their surroundings. Vulnerabilities in these devices and networks could expose homeowners to privacy breaches and cyber threats, eroding trust in smart home technologies. Regulatory barriers, such as zoning restrictions and permitting requirements, may impede the deployment of sustainable technologies in residential and commercial buildings, stifling innovation and investment.

Moreover, consumer perceptions of renewable energy and IoT technologies as unproven or risky investments contribute to resistance to adoption, particularly among those unfamiliar with the technology or unsure of its reliability. To address these challenges, education and outreach efforts are crucial to increasing awareness and understanding of sustainable IoT solutions. Collaboration among stakeholders, including governments, industry players, and advocacy groups, is essential to promoting investment, innovation, and adoption of these systems. Policy interventions, financial incentives, and technical standards can help overcome barriers and facilitate the transition to smart, sustainable living solutions.

9.1 Charting a Course for Future Sustainability

To propel the adoption of sustainable IoT home automation systems, various avenues for research, innovation, and collaboration can be explored. Efforts should focus on driving down costs associated with renewable energy technologies, energy-efficient devices, and IoT infrastructure, leveraging technological advancements and economies of scale. Financial incentives from governments and utilities, like rebates and tax credits, can help defray upfront expenses for homeowners and building owners. Standardization efforts led by industry consortia should aim to establish common frameworks for IoT devices and communication protocols, facilitating seamless integration and interoperability across different platforms.

Advancements in IoT technologies, including sensors, edge computing, and security protocols, are imperative to enhance performance, reliability, and data privacy. Edge computing capabilities can enable real-time data processing at the device level, reducing reliance on cloud services and improving responsiveness. Innovations in data encryption and privacy-preserving techniques can bolster security while ensuring user confidentiality. Human-centered design principles should guide the development of IoT systems, prioritizing user experience and accessibility.

Educational initiatives are crucial to raise awareness and empower consumers with the knowledge needed to adopt sustainable technologies effectively. Supportive policies and regulations from governments should incentivize innovation and sustainability in the built environment while addressing barriers to adoption and ensuring consumer protection. Additionally, research and development efforts should focus on improving the efficiency and affordability of renewable energy technologies, integrating them with smart grids and demand response programs to enable energy sharing and enhance resilience.

By fostering collaboration and investment in these areas, stakeholders can overcome existing challenges and unlock the full potential of sustainable IoT home automation systems, driving energy efficiency, environmental sustainability, and enhancing quality of life for residents and communities.

10 Conclusion

The advancement and acceptance of sustainable IoT home automation systems offer substantial potential for enhancing energy efficiency, environmental sustainability, and user satisfaction in both residential and commercial settings. By integrating renewable energy sources, energy-efficient devices, and IoT technologies, these systems offer notable environmental advantages, such as reducing carbon emissions, conserving resources, and mitigating pollution. Through resource optimization, demand response strategies, and IoT automation, sustainable IoT home automation systems can yield significant energy savings, cost reductions, and decreased reliance on fossil fuels. Users experience improved comfort, control, and contentment with personalized settings, real-time feedback, and remote management options. However, despite these benefits, several obstacles impede the widespread adoption of sustainable IoT home automation systems. Challenges include high initial costs, technical complexities, interoperability issues, data privacy concerns, regulatory barriers, and limited awareness and education. To overcome these hurdles and drive adoption, stakeholders must concentrate on cost reduction, standardization, technological advancements, data privacy and security, user-centric design and education, policy and regulatory support, advancements in renewable energy technologies, and integration with smart grids and microgrids. Collaboration among developers, policymakers, industry players, and consumers is crucial to fostering innovation, investment, and adoption of sustainable IoT home automation systems. A holistic approach that addresses technical, economic, regulatory, and societal factors while prioritizing user needs, environmental sustainability, and energy resilience is essential. These systems present a transformative opportunity to create smarter, more efficient, and more sustainable buildings and communities. By addressing key challenges, leveraging technological advancements, and promoting collaboration, stakeholders can fully unlock the potential of sustainable IoT solutions to promote a greener, healthier, and more resilient future.

At the juncture of technology and sustainability, it is imperative to prioritize environmental stewardship in IoT solutions development and deployment. The rapid expansion of IoT devices presents opportunities and challenges alike. Let's shift our focus towards sustainability, recognizing the environmental impacts of every decision made in IoT technology. Innovation should not compromise environmental integrity; thus, energy efficiency, renewable energy integration, and eco-friendly materials should be central in product design and manufacturing. Collective action is required to address environmental challenges. Collaboration across sectors—industry, academia, government, and civil society—is essential to sharing knowledge and best practices for developing sustainable IoT solutions. Governments must shape regulatory frameworks that promote energy efficiency and renewable energy adoption. Educating and empowering consumers to demand eco-friendly products and services can drive sustainability. Establishing metrics for measuring environmental impact is crucial for accountability and continuous improvement. Stakeholders must lead by example, demonstrating commitment to environmental sustainability through actions and initiatives. By prioritizing sustainability, we can harness technology's transformative power to create a greener, more resilient, and equitable world. Together, let's build a sustainable IoT ecosystem that fosters innovation, prosperity, and planetary well-being.

References

1. Almalki, F.A., Alsamhi, S.H., Sahal, R., Hassan, J., Hawbani, A., Rajput, N.S., Breslin, J.: Green IoT for eco-friendly and sustainable smart cities: future directions and opportunities. Mob. Netw. Appl. **28**(1), 178–202 (2023)
2. Saini, P., Gupta, L., Malhotra, S., Dias, I.: The role of sustainable IoT in presentday Life. In WSN and IoT, pp. 35–63. CRC Press (2024)
3. Stolojescu-Crisan, C., Crisan, C., Butunoi, B.P.: An IoT-based smart home automation system. Sensors **21**(11), 3784 (2021)
4. Vishwakarma, S.K., Upadhyaya, P., Kumari, B., Mishra, A.K.: Smart energy efficient home automation system using IoT. In 2019 4th International Conference on Internet of Things: Smart Innovation and Usages (IoT-SIU), pp. 1–4. IEEE (2019)
5. Urban, B., Roth, K., Harbor, C.D.: Energy savings from five home automation technologies: A scoping study of technical potential. Boston, MA, Fraunhofer USA Center for Sustainable Energy Systems (2016)
6. Greenblatt, J.B., Shaheen, S.: Automated vehicles, on-demand mobility, and environmental impacts. Curr. Sustain. Renew. Energy Rep. **2**, 74–81 (2015)
7. Nižetić, S., Šolić, P., Gonzalez-De, D.L.D.I., Patrono, L.: Internet of Things (IoT): opportunities, issues and challenges towards a smart and sustainable future. J. Clean. Prod. **274**, 122877 (2020)
8. Bibri, S.E.: The IoT for smart sustainable cities of the future: an analytical framework for sensor-based big data applications for environmental sustainability. Sustain. Cities Soc. **38**, 230–253 (2018)
9. Mahmood, Y., Kama, N., Azmi, A., Ya'acob, S.: An IoT based home automation integrated approach: impact on society in sustainable development perspective. Int. J. Adv. Comput. Sci. Appl. **11**(1), 240–250. (Figure 1) (2020)
10. Shah, S.K.A., Mahmood, W.: Smart home automation using IOT and its low cost implementation. Int. J. Eng. Manuf **10**(5), 28–36 (2020)
11. Harsha, S.S., Reddy, S.C., Mary, S.P.: Enhanced home automation system using internet of things. In: 2017 International Conference on I-SMAC (IoT in Social, Mobile, Analytics and Cloud) (I-SMAC), pp. 89–93. IEEE (2017)
12. Srisruthi, S., Swarna, N., Ros, G.S., Elizabeth, E.: Sustainable agriculture using eco-friendly and energy efficient sensor technology. In: 2016 IEEE International Conference on Recent Trends in Electronics, Information & Communication Technology (RTEICT), pp. 1442–1446. IEEE (2016)
13. Wang, X., Mao, X., Khodaei, H.: A multi-objective home energy management system based on internet of things and optimization algorithms. J. Build. Eng. **33**, 101603 (2021)
14. Godina, R., Rodrigues, E.M., Pouresmaeil, E., Matias, J.C., Catalão, J.P.: Model predictive control home energy management and optimization strategy with demand response. Appl. Sci. **8**(3), 408 (2018)

15. Liu, Y., Zhang, D., Gooi, H.B.: Optimization strategy based on deep reinforcement learning for home energy management. CSEE J. Power Energy Syst. **6**(3), 572–582 (2020)
16. Prakashraj, K., Vijayakumar, G., Saravanan, S., Saranraj, S.: IoT based energy monitoring and management system for smart home using renewable energy resources. Int. Res. J. Eng. Technol. **7**(2), 17901797 (2020)
17. Tipantuna, C., Hesselbach, X.: IoT-enabled proposal for adaptive self-powered renewable energy management in home systems. IEEE Access **9**, 64808–64827 (2021)
18. Hasan, M., Talukder, T.I., Saima, F.T.Z., Joy, M.N.U., Das, A., Sheham, M.N.H.: Smart home automation system powered by renewable energy. In: 2022 IEEE International Conference on Distributed Computing and Electrical Circuits and Electronics (ICDCECE), pp. 1–7. IEEE (2022)
19. Bhattarai, D., Singh, A.K., Newpaney, S., Pyakurel, P.: Design and prototype implementation of a renewable energy-powered home with home automation system using internet of things (IoT). Adv. Greener Energy Technol. 489504 (2020)
20. Shabnam, F., Islam, T.U., Saha, S., Ishraque, H.: IoT based smart home automation and demand based optimum energy harvesting and management technique. In: 2020 IEEE Region 10 Symposium (TENSYMP), pp. 1800–1803. IEEE (2020)
21. Molla, T., Khan, B., Singh, P.: A comprehensive analysis of smart home energy management system optimization techniques. J. Auton. Intell. **1**(1), 15–21 (2018)
22. Al-Ali, A.R., Zualkernan, I.A., Rashid, M., Gupta, R., AliKarar, M.: A smart home energy management system using IoT and big data analytics approach. IEEE Trans. Consum. Electron. **63**(4), 426–434 (2017)
23. Ain, Q.U., Iqbal, S., Khan, S.A., Malik, A.W., Ahmad, I., Javaid, N.: IoT operating system based fuzzy inference system for home energy management system in smart buildings. Sensors **18**(9), 2802 (2018)
24. Kim, J.: HEMS (home energy management system) base on the IoT smart home. Contemp. Eng. Sci. **9**(1), 21–28 (2016)

EEG-Based BCI Technology in IoT Enabled Smart Home Environment: An In-Depth Comparative Analysis on Human–Computer Interaction Techniques

Dibas Kumar De, Soumya Ranjan Samal, Aryan Choudhry, and Hitesh Mohapatra

Abstract The advent of smart home technology has revolutionized the way individuals communicate with their living spaces, offering efficiency, convenience, and comfort. The integration of brain-computer interface technology within smart home environments presents a promising avenue for transforming human–computer interaction paradigms. This review paper synthesizes current research findings on optimizing smart home user interfaces through humancomputer interaction utilizing Electroencephalography based BCI technology. EEG-based BCIs offer a novel approach to interface design by directly interpreting users' neural signals, thereby enabling seamless interaction with smart home devices. By leveraging EEG signals to interpret users' cognitive states and intentions, BCI offers a direct pathway for intuitive communication between humans and devices, bypassing traditional input modalities. The paper examines a few explicit key components such as signal acquisition, feature extraction, feature selection, classification algorithms, and system integration. Furthermore, the review evaluates the effectiveness, challenges, and future prospects of EEG-based BCIs in optimizing human–computer interface within smart home ecosystems. Insights from this review contribute to the understanding of how EEG-based BCIs can revolutionize user interaction paradigms, leading to more intuitive, efficient, and personalized smart home environments. This work presents a comprehensive study on the proposed topic by consolidating useful information from various sources and exhibiting it in a single paper to provide quality data to the novice researchers to help them in this field of research.

Keywords Smart city · IoT · Elderly · PWD · Smart environment · Social inclusion

---

D. K. De · S. R. Samal · A. Choudhry · H. Mohapatra (✉)
School of Computer Engineering, KIIT Deemed to be University, Bhubaneswar, Odisha, India
e-mail: hiteshmahapatra.fcs@kiit.ac.in

B. K. Mishra and A. Rocha (eds.), *Future of AI, IoT, and Sustainability*, Information Systems Engineering and Management 23,
https://doi.org/10.1007/978-3-031-76286-4_6

1 Introduction

In this dynamically evolving era of modernized living, the incorporation of technology in our homes has reshaped the way we engage with our surrounding environment. As stated by some of the popular researchers, a home must have a minimum of multiple sensors and actuators, user interfaces (including visual displays and voice control), building services (like lighting, heating and ventilation) and various network appliances with a view to qualify as a smart home [1]. Smart homes are reforming the residential life of human beings by harnessing sophisticated computing capabilities to enhance comfort, convenience and efficiency. The internal in-house networks are combined with the external networks (mobile phones and internet) [2]. Within this framework, a smart home is based on integrated communication services and centered around an automated structure by virtue of existing building infrastructure [2]. Scholars generally agree that a smart home is fragmented into three components that are "Home Automation", "Internal Networks" and "Intelligent Controls" [3]. HumanComputer Interaction (HCI), a multidisciplinary field which studies the seamless interaction among humans and machines which plays a vital role in transforming the user experience within these smart home environments and in return it offers a bridge between human needs and the potential of intelligent devices. Residents may easily control, monitor, and customize their living spaces according to their preferences due to the dynamic ecosystem created by the collaboration of humans and machines (Fig. 1).

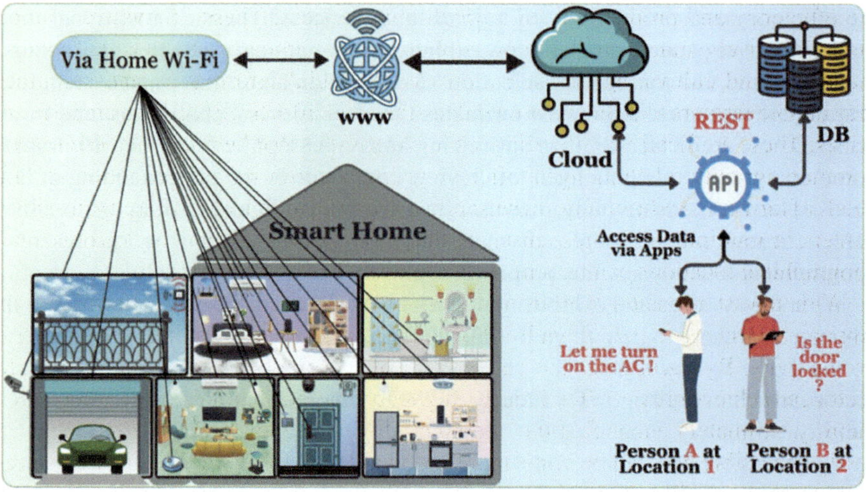

Fig. 1 Ubiquitous computing: IOT model level 5

1.1 Ubiquitous Computing

As computer technology progresses, Mark Weiser "The Father of Ubiquitous Computing" proposed that UbiComp has evolved from an innovation to a tangible reality in our everyday lives [4]. Recently, existing computing hardware (HW) embedded devices integrated with AI powered software (SW) are intended for enhancing human lives [5]. However, Ubicomp's internal modules must interact with themselves prior to engaging with end users [5]. This kind of conception raises the significance of Human–Computer interaction in smart home environments [6]. Within this aspect, a smart home evolves into a Ubicomp-focused ecosystem fabricated of cooperative autonomous spaces each of which facilitates a service area provided with HW/SW to offer a range of services overseen by agents [5]. One of the biggest benefits of Ubicomp-based smart home is its ability to execute numerous services with multiple configurations [5]. Nevertheless, the most challenging task is figuring out the best configuration [5]. From the perspective of human-centric requirements, a home is typically thought of as offering services which fulfills "comfort", "security" and "convenience", which distinguishes a smart home from other UbiComp-based environments [5]. In specific terms, we illustrate "comfort" to be the Quality of services (QoS) and the method of providing these services are carried out by "convenience", which establishes a connection between the end user and the environment where these services are provided, and "security" be inspecting the issues of information privacy and security [5]. Additionally, every service interaction with end users needs to be adjusted with changing contexts of the environment [5] (Fig. 2).

1.2 Voice Recognition and Orientation

In contemporary Human–Computer Interaction (HCI), speech-based interaction stands out as a highly favored form, allowing individuals to engage in verbal communication with machines through technologies like speech recognition and speech synthesis [7]. Voice interaction has been recognized as one of the most preferred HCI techniques by virtue of its efficiency and humanized traits [8]. Intelligent voice technology is amplifying quickly and substituting people's earlier lifestyles due to advancements in Internet of things (IOT) and deep learning along with improved processing capability [9]. The necessity for human–machine speech interaction goes beyond recognition precision and a greater focus on user's experience elements [7]. Voice orientation primarily aims to the direction the head is looking at the time the speech signal is released [7]. Voice orientation and recognition is not just a mechanical speech recognition technique but it also inculcates the knowledge of people's states resulting from the understanding of voice orientation, which in further enhances the experience and enjoyment of speech-based HCI [7]. Despite being beneficial, voice recognition cannot always be determined only due it's acoustic characteristics [7].

Fig. 2 Smart bedroom scenario via voice recognition

The two major drawbacks in voice recognition are firstly, "The Multi path Effect", which does not define the path of propagation and source of the voice signal and secondly, there is no specific spectrum for frequencies of various speech recognition [10]. Furthermore, we must execute comparatively better identification accuracy using minimal and basic equipment's and reasonable hardware in a field of real-world occurrences [7].

1.3 Brain Computer Interface (BCI)

Brain-Computer interfaces (BCIs) have evolved as a valuable tool to facilitate daily life activities of elderly people and those with physical disabilities, over the past few decades [11]. BCIs establishes a direct communication pathway among the individual's brain impulses and various external kinds of actuation devices [11]. Lately, due to the various developments in augmented reality (AR) and Internet of things (IOT), smart homes have become an alternative option to enhance the daily activities of human beings [12] (Fig. 3).

The number of members engaged in BCI system testing were young even though the target were elderly. Hence, it was found that some evoked potentials such as P300 and steady-state visual evoked potential (SSVEP) was stimulated more in younger people than the elderly [13]. Regardless, the BCI system may be user-friendly for the youngsters but the elderly who are less acquainted with electronic devices may

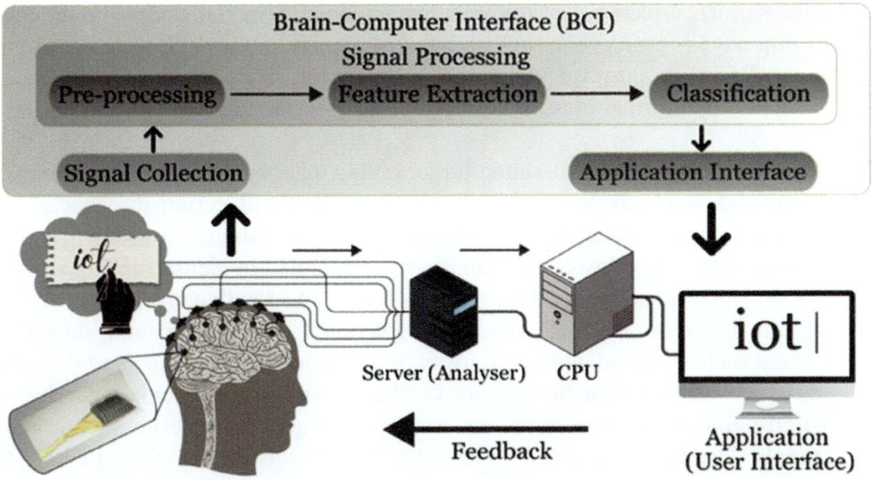

Fig. 3 Outline for BCI methodology

find it difficult to operate [13]. Therefore, by the integration of innovative electroencephalography (EEG)-based BCI and IOT, we can achieve a feasible system for people who are above the age of 65 [11].

1.4 Recognition Activities in an Intelligent Home

In an intelligent home, to facilitate the identifying actions one must obey three major steps:

- Collect granular-level sensor data (collecting data from the sensors).
- Analyze data gathered (processing the collected data); and
- Reasoning learning or utilizing techniques to deduce activities from the analyzed data (activity recognition).

1.4.1 Acquiring Sensor Data

The initial phase of data gathering involves employing sensors and actuators, compact and cost-effective devices positioned throughout the household, capable of detecting, observing, and relaying various human activities. There are a restricted number of sensors on the market tailored specifically for recognizing activities. Depending on their characteristics such as type, intended function, input and output signals, as well as technical setup, sensors can be categorized as follows:

1. Dense sensors are further categorized into two broad classes as: vision-based sensors or obtrusive sensors and non-obtrusive sensors. Vision-based sensors,

which utilize video cameras for activity recognition, fall under the category of obtrusive sensors, requiring no human intervention and thus are commonly favored for this purpose. Additionally, non-obtrusive sensors can be categorized into two further classes: dense-sensing environment sensors and wearable sensors.

2. Wearable sensors are well-suited for tasks like tracking pulse, skin temperature, movement, and body position. However, they come with certain drawbacks such as limited battery life due to continuous operation and reliance on user willingness to wear them. As a result, they are being supplanted by dense-sensing sensors, which gather data without requiring physical contact with end-users.
3. Within smart homes, the volume of generated data is constant, shaped by factors such as the quantity of sensors, the number of occupants, and the activities conducted within the premises. These sensors utilize either wireless or wired communication methods, necessitating Unique IDs, timestamps, and status signals. The precision of sensor data directly impacts the noise level within the smart home environment. Therefore, employing temporal-ordered random data processing becomes essential for accurate activity recognition.

1.4.2 Analysis and Processing of Data

Data analysis plays a crucial role in activity recognition within smart homes, primarily conducted using logical approaches and machine learning algorithms. However, as smart homes become increasingly complex, the collected data also becomes noisier, necessitating more extensive processing before it can be subjected to further analysis. Filtering raw data is essential to eliminate anomalies and remove outliers. Techniques such as median filters, low-pass filters, Kalman filters, Bayes, and particle filters can be deployed to refine the raw data. Following filtering and the removal of data gaps, the subsequent stage of data analysis involves formatting the data to align with established algorithms. Data segmentation also plays a vital role since sensor data collected in smart homes is recorded at frequent time intervals. During the segmentation process, the data collected is divided into three main smaller segments: activity-based, temporal-based and sensor-based.

1.4.3 Activity Recognition

In the final phase depicted in Fig. 4, activity recognition relies on two primary approaches: knowledge-driven, exemplified in [14] and [15], and data-driven, demonstrated in [16]. Knowledge-driven methods leverage existing domain knowledge to characterize ongoing activities, encompassing knowledge acquisition, formal modeling with presentation of knowledge [17, 18]. Within the domain of knowledge-driven models with the task involving logical reasoning such as abduction, induction, and deduction play essential roles in activity recognition or prediction [2]. These models display semantic clarity, logical elegance, and immediate and smooth applicability for users, effectively providing solutions to the cold-start problems [19]. Nevertheless, they have limitations as static methods, offering less expertise in managing uncertainty and real time data. In contrast to driven by knowledge approaches, models driven by data extract insights from Pre-existing datasets containing user-generated behaviors, leveraging data mining and ml algorithms for learning process [2]. These models employ statistical or probabilistic methods to address temporal issues and data uncertainty. The models utilize probabilistic methods to address temporal issues and data uncertainty. Further classification of data-driven approaches reveals two main types: generative and discriminative [20]. Within the generative approach, comprehensive representation of the input data is created using a probabilistic model [21]. For example, Conditional Random Fields (CRFs) yield satisfactory results for activity recognition as they integrate probability concepts [22]. Conversely, a discriminative approach utilizes past submissions to compile datasets containing both correct and incorrect data [2]. As an illustration, the nearest-neighbor algorithm (KNN), which relies on an extensive array of training samples, expands exponentially in accordance with the targeted level of accuracy. Discriminative approaches are crucial in tackling the problems of classification, unlike the problem of representation in case of approach based on generation. To exemplify, one such approach is the nearest-neighbor method, which scrutinizes the dataset based on training groups together sequences that intimately match each other. Another instance of this method is decision trees, which are utilized to acquire descriptions of logical activities derived from sophisticated sensor data readings [2].

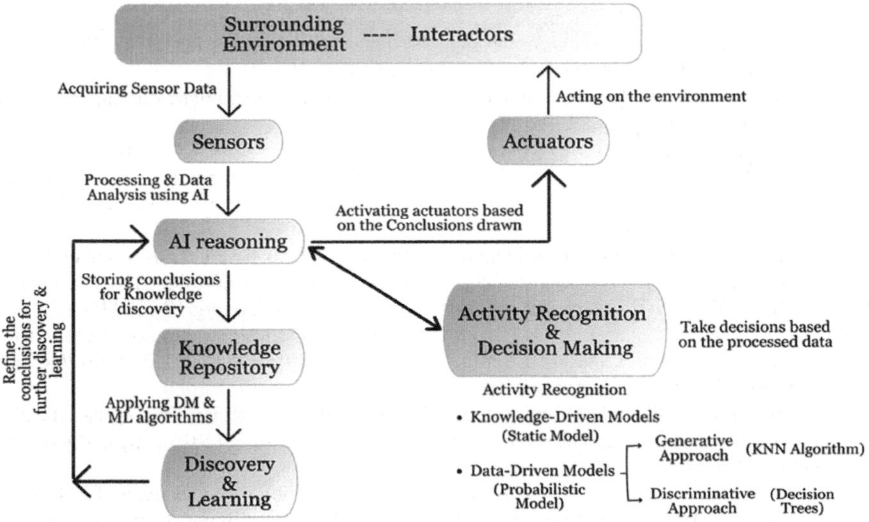

Fig. 4 Data-flow illustration for activity recognition

2 Literature Review

A brain-computer interface (BCI) is integration of brain and an external device which transmits brain signals in order to implement various external tasks without any foreign nerves and muscles [23]. BCI is primarily used by people with acute movement disorders, specifically patients with Motor neurine diseases, midbrain strokes, or other neuromuscular diseases [24]. BCI-based applications can be utilized to operate wheelchairs, orthic, assistive devices and multiple computer applications. In addition, BCI-based home automation control system was found advantageous in the domain of Artificial Intelligence (AI). Moreover, BCI-controlled home automation interfaces showcase independence with respect to illumination and audio spectrums [25]. Additionally, it allows a home automation interfaces to be controlled directly by the user's desired actions. In order to avoid surgical techniques, there are broadly four noninvasive measurement techniques which are implemented for collecting brain signals: electroencephalogram (EEG) [26], magnetoencephalography (MEG) [27], functional magnetic resonance imaging (fMRI) [28], and functional near-infrared spectroscopy (fNIRS) [29]. BCI-based control system is found beneficial over other type of brain techniques due to their good portability, temporal resolution, and less expensive.

Electroencephalograph (EEG) is the most preferred technology for reading signals from the brain. It's a kind of non-invasive BCI technique which involves intercommunication between the brain and an external equipment without any direct physical contact with brain tissue or implantation. It detects the potential fluctuation within the neurons whenever there are any external stimuli; basically, the variations in signals affect the location of synapses generating a potential difference which is then recorded [30]. The functioning of cerebral activity is most effectively observed in those areas where it is augmented [26]. The obstruction formed between the sensors and neurons often generates frequencies beyond 40 Hz making it challenging for measuring accurate EEG signals [31]. Among brain activity measurement techniques, EEG stands out for its incomparable combination of affordability, portability, and compatibility. Moreover, it offers exceptional temporal resolution, capturing brain activity with millisecond precision. However, various research indicates that the EEG signals obtained after pre-processing are quite noisy due to their low signal-to-noise ratio, low spatial resolution, and several external factors such as artifacts and interfering frequencies. The EEG activity signals often exhibit a relatively low strength, typically measured within the theta and gamma frequency ranges. Considering EEG electrodes are positioned according to a standardized system on the scalp's surface which results in a low signal-to-noise ratio. Furthermore, it was observed that an extremely higher signal-to-noise ratio is obtained in the case of ECoG (Electrocorticography) since the sensor electrodes are explicitly placed on the brain. The noise must be processed to derive some meaningful insights from the signal [32, 33]. There are two major techniques for removal of noise from signals; one is filtering and other is artifact removal [30]. Upon completion of the signal enhancement stage, the resulting noise-free EEG data undergoes feature extraction to isolate relevant characteristics of brain activity.

Feature extraction from EEG signals is achieved through diverse techniques like Autoregressive models (AAR) for capturing signal dynamics, Principal Component Analysis (PCA) for dimensionality reduction, and Wavelet Packet Decomposition (WPD) for time–frequency analysis [30, 32, 34, 35].

After the feature extraction phase, the brainwaves obtained are sorted into diverse classes using multiple classifiers. These classifiers are linear classifiers, Artificial Neural Networks (ANN) based classifiers, non-linear Bayesian classifiers and nearest neighbor classifiers [30, 36]. As specified by [37], Support vector Machine (SVM) appears as a suitable algorithm for classifying Mind-control system powered by EEG data for increasing classification accuracy, as referred to by several research papers in their analysis. Multiple references suggest that a Gaussian classifier offers higher suitability for high-dimensional EEG-based robot controllers [38]. By implementing this classifier, along with threshold method by as proposed by [38], it not only offers reliable classification accuracy for BCIs but also filters out unrecognizable Motor Imagery (MI) tasks, reducing false positives in classification outcomes [30]. A study on brain-computer interfaces using EEG [39] found success with k-Nearest Neighbor (k-NN) and Linear Discriminant Analysis (LDA) for classifying brain signals. The system achieved impressive accuracy, reaching up to 82%. To extract key characteristics from the brain signals, the researchers used techniques called Mean Derivative (MD) and Hilbert Transformation (HT). These findings encourage a potential interest in exploring alternative classification algorithms to enhance data utilization [30]. Another study [40] compared the performance of three classification algorithms: Linear Discriminant Analysis (LDA), Support Vector Machine (SVM), and Gaussian Naive Bayes (GNB). This comparison was conducted for a Brain-Computer Interface (BCI) controlled by motor imagery (MI). The study [40] found that the Gaussian Naive Bayes (GNB) method achieved higher classification accuracy compared to the other two (LDA and SVM) and argued that GNB is particularly well-suited for analyzing EEG data. Additionally, the researchers proposed that GNB's effectiveness could be further explored for developing real-time Brain-Computer Interfaces (BCIs) based on motor imagery (MI).

EEG and Brain-Computer Interface (BCI) are integrated along with smart home environments to facilitate a more innovative, efficient interface for controlling devices and enhanced functionalities for an interactive user experience. End users can control smart home devices with the help of brain signals which in return reduces the need of physical interaction or voice commands. This integration offers various benefits such as providing assistance to people with disabilities and hands-free operation for tasks in physically constrained environments. As suggested in [41], BCI technology acts as an useful device for HCI, it is capable of capturing ambient properties of human activities rather than performing active operations. Unlike traditional interfaces, EEG technology allows for user interaction without placing a burden on the user's mental focus. This makes it a valuable tool for various interactive devices. Furthermore, EEG driven by BCI are directed to encapsulate neural activities into smart wearables. This wearable device allows user's brain activities to be monitored even while free roaming, enabling the prototyping of ubiquitous computing. The prototype is demonstrated using EEG driven mobile BCI, which can be used as a novel form of human–computer interaction device. Highlighting the security benefits of eye gaze technology, we created a login system where users can authenticate by looking at the screen from a close distance. The functionality is achieved through the identification of characteristic EEG patterns elicited by user visual attention to the computer terminal's blinking display. Building on similar concepts, research by [41, 42] explores eye sensors as an alternative for managing cell phone notifications. Their system leverages a user's gaze direction to infer their intent. This effectively merges eye tracking technology with Brain-Computer Interface (BCI) control. In simpler terms, imagine interacting with your phone by simply looking where you want to click—a truly intuitive approach that eliminates the need for physical touches or gestures.

As specified in [43], a potential framework for ubiquitous brain-aware computing (UBAC), characterizes the usage of wearable, ubiquitous BCI technology to facilitate cognitive state information to contextually aware and other HCI applications. The above-mentioned paper analyzes the feasibility of using a cost-effective, single electrode, wireless-enabled, non-intrusive, gel-free, and battery-operated EEG recording device for accurate identification of cognitive states. The preliminary outcomes of the above referred paper indicates that by using a straightforward single-electrode system, we can potentially distinguish among various categories of cognitive and emotional states. For instance, the states associated with reading "uninteresting" material, reading "interesting" material, listening to music, and engaging in relaxation is defined under this category. Numerous smart appliances and context-aware apps can become "smarter" if they have the access to mental models of their users. In future, one may imagine some certain brain-state information can be projected into real world enabling more seamless and smooth integration of people with their surroundings. A key feature of this data process decoupled, layered architecture is that it segregates different modules at different levels of processing (which includes feature extraction, selection, classification, high-level interpretation) [30, 37, 43]; developers working within one particular layer need not have in-depth knowledge regarding how other layers or processes (even within the same layer are implemented. This architecture is similar in nature to TCP/IP-based BCI platform BCI2000 [44].

According to a study [45], persons with disabilities and elderly people face difficulties in operating home appliances, hence this challenge can be tackled by using Neuro-control interface. Brain-computer interfaces (BCIs) employ non-invasive electrodes placed on the scalp to establish a communication channel between the external devices and the brain. Users connect and interact with the Brain sense headset through the Neuro view software. The electrodes detect brainwave activity. This information is then processed and transformed into a specific type of signal (blink wave) suitable for the Arduino board. Finally, the Bluetooth module wirelessly transmits this signal to the Arduino board. Once received by the Arduino board, the digital signal is processed to understand the user's intent. This information is then used to activate the corresponding appliances. An Arduino-based BCI-controlled home automation system for the physically disabled is presented in this paper. This paper suggests a BCI system specifically designed to empower individuals with paralysis or disabilities. The system allows for independent control of easy-to-build, cost-effective appliances. This scalable design can be adapted to manage any number of devices and has broad applications in various fields, including automobiles, industrial settings, remote control operations, and beyond.

The paper proposed in [46] discusses the development of a Mind Controlled Robot utilizing BCI technology, which analyzes brain waves using LabVIEW software technology facilitates a direct link between the brain and physical devices. By analyzing different brain activity patterns instantaneously, BCIs allow for the control of mobile robots. It focuses on analyzing brain wave signals, which represent the complex interactions between neurons in the human brain and manifest them as thoughts and emotional states. As human thoughts evolve, these patterns change, resulting in the production of different electrical waves which are further transmitted to the robotic hardware system (often used with Arduino) by using Bluetooth transmission protocol.

The research performed in [47] involves the decoding and interpreting of EEG signals to recognize Hook and Span hand gestures. This model utilizes various signal processing methods alongside supervised machine learning algorithms like SVM, Decision Tree, Adaboost, and Random Forest to achieve classification. The system begins by processing the EEG signal. It first removes unwanted noise using band-pass (14–30 Hz) and notch (49–51 Hz) filters. Then, it extracts relevant features from the cleaned signal and finally classifies it as either a hook or span hand gesture. The objective of this work is to offer assistance to individuals with nerve damage by equipping users with the ability to interact with the automated features of their wheelchairs. The significance in the above research lies in it's specific focus on operating wheelchair gear control and release, just by using hand gestures.

Conventionally, the EEG system driven by BCI is segmented into four different patterns, that are Steady-state visual evoked potential (SSVEP), Motor Imagery (MI), P300 potentials, and slow cortical potentials. Each signal pattern has its own pros and cons [48]. While reference [49] demonstrated controlling a system with MI signals based on hand grasps, it suffered from a restricted set of available commands. Research by [50] investigated eye blinks and eye movements for controlling a home lighting system. They concluded that eye blinks offered greater accuracy. Eye movements, they found, could be influenced by involuntary factors or slight head movements, introducing errors and potentially causing unintended light activations. When MI signals are compared with eye-blink signals, it is observed that P300 and SSVEP signals exhibit exceptional accuracy and rapid response times. As a result of which, the majority of the existing home automation systems based on BCI technology rely on P300, alpha rhythm and SSVEP signals to achieve faster and more precise control. Both P300 and SSVEP signals require external triggers for operation. P300 signals are particularly well-suited for recognizing a wider variety of triggers (typically more than six). In contrast, SSVEP signals demonstrate optimal performance when dealing with a limited number of triggers (under six). A person whose entire body is completely paralyzed can still communicate using eye signals to execute gestures through the hook and span method described in the aforementioned paper [47].

In accordance with the study made in [51] a multimodal BCI system for home automation control was proposed. This study investigated the use of electrodes to record two specific brain signals. Participants focused solely on a stimulus and used eye blinks to interact. The user sees a specific visual pattern (stimulus) that triggers the desired commands through SSVEP brain signals. Single eye blinks indicate selection confirmation, while two consecutive eye blinks signify denial and prompts reselection, thereby calibrating the SSVEP command. The study concludes that the proposed model provides 38 navigation commands with a duration of 2 s with an accuracy of 96.92%, leveraging a single bipolar EEG channel for data acquisition. His research proposes a groundbreaking brain-computer interface (BCI) system specifically designed for automation of home applications. The system utilizes eye blink signals, SSVEP, aiming to empower individuals with disabilities.

The use of EEG for UAV control is limited due to the demanding nature of drone piloting, which requires precise control and sustained mental focus. While EEG shows promise for UAV control, using it alone might be challenging due to the limitations in accurately translating complex user intentions into precise flight commands. However, a hybrid BCI system that incorporates EOG signals alongside EEG can address this limitation by offering a more robust and intuitive control interface [51, 52]. By implementing an obstacle avoidance algorithm, we can effectively prevent collisions and navigate around dead ends in any given environment. The integration of eye-tracking zones empowers users to navigate more effectively by leveraging real-time video from the UAV camera. By fixating their gaze on specific areas of the live feed, users can direct the drone's movement. This system incorporates an additional control layer through visual attention. Users can control the drone's ascent, descent, and forward/backward velocity by directing their gaze on the live video feed. Meanwhile, motor imagery (MI) signals are employed for precise control over the drone's tilt (roll), sideways movement (pitch), and turning (yaw). A novel application of a hybrid autogenous BCI was demonstrated in [53] for controlling a drone swarm. The control system leveraged different brain signals: visual imagery (VI) for specifying high-level goals, motor imagery (MI) for directional control, and speech imagery for commanding specific swarm configurations.

The study of attention, a critical aspect of brain function, represents a highly valuable domain in the field of neuroscience. It has a significant impact on some crucial cognitive processes like learning, driving vehicles, etc. Artificial Intelligence (AI) has accelerated the growth of brain-computer interfaces (BCIs), enhanced interpretation and understanding of brain activity. Another study [54], detected and measured the cognitive load of a person by using BCI and EEG Signals. This model could help ensure the alertness of drivers and pilots by monitoring their brain activity and identifying potential fatigue or distraction. This could lead to safer roads and skies. This paper explores the use of electroencephalography (EEG) and machine learning to classify different mental attention states in humans. This research focuses on its potential application in diagnosing attention deficit hyperactivity disorder (ADHD), where attention difficulties are a core symptom. The study recruited five volunteers to assess their mental focus levels. Researchers employed a technique called Short-Time Fourier Transform (STFT) to extract relevant features from their brain activity data

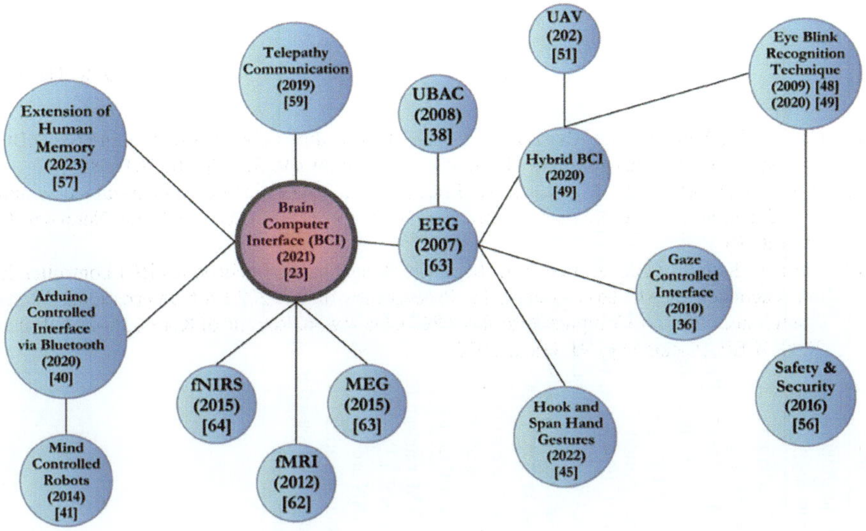

Fig. 5 Visual representation depicting different studies in the field of BCI

recorded through EEG. These features were then fed into a Support Vector Machine (SVM) classification algorithm, which categorized the volunteers' mental states into three main categories: focused, unfocused, and drowsy [55]. As more features are included during model training, the accuracy of predicting human mental states can be significantly improved. This allows for more precise classification of focus levels, leading to potentially better applications. More accurate prediction of drivers' mental states could reduce accidents, while monitoring students' brain activity could assess their engagement during learning [54] (Fig. 5).

Employing EEG technology for security purposes will unlock innovative methods in enhancing safety of individuals. Addressing occupational safety through strategies for managing IoT networks is yet a persistent challenge. A paper [56] suggested the use of UAVs to monitor mine workers' exposure to harmful conditions. Another paper [57] proposed a smart helmet equipped with EEG sensors, enabling constant monitoring of the user's health condition via IMU and EEG sensors [58]. To promote worker well-being, sensors monitor for signs of fatigue or stress and send alerts to connected devices or apps. Exploration teams working in remote locations, often beyond the reach of traditional communication, could benefit from EEG technology. This allows for emergency alerts to be sent directly from a worker's brain activity. However, whether utilized independently or in conjunction with other physiological indicators, EEG presents an opportunity for enhancing security systems, contributing to the detection of hidden targets that would be difficult to find using traditional methods [30] (Table 1).

Following the idea of Stephen Hawking to expand human memory by transferring the mind to a computer system [60], advancements in Brain-Computer Interfaces

Table 1 Summarizing the objectives and models used in different studies of BCI along with their citations

Citations	Title	Objectives	Models/algorithm/ technology used	EEG control signals
[30]	An overview of the use and applications of brain-computer interfaces based on electroencephalographs (EEG)	• This paper aims at demonstrating various methods for collecting brain signals • It provides a clear and concise examination of each approach, including the benefits and drawbacks of each	• EEG-based BCI technology • fMRI, MEG, fNIRS ECoG methods are used for signal acquisition • FFT, ICA, PCA, WT, AAR are mostly used for feature selection • SVM, k-NN, ANN, LDA are some the classification algorithms used	• Desynchronization/ synchronization (ERD/ERS) • Steady-State Visual Evoked Potentials (SSVEP) • P300 • Visual Evoked Potentials (VEP) • Slow Cortical Potentials (SCP)
[59]	A review of classification algorithms for EEG-based brain–computer interfaces	• This research paper provides an overview of some of the classification algorithms used in EEG based BCI systems • It outlines their key attributes, and compares their performance based on the existing literature	• SVM, random forest, k-NN, LDA and ANN are few classification algorithms used in this paper	• SSVEP • P300 • ERD/ERS

(continued)

Table 1 (continued)

Citations	Title	Objectives	Models/algorithm/ technology used	EEG control signals
[60]	EEG-based BCI control schemes for lower-limb assistive-robots	• The objective of this research is to explore applications of EEG based BCI for mobility rehabilitation and locomotion, highlighting advance-ments and challenges in controlling wearable lowerlimb devices and assistive-robots	• TCP/IP protocol suite is used for transmission purposes • SVM, LDA, ANN and Gaussian classifiers are used for classification	• P300 • SCP • VEP • ERD/ERS • MI
[39]	Quaternion-based signal processing for electroencephalographic signals-based motor imagery classification	• The purpose of this work is to extract features from EEG data and use the QSA approach to represent motor commands	• k-NN, LDA, SVM and DT classification algorithms are used • To represent the signals, features are extracted using Mean Derivative (MD) and Hilbert Transformation (HT)	• SSVEP • P300

(continued)

Table 1 (continued)

Citations	Title	Objectives	Models/algorithm/ technology used	EEG control signals
[43]	Towards ubiquitous brain-aware computing: a preliminary EEG study	• This paper suggests a potential framework for ubiquitous brainaware computing (UBAC), characterizing the usage of wearable, ubiquitous BCI technology to facilitate the identification of cognitive states and other HCI applications	• Ubiquitous Brain-aware Computing (UBAC) framework is implemented • Fisher Linear Discriminant Analysis (FLDA) is performed for task classification	• ERP/ERD/ERS • P300
[45]	An arduino home automation system with brain-computer interface for people with physical disabilities	• This paper aims at utilizing bluetooth (for communication) and arduino (for controlling) to interpret brain signals and convert them into real life actions by integrating BCI with IoT so as to facilitate home automation for paralyzed and disabled ones	• MATLAB software is used to visualize and sense the brain signals	• Eye Blink

(continued)

Table 1 (continued)

Citations	Title	Objectives	Models/algorithm/technology used	EEG control signals
[47]	An electroencephalogram based detection of hook and span hand gestures	• This work analyzes and classifies the EEG signals to recognize hook and span hand gestures. Its objective is to offer assistance to individuals with nerve damage by enabling them to independently operate automated wheelchair controls	• SVM, decision tree, adaboost, and random forest classification algorithms are implemented • A notch filter of 4951 Hz and a band pass filter of 1430 Hz is used during preprocessment	• SSVEP • Alpha rhythm
[50]	Eye blinks detection from EEG data for home lighting activation systems	• This study investigates an optimal eye activity to trigger home lighting systems and subsequently identifies this activity from EEG signals	• The technique of Short-time Fourier transform (STFT) is utilized to extract features • Convolutional neural network (CNN) is used as a classification algorithm	• Motor Imagery (MI) • Eye blink • Eye movement • SSVEP • P300 • Alpha rhythm

(continued)

Table 1 (continued)

Citations	Title	Objectives	Models/algorithm/ technology used	EEG control signals
[51]	A hybrid brain-computer interface system for home automation control based on bipolar channel using eye-blink signals and steady-state visually evoked potential	• The aim of this study was to create and authenticate a hybrid BCI system designed for controlling home automation and leveraging SSVEP and eye blink signals, with potential benefits for individuals with disabilities (esp. fullbody paralysis)	• Short-time Fourier transform (STFT) is applied for feature extraction • Convolutional neural network (CNN) is used as a classification algorithm	• Eye blink • SSVEP
[53]	Design of an endogenous BCI based drone swarm control system using EEG	• This study focuses on an endogenous model with an emphasis on EEG-based Unmanned Aerial Vehicle (UAV) deployment and drone swarm control	• Common spatial pattern (CSP) is used for extracting spacial features • CSP-LDA algorithm is used to classify the model • Robotic Operating System (ROS) is used to interact with MATLAB for controlling the UAV device remotely	• Motor Imagery (MI) • Visual Imagery (VI) • Speech Imagery (SI)

(continued)

Table 1 (continued)

Citations	Title	Objectives	Models/algorithm/ technology used	EEG control signals
[46]	Bluetooth based brain computer interface system for mind-controlled robot	• The paper discusses the development of a mind controlled robot utilizing BCI technology, which analyses brain waves using LabVIEW software. It facilitates people with severe organ disorders and help them improve their quality of life	• XBee and XBee-PRO modules along with Zigbee are used to transmit data to robot module • Think-gear is another technology that is used to empower and enable devices to interact with the user's brainwaves	• P300 • SSVEP • Eye blink

(continued)

Table 1 (continued)

Citations	Title	Objectives	Models/algorithm/ technology used	EEG control signals
[54]	Human attention detection using artificial intelligence via brain-computer interface for health care monitoring	• This paper focusses on detecting and measuring the attention of a person by using EEG based BCI to avoid problems associated with lack of consciousness of a human mind (ADHD)	• Fast Fourier transform (FFT) is used in preprocessment • k-NN classification algorithm is used	• SSVEP
[57]	An intelligent safety helmet that uses IMU and EEG sensors to identify worker tiredness	• This study describes the use of a Smart Safety Helmet (SSH) to monitor miners' head motions and brain activity with the aim to detect abnormal behaviors	• Inertial Measurement Unit (IMU) is used to recognize human head gestures	• P300

(BCI) suggest a potential future where scientists can accurately extract and store sensitive brain information in external physical memory devices, such as portable flash drives. It's working can be beneficial in many possible ways for instance, a counseling psychologist is equipped with precise insights acquired through a BCI device regarding an individual's behaviors and traits. Naturally, such an expert would be anticipated to offer insightful guidance and conclusions, significantly impacting the individual undergoing counseling. This scientific endeavor demands rigorous multidisciplinary research to gain maximum excellence. A prominent researcher suggests that Brain Computer Interfaces (BCI) combined with computer-brain interfaces (CBI) [61] could enable a form of telepathic communication, bypassing physical interaction and sensory channels. While Brain-Computer Interfaces (BCI) combined with other Brain Computer Interfaces (CBIs) create a "brain-brain interface" technology still in its early stages [62], further research is needed. Integrating BCI with the Internet of Things (IoT) and other communication methods like mind-to-mind and mind-tomachine interfaces holds promise. Exploring these possibilities could unlock new functionalities in human–machine-human communication [63].

3 Proposed Review and Comparative Analysis

A variety of comprehensive review papers have been published on EEG-based BCI in smart home environment [64–66]. But very few of these papers have intensively focused on the processes involved in extracting, pre-processing, and classifying the brain signals with respect to their applications. In this paper, we perform an explicit review on various pre-processing, feature extraction and classification techniques on EEG signals over the last few decades and provide an overview of integrating smart home environment with EEG-based brain computer interface (BCI) along with their features. This paper derives from an extensive study of diverse models proposed across a wide range of papers and perform an in-depth comparative analysis of these models based on different process criteria and demonstrate them in a tabular structure. The annual trajectory of research publications with reference to EEG-based BCI in smart home environment highlights the encouraging outcomes of BCI systems in the domain of IoT. Moreover, the limited presence of review papers signifies a deficiency of comprehensive analyses that could aid novice researchers in gaining deeper insights into the domain and better understand the field of research while keeping up with the new and emerging trends and advancements in the field.

EEG SIGNAL PROCESSING METHODOLOGIES

The implementation of EEG paradigms involves a series of sequential steps for deployment. According to a study [67], there exist five steps in this process, these include: we start by acquiring and segmenting signals. Pre-processing cleans them, then features are extracted and fed to models for predictions. These predictions are then put to use.

3.1 Signal Acquisition and Partitioning

Understanding the brain's structure is crucial for acquiring meaningful EEG signals. The brain is divided into two main regions: the cerebral cortex and subcortical regions. While subcortical areas manage vital functions like body temperature and reflexes, the cerebral cortex is responsible for higher functions like sensory processing and decision-making. Further divided into lobes with specialized functionalities [68], the cerebral cortex offers a map for interpreting EEG signals based on their origin. This knowledge is essential for developing effective Brain-Computer Interfaces (BCI).

1. Frontal lobe: It spans across the frontal and upper regions of the cortex particularly in front of the brain's central sulcus. The prefrontal lobe, frontal motion area, and primary motion area all constitute this region. Its functions encompass a wide range, such as cognition, decision-making, planning, speech, memory, judgement, problem-solving, social skills, consciousness, personality traits, emotional and behavioral control, intellectual capabilities, and self-awareness. It serves as the highest-level sensory center.
2. Parietal lobe: Located in the back upper area of the cerebral cortex, specifically behind the central sulcus and bordering the front of the occipital fissure. It is responsible for a wide range of functions, including sensation, touch, pain, temperature, taste, pressure, reading comprehension, spelling, differentiating leftright, spatial orientation, visual attention, perception, language interpretation and word recognition. This region is also associated with logical and mathematical reasoning.
3. Temporal lobe: It is located at the bottom central region of the cortex, just behind the temples particularly beneath the lateral fissure, sandwiched between the frontal lobe at the front and the occipital lobe at the back, the parietal lobe occupies the upper region of the brain. Its functions include smell, hearing (interpreting auditory signals from the ears), object categorization, recognizing faces, and generating emotions.
4. Occipital lobe: It is situated at the lower, posterior region of the cortex particularly behind the occipital sulcus. Its primary functions include visual perception, color identification, tracking object motion, perception of behavior and abstract concepts [30].

The most widely used system for non-invasive electrode placement is the 10–20 system, endorsed by the American Electroencephalographic Society. This system ensures standardized electrode positioning across individuals, facilitating the gathering and processing of EEG data in a consistent and comparable manner. As illustrated in Fig. 6, the 10–20 system relies on specific distances for electrode placement. These distances are derived from measurements of the head: 10% of the total distance from the nasion (bridge of the nose) to the vertex (top of the head) and 10% of the total distance from the inion (bony bump at the back of the head) to the preauricular point (just in front of the ear). The remaining electrode positions are spaced at intervals of 20% along these total lengths. Following this approach, the 10–20 system assigns

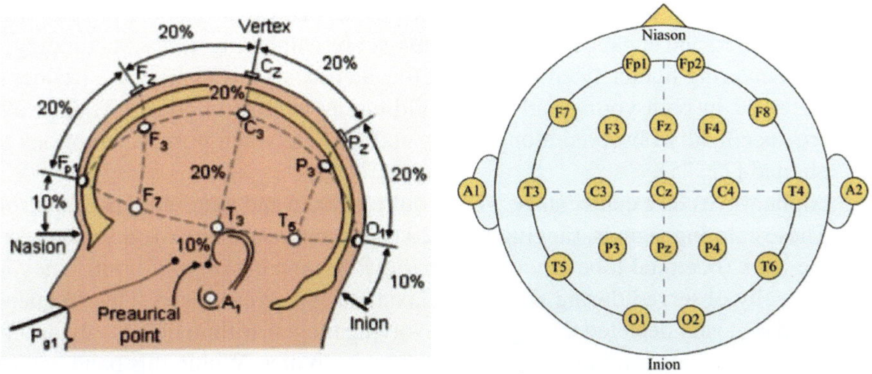

Fig. 6 International 10–20 EEG electrode placement system on the scalp and letternumber-denotation

a total of 21 specific electrode positions across the scalp. Each position is denoted using a letter-number combination. The letter corresponds to the underlying cortical region the electrode is closest to (Fp: pre-frontal, F: frontal, C: central, etc.), while the number indicates the hemisphere (odd for left, even for right). Numbers also specify position within a hemisphere, which is important because brain functions can differ between the left and right sides. The "z" designation is used for midline positions, particularly for baseline measurements [69, 70].

The brain is a constantly buzzing hive of electrical activity! These electrical signals can be categorized into distinct frequency bands based on their speed (measured in Hertz, Hz). The five main bands are delta (slowest, 0.5–3.5 Hz), theta, alpha, beta (fastest), and gamma. There's also a sub-band within alpha called mu (7–11 Hz). The strength of these signals, measured in microvolts (μV), is typically quite weak, ranging from 0.5 to 100 μV [70–72]. The brain emits a variety of rhythmic signals across different frequency ranges, as discussed below:

1. Delta Wave (0.5–3.5 Hz): These waves usually manifest in the frontal cortex of the brain with amplitudes ranging from 20 to 200 μV. They are commonly observed during the state of unconsciousness due to oxygen deficiency, deep sleep without dreams, or under anesthesia. When we enter the deepest stages of sleep, our brain activity slows down, and the electrical waves measured by EEG machines become larger in amplitude. Variatioin the depth are caused by several neurological and physiological conditions. Disrupted sleep, often associated with depression, anxiety, and ADHD are frequently observed in adults and, posteriorly, in children. These waves disappear in adults who tend to be in awake or alert situation [73, 74].
2. Theta waves, pulsing between 4 and 7 cycles per second (Hz), are most active in the brain's parietal and temporal regions. These waves are relatively strong, with electrical activity reaching between 100 and 150 microvolts (μV). They are associated to a state of relaxation and cognitive tasks involving working

memory. Theta waves are like the brain's in-between gear. They're present when you're alert and engaged, actively learning or remembering things, meditating, or even entering lighter sleep stages (not the deepest snooze!). They are frequently observed in both young and older children, as well as in adults. Theta waves are specifically observed along the frontal midline when positive emotions are elicited [72, 75].

3. Alpha waves take center stage when you're relaxed and your mind is wandering. These calming signals, ranging from 20 to 100 microvolts (μV), are strongest in the back (occipital lobe) and upper (parietal lobe) parts of your brain. They are typically observed during states of relaxation with eyes closed. This frequency band is mostly detected during tasks involving mental arithmetic, visualization of images, and performing short-term memory activities. Within this band, various small and large alpha waves are accounted, signifying the dynamic involvement and disengagement in the specific task at hand. Alpha waves (relaxation) fade with sight, sound, or mental effort. Moreover, these waves get suppressed when the eyes are open or when the individual experiences drowsiness and sleepiness. Interestingly, they exhibit higher oscillatory energy as compared to beta and gamma waves during both positive and negative emotional states [76, 77].
4. Mu Wave (7–11 Hz): These waves are typically detected in the sensorimotor cortex, particularly in the central region of the brain with amplitudes roughly ranging from 30 to 120 μV. They are often associated with states of relaxation and immobility, and can be suppressed during voluntary movement or motor tasks. This rhythm is specifically studied in the context of motor planning, imitation, and mirror neuron activity and is thought to play an essential role in social cognition, modulating motor functions and understanding others' actions. These waves are of key interest in various fields, including neuroscience, psychology, and neuro-rehabilitation, for their potential involvement in motor control and social interaction research [70].
5. Beta Wave (13–30 Hz): They are typically observed primarily in the frontal lobe of the brain; however, during contemplation, these waves may arise in various brain regions with amplitudes ranging from 5 to 20 μV. This rhythm is associated with high mental activity, active cognition processing, focused mindset, alertness in brain activity, and the perception of an individual. These waves signify a re-synchronization that occurs in response to an external stimulus and serve as a measure of movement preparation. They exhibit symmetrical distribution on both sides of the brain. The beta rhythm eventually disappears as emotional activity increases, whereas alpha waves predominate in the cerebral cortex when the body is at rest. The amplitude of alpha waves diminishes and the frequency of beta waves increases in conditions when the central nervous system is tense, stressed, or straining, resulting in a gradual shift from alpha to beta waves. When the cerebral cortex seems to be in the beta state, an excited state is detected [75, 78].
6. Gamma Wave (>30 Hz): The EEG signal originates from widespread cortical networks, encompassing both sensory and non-sensory processing areas. These electrical waves exhibit low amplitudes, usually below 2 μV. Among all EEG rhythms, this frequency band exhibits the highest value, corresponding to a state

of peak focus. These waves, prominent during multimodal sensory processing integration, and feedback within the brainstem. Additionally, they are associated with states of intense focus and represent the optimal brain frequency for cognitive performance.

EEG acquisition can be achieved through wired or wireless methods [79]. Wired systems utilize an elastic cap with electrodes (1–256) attached to the scalp, often requiring conductive gels for optimal signal quality [70]. However, this approach can be time-consuming and cumbersome. Dry electrodes offer a faster alternative but suffer from reduced signal fidelity (30% lower information rate) compared to gel-based systems [79]. While wired EEG is well-established, limitations include complex setup and restricted user mobility due to cables [79]. To address these issues, wireless EEG systems have gained popularity due to their non-invasive nature and unrestricted movement [79]. Headset selection hinges on application; research-oriented models offer higher channel counts and resolution for advanced studies, while lower-cost options cater to non-research purposes like meditation or home therapy [70, 79]. Ultimately, the optimal acquisition method depends on the specific BCI application.

3.2 EEG Signal Preprocessing

EEG preprocessing tackles unwanted noise from muscles, blinks, or even the environment, leaving a clearer picture of the brain's electrical activity. EEG signals, despite originating from the brain itself, are easily corrupted by electrical noise from both biological sources (muscle activity, eye blinks) and environmental sources (power lines). This noise can originate from the electrodes or from the body itself and is commonly referred to as "artifact" or "interference". EEG recordings, while offering a window into brain activity, are vulnerable to electrical noise from the environment. This includes the constant hum of power lines (50/60 Hz), flickering lights, and even the buzz of nearby electronics like computers, phones, and tablets. EEG electrodes may capture unwanted physiological signals, during signal recording leading to the rise of physiological noise which derives from various bodily activities, including movement, other bio-electrical potentials, and fluctuations in skin resistance. ey sources of physiological noise include activity from electrooculography (EOG, related to eye movements), electrocardiography (ECG, heart signals), scalp electromyography (EMG, muscle activity), ballistocardiography (motion related to heartbeats), and breathing. Physiological artifacts pose a challenge in EEG-based state recognition. Eye movements (EOG) and muscle activity (EMG) generate electrical signals that contaminate the EEG. Cardiac activity, particularly the QRS complex (ECG), can also introduce noise via volume conduction through cerebrospinal fluid. Additionally, motion artifacts arise from cable movement and electrode displacement during subject movement. While artifact removal is common, caution is necessary, as some artifacts may contain valuable information about brain states relevant to recognition algorithms. Therefore, this pre-processing phase is crucial to prevent and minimize the noise contamination in raw EEG data that could potentially effect subsequent classification phase. There are two main techniques through which we can reduce the noise or artifacts present in the acquired EEG signals. These techniques are discussed as follows:

3.2.1 Filtering

Extracting meaningful information from raw EEG necessitates amplification. The digitization process amplifies these weak signals by a factor of 500–2000. Various pre-processing techniques are employed while interpreting the raw EEG data. One of the techniques to do so is filtering, which reduces artifacts in the collected EEG signals by applying various types of filters on them. The three frequently used filters in the domain of EEG include: low-frequency filters, high-frequency filters, and notch filters. The initial filtering stage employs two distinct filters to isolate the desired frequency band (1–50/60 Hz). These filters, known as low-pass and high-pass filters, respectively remove electrical noise below 1 Hz and above 50/60 Hz. For even more targeted noise removal, notch filters are employed. Unlike low-pass and high-pass filters that eliminate a range of frequencies, notch filters focus on a very specific frequency and suppress it. In EEG recordings, this technique is commonly used to eliminate the electrical noise from power lines, which typically oscillate in the range of 50–60 Hz depending on the regional electricity standard [71]. EEG signal processing often utilizes frequency domain filters to refine the bandwidth of interest. Specific filter types, like Butterworth, Chebyshev, and inverse Chebyshev, are favored due to their desirable characteristics in this application [80]. There exist some unique characteristics in each of them that require thorough analysis. The filtering toolbox proves to be a powerful weapon against EEG artifacts. Various filter types, including those mentioned previously, effectively remove unwanted noise and enhance the signal of interest [78]. However, caution must be exercised when applying filters. Generally, in the time domain, the EEG signal's waveform and structure is distorted using filters. Hence, filtering should be used sparingly to prevent loss of EEG signal information. This processing step also tackles low-frequency noise, like the hum from power lines, that sneaks into EEG recordings from external sources [81].

3.2.2 Artifact Removal

While band-pass filtering effectively mitigates EMG and EOG artifacts by removing frequencies outside the desired EEG range (typically 1–50/60 Hz), other sources like blinks and facial movements can remain prominent due to their spectral overlap with neural activity. These residual artifacts necessitate further processing techniques for optimal signal quality [82]. EEG recordings are susceptible to contamination from two primary artifact categories: extrinsic and intrinsic. Extrinsic artifacts originate from external sources, including power line interference (DC line noise) and electromagnetic interference from wireless signals. Band-pass filtering proves to be a valuable tool in mitigating these artifacts by attenuating frequencies associated with device drift and power line noise (typically 50/60 Hz depending on the region) [67]. Intrinsic artifacts, in contrast, arise from physiological processes within the subject's brain. These include electrical activity generated by eye movements (EOG), blinking, and muscle contractions (EMG) [83]. Addressing these intrinsic artifacts requires more sophisticated techniques compared to band-pass filtering. Common approaches include adaptive filtering, which dynamically adjusts to suppress noise, spatial filtering that utilizes electrode placement to isolate desired brain regions, and blind source separation techniques like independent component analysis (ICA) to separate the underlying brain signals from the contaminating artifacts [83]. Preprocessing is a complex process aimed at eliminating unwanted components present in the EEG signal. Effective pre-processing enhances signal quality, thereby improving feature distinguishability and classification performance. Nevertheless, pre-processing helps to distinguish between various signals and their sources. Tables 2 and 3 presents some of the popular methods of artifact removal used in pre-processing of EEG signals, along with their characteristics, advantages, limitations, literature usage statistics (in %) for last one decade and many other attributes [84].

3.3 EEG Feature Extraction Methods

After acquiring the noise-free signals from the signal enhancement phase (i.e., preprocessing and artifact removal), essential features were extracted from the brain signals which will be fed to the classifier in the next phase. Its main distinguishing characteristics include its structure, frequency, rhythmic variations, phase variations, power, latency, and reactivity [103]. This prompts the question: how does one determine which feature(s) to select for a specific task? Considering all the features are both computationally expensive and time consuming. Therefore, the objective of this phase is to capture one of the three types of features, they are: (a) time domain features (describing the shape and peak values of the waveforms), (b) frequency domain features (describing the spectrum of the EEG data during rhythm variations), (c) spatial domain features.

Time-domain features incorporate various metrics such as Hjorth features, higherorder crossing (HOC), event-related potential (ERP), principal component analysis (PCA), and Higuchi's fractal dimensions (FD), independent component analysis (ICA), which serve as an indicator of self-similarity and signal complexity within the domain. This kind of analysis is particularly relevant for applications that primarily deal with Event-Related Potentials (ERPs) and the shape and latencies of waveforms [104–106].

In order to extract and assess frequency domain characteristics, frequency domain analysis algorithms convert time-domain EEG data into frequency-domain signals. Typically, the EEG signal is split up into several sub-bands from which features are

Table 2 Comparative analysis on various EEG-based preprocessing methods

Citations	Preprocessing algorithms/ artifact removal techniques	Characteristics	Advantages	Limitations
[85–87]	Independent component analysis (ICA)	Without the use of reference channels, it analyzes the properties of the data to extract artifacts from EEG signals. Effective in eliminating ocular artifacts, it separates multi-channel EEG data into discrete components that are both temporally unique and spatially fixed	It breaks down signals into temporally autonomous and spatially consistent elements, which efficiently distinguishes artifacts from noise components	The success of ICA depends on specific conditions where one signal surpasses the others in magnitude. The effectiveness of the corrected signals is heavily influenced by the quality of the artifacts
[88–91]	Common Average Reference (CAR)	It is employed to create a reference for individual channels by averaging all recordings across each electrode, enhancing the signal-to-noise ratio quality	It surpasses traditional electrical referencing methods, achieving a noise reduction of over 30%	Calculating the average may pose challenges with incomplete head coverage, and limited sample density
[92–95]	Surface Laplacian (SL)	It provides high spatial resolution when viewing EEG data. Without requiring precise information regarding volume conduction, it calculates the current density entering or leaving the scalp through the skull while considering the external shape of the volume conductor	Since its estimates are not influenced by references, they produce identical SL estimates regardless of the EEG recording reference scheme. It improves the EEG signal's spatial resolution without demanding further functional neuroanatomy assumptions	Spline patterns and artifacts influence it

(continued)

Table 2 (continued)

Citations	Preprocessing algorithms/ artifact removal techniques	Characteristics	Advantages	Limitations
[96–99]	Principal Component Analysis (PCA)	It identifies patterns within data by rotating the coordinate axes, thereby aligning them with linear combinations of sets of time points, collectively representing patterns within the signal. This rotation aims to maximize the variance along the first axis while keeping the axes orthogonal	It contributes to reducing feature dimensions and facilitates data classification through ranking	It does not entirely eliminate noise, but it can diminish its presence. Compared to ICA, PCA compresses data and enables data separation
[100–102]	Common Spatial Patterns (CSP)	It utilizes spatial filters to differentiate between various classes of EEG signals, such as those associated with different types of motor activity	It does not necessitate pre-selection of specific frequency bands or prior knowledge about them	It relies on a large number of electrodes, and alterations in electrode placement can impact classification accuracies

Table 3 Comparative analysis on various EEG-based preprocessing methods on some additional parameters

Preprocessing algorithms/ artifact removal techniques	Automatic	Single channel usability	Percentage of literature usage (2010–2020)
ICA	NO	YES	26.8
CAR	YES	NO	5.0
SL	YES	NO	0.4
PCA	NO	YES	50.1
CSPS	YES	NO	17.7

extracted for analysis. These features include logarithmic energy spectrum, power spectral density (PSD), differential entropy (DE), higher-order spectrum (HOS) [107, 108]. The fast Fourier transform (FFT) is widely recognized as the most popular frequency-domain method. AR (Auto recessive) models serves as a replacement for Fourier-based techniques for analysing frequency spectrum of signal [109, 110]. Frequency-Time domain analysis techniques are founded on the assumption that signals based on EEG are linear and quasi-stationary, meaning the frequency content of the EEG signals remains constant throughout the analysis window. Conventional timedomain analysis techniques, which depend on features such as duration, amplitude, variance and auto-correlation, not suitable for analyzing signals that are non stationary such as EEG based signals in adults. Examining non-stationary signals necessitates insight into energy distribution across frequencies and fluctuations in frequency over time, aspects not addressed by either time-domain or frequency-domain analysis techniques. Frequency-domain representation sacrifices temporal information during spectrum construction, thereby imposing constraints. In response to these limitations, alternative tools have emerged for representing signals, referred to as time–frequency domain analysis techniques. This approach integrates information from both the time and frequency domains, facilitating localized analysis within the frequency-Time domain.

Results of which Features in the frequency domain are adept at capturing timevarying and non-stationary signals, enabling the characterization of diverse emotional states [71]. Among time–frequency analysis methods, the Wavelet transform emerges as the predominant approach. The Short-time Fourier transform (STFT) [72], Hilbert Huang transform (HHT) [122], and wavelet packet transform (WPT) [117] represent key methods in time–frequency domain analysis. The time–frequency domain capitalizes on variations in both time and frequency, delivering detailed descriptions of neural activities. To accomplish this, techniques such as wavelet transform (WT) and wavelet packet decomposition (WPD) are utilized [123]. Spatial-frequency domain analysis techniques are employed to analyze brain activity across various brain regions. Spatially localized features can be used when information from specific electrode positions needs to be isolated, extracted, and utilized exclusively [78]. Table 4 illustrates some of the commonly used algorithms in the feature extraction method along with their characteristics, advantages, limitations, literature usage statistics (in %) over the last decade, and several other attributes.

3.4 EEG Feature Selection and Dimensionality Reduction Methods

The process of feature selection is one of the most crucial phases as it identifies the signal characteristics that most accurately characterize the EEG parameters to be classified. The feature vector in BCI systems typically has a large dimensionality [124]. Feature selection aims to reduce the number of input variables for the classifier, distinct from dimensionality reduction, which merges features to decrease their amount. Feature selection methods do not alter the fundamental properties but rather exclude some based on specific criteria of usefulness. These methods strive to achieve optimal results by processing minimal data. They eliminate features that are unecessary, reduntant and do not participate in the classification for simpler, faster, and more effective classification models. Moreover, feature selection methods help

Table 4 Comparative analysis on various EEG-based feature extraction methods

Citations	Feature extraction algorithms	Characteristics	Advantages	Limitations	Feature extraction domain	Literature's usage statistics % (2010–2020)
[110–113]	Independent Component Analysis (ICA)	ICA functions as an algorithm for feature extraction as well as a signal enhancement technique. It isolates independent elements by leveraging the principle of statistical independence	Offers computational efficiency Demonstrates exceptional performance with large datasets Dissects signals into temporally independent and spatially fixed components	Not suitable for situations with low clarity Demands additional calculations to be made for the breakdown	Time In the frequency domain, there is exists a FastI-CA	11.3
	Principal Component Analysis (PCA)	Even though dimensionality reduction is one of the functions for the PCA technique, feature extraction is its primary objective. It diminishes signal extent by generating new, uncorrelated variables	An effective tool for data analysis and dimensionality reduction with minimal information loss	Struggles to handle complex manifold structures in data processing Relies on data being linear and continuous	Time	19.7

(continued)

Table 4 (continued)

Citations	Feature extraction algorithms	Characteristics	Advantages	Limitations	Feature extraction domain	Literature's usage statistics % (2010–2020)
[114–116]	Wavelet Transform (WT)	The WT approach uses discrete or continuous wavelets, which are safe and straightforward building blocks, to represent the actual EEG signal	Able to analyze signals with discontinuities by adjusting window size dynamically Able to extract features such as energy, distances, or clusters	Lack of a well-defined approach for applying wavelet transform to widespread noise Heisenberg uncertainty principle limits the performance	Time frequency	26
[109, 110]	Auto Regressive (AR)	It utilizes a parametric approach to estimate the power spectrum density (PSD) of EEG signals. The process of PSD estimation includes determining the parameters or coefficients of the linear system under examination	Needs shorter data record durations. Enhances frequency resolution and minimizes spectral loss issues	Not suitable for nonstationary signals Challenges arise in determining the model properties of EEG signals	Frequency	1.6

(continued)

Table 4 (continued)

Citations	Feature extraction algorithms	Characteristics	Advantages	Limitations	Feature extraction domain	Literature's usage statistics % (2010–2020)
[117–119]	Wavelet Packet Decomposition (WPD)	To demonstrate the breakdown procedure, it makes use of a sub-band tree structure facilitated by a complete binary tree It independently breaks down the original signals in an orthogonal manner, adhering to the conservation of energy principle Feature extraction involves extracting the energy distribution	Capable of analyzing non-stationary signals	Extended computation duration	Time frequency	1.6
[109, 110]	Fast Fourier Transformations (FFT)	Power spectral density (PSD) estimation is used to evaluate and calculate features of EEG signals in order to represent EEG sample signals selectively	Effective technique for frequency analysis	Highly sensitive to noise, resulting in decreased performance Only suitable for stationary signals and linear random processes	Frequency	2.2

(continued)

Table 4 (continued)

Citations	Feature extraction algorithms	Characteristics	Advantages	Limitations	Feature extraction domain	Literature's usage statistics % (2010–2020)
[104–106]	Hjorth features	These statistical indicators use normalized slope descriptors. They include activity, which measures the variance of a time function; mobility and complexity, which evaluates how close the signal is to a pure sine wave in relation to the frequency change	Suitable for real-time analysis due to their low computational expense	Potential bias may occur when calculating signal parameters statistically	Time	17
[107, 108, 120, 121]	Differential Entropy (DE)	Entropy indicates the scattering within data, while DE can capture spatial variations in signals	Entropy and related measures show how information moves around inside the brain	Entropybased measures can be sensitive to noise in the data, potentially leading to inaccurate assessments of information flow within the cortex	Time–spatial	4.9

to reduce the risk of over-fitting in standard datasets, flexible models, or datasets with numerous features but lacks sufficient observations. Feature selection methods can be classified into two categories based on the number of variables, these include: (1) Univariate and (2) Multi-variate. Uni-variate methods evaluate the input features individually, whereas multi-variate methods analyze the entire group of characteristics simultaneously. Another type of classification categorizes feature selection methodology into filtering, wrapper, and built-in algorithms.

Filtering methods access features based on the inherent characteristics of the data. Moreover, majority of these methods are uni-variate, as each attribute is independently assessed. These techniques are less computationally intensive, making them suitable for huge datasets. Depending on the type of classifier, wrapping techniques select additional features based on their ability to affect previously chosen characteristics, with only attributes that raise accuracy are maintained. Built-in techniques operate internally within the classification methods, like those used in deep learning techniques. It generally requires less computational resources as compared to wrapper techniques. Some of the popular feature selection algorithms include Stepwise discriminant analysis (SDA), Genetic Algorithms (GA), Fisher score and many more as described in Table 5 [78]. Genetic algorithms employ evolutionary methods to lower the size of the feature vector, retaining only the most informative features [125–127]. By introducing the stepwise function, the stepwise discriminant analysis algorithm enhances the statistical tool used for discriminant analysis [128]. Another feature selection technique that employs statistical metrics to determine the relationship between each feature and the output classes is Fisher score. Table 5 presents some of the widely used feature selection algorithms along with their literature usage statistics (in %) in last one decade [129].

3.5 EEG Signal Classification Methods

After acquiring the features, they can undergo post-processing analysis to derive meaningful insights. Typically, this involves training a classifier on these features. The classifier establishes a mapping between the EEG feature vectors and a classification scheme it acquires during training. Small variations can significantly influence the intricate patterns of EEG signals related to human cognition, highlighting the need for a highly efficient and resilient classifier. In a BCI system, this phase takes the feature vector, a summary of brain activity extracted earlier, and aims to decode the user's intentions based on those features. This objective can be accomplished through classification and regression techniques, with classification algorithms currently being the most preferred choice. Regression analysis predicts user intents by using features obtained from EEG signals as distinct variables. Conversely, classification algorithms define boundaries between various targets in the feature space by using the retrieved features as distinct variables [130]. According to a study [131], the selection of classifiers is influenced by three key factors: (a) the likelihood of features being affected by noise, (b) feature dimensionality, and (c) whether

Table 5 Comparative analysis on various EEG-based feature selection and dimensionality reduction methods

Citations	Feature selection algorithms	Properties	Advantages	Limitations	Percentage of literature usage (2010–2021)
[78]	Uni-variate	This method evaluates the input features one-byone	It is simple to implement and computationally efficient	It ignores interactions between features and could discard some relevant features	6.3
[78]	Multivariate	This method examines the entire group of features simultaneously	It considers relation-ships between features and can be more effective than uni-variate methods	It can be computationally expensive and may be sensitive to irrelevant features	6.3
[125–127]	Genetic algorithm	This method uses evolutionary techniques to reduce dimensionality of a feature vector, retaining only the most informative features	It can handle complex feature interactions and works well for non-linear relationships	It can be computationally expensive and parameter tuning can be challenging	32.3
[104, 129]	min-Redundancy Max Relevance (mRMR)	This algorithm assesses the common data between every feature and every class at the output for comparison	It reduces computational cost by selecting a small subset of nonredundant and relevant features. This improves model performance and enhances interpretability	It may not be suitable for problems with highly non-linear relationships between features and target	11.5

(continued)

Table 5 (continued)

Citations	Feature selection algorithms	Properties	Advantages	Limitations	Percentage of literature usage (2010–2021)
[128]	Stepwise Discriminant Analysis (SDA)	This algorithm enhances the statistical tool for discriminant analysis by integrating it with the stepwise function	It is efficient for feature selection in classification and considers class labels	It may be biased towards features that separate initial classes well and not suitable for highdimensional data	17.7
[129]	Fisher score	It is a feature selection technique that employs statistical metrics to determine the relationship between each feature and the output classes	It is simple to implement and works well for linear relationships	It ignores feature interactions and may not be suitable for non-linear relationships	7.3
[78, 129]	Wrapper method	Only features that improve accuracy are kept in this strategy, which depends on the type of classifier that selects new features based on their influence on previously selected features	It can be very effective in finding optimal subsets and handle complex interactions	They can be computationally very expensive, especially for high dimensional data	15.6
[78]	Built-in methods	This method operates internally within the classifier algorithms, such as those used in deep learning	They are efficient, promote sparsity, and some can be interpretable. Ex: Lasso regression algorithm	Their performance depends on the specific algorithm and may not be suitable for all data types	3.1

the features exhibit non-stationary characteristics. A research proposed [103], categorizes EEG based BCI classifiers into five major groups: (a) linear classifiers, (b) nearest neighbor classifiers (NN), (c) hybrid classifiers (or ensemble learners), (d) non-linear Bayesian classifiers, and (e) artificial neural networks (ANN), as described in Table 6, specifying their advantages and disadvantages along with the citations. Among these, linear classifiers are not commonly utilized in practical applications due to their limited learning capability. Classification methods transform the data extracted into specific motor oriented tasks, such as foot movements, hand gestures, word production, and so on in MI based BCIs [132]. Accurate classification of the same brain activity can be improved by combining distinct signal properties from different sources or devices. For example, fingertapping and hand/arm movement have been successfully identified using a combination of fNIRS and EEG techniques [133]. ML and DL techniques are utilized to recognize and detect EEG-based BCI signals. ML algorithms are typically categorized into three classes based on their outcomes, these include: supervised, unsupervised, and reinforcement learning [132]. Additionally, DL methodologies signified enhancements in classification accuracy. Table 6 showcases a range of classification algorithms, detailing their categories, characteristics, advantages, limitations, literature usage statistics (in %) in last one decade, examples and various other attributes.

Some of the well-known classification algorithms include linear discriminant analysis (LDA), multilayer perceptron (MLP), convolutional neural network (CNN) [134], k-nearest neighbors (k-NN), support vector machine (SVM) [132], and decision trees [135]. Notably, the SVM classifier demonstrates superior performance compared to other classification algorithms like k-NN, MLP, LDA, etc. A comprehensive comparative analysis [114] was conducted to study three classification models for their accuracies, namely: k-NN, MLP, and SVM classifiers, with the objective to identify the optimal features based on wavelets. In general, all the three models performed very well, achieving accuracy above 98%. SVM took the top spot with 98.75% accuracy for approximate coefficients (A4) related to delta waves (0.53–3.06 Hz) while the Multi-Layer Model (MLM) reached 98.57% accuracy for detailed coefficients (D4) associated with theta waves (3.06–6.12 Hz). Moreover, both of these classifiers come under the category of non-linear classification method. During the assessment of classification accuracies for SVM, backpropagation neural network (BPNN), and SVM, it was observed that both LDA and SVM classifiers exhibited comparable high accuracies, surpassing those of BPNN [136]. Deep Learning models (DL) bypass the need for an explicit feature extraction phase as they can learn directly from EEG data over a period of time. [137–139]. Deep learning streamlines the process by automatically identifying relevant features and incorporating them directly into the classification model [67], removing the separate feature extraction and selection stage. However, deep learning models exhibit high computational complexity, necessitating extensive training period.

In EEG-based BCI research, popular DL models include Recurrent Neural Networks (RNNs), Convolutional Neural Networks (CNNs), General Adversarial Networks (GANs), and Auto-encoders. Among these, CNNs are frequently favored and often demonstrate superior performance compared to other algorithms [36].

Table 6 Comparative analysis on various general classification methods

Types of classifier	Characteristics	Advantages	Limitations	Classification algorithms (with citations)	Percentage of literature usage (2010–2021)
Linear classifiers	These algorithms create decision boundaries, like straight lines or flat planes, using simple equations to divide data points into different categories	These algorithms exhibit a desirable balance between classification accuracy and the ability to generalize to unseen data	These algorithms often exhibit sub optimal performance when processing complex nonlinear EEG signal data	Discriminant analysis, Bayesian Linear, Graph Regularized Sparse Linear Regularized GRSLR [32], Linear Discriminant Analysis LDA, Support Vector Machine SVM [125, 126], Long Short-term Memory Recurrent Neural Network LSTM-RNN [65–68], Multilayer Perceptron MLP [127]	5.5 1.4 30.3 0.02
Nearest Neighbor Classifiers (NN)	These algorithms make predictions by comparing new data to known categories based on similarity	kNN shines in asynchronous BCI due to its effectiveness with lowdimensional data, a common characteristic of BCI feature vectors	High dimensionality and noisy features can weaken the effectiveness of kNN for classification tasks	Mahalanobis Distance [133], k-Nearest Neighbors kNN [132]	4.5 0.1

(continued)

Table 6 (continued)

Types of classifier	Characteristics	Advantages	Limitations	Classification algorithms (with citations)	Percentage of literature usage (2010–2021)
Nonlinear-Bayesian classifiers	Unlike some classifiers, generative models can create flexible decision boundaries to separate data classes	Generative classifiers effectively identify and discard uncertain samples	In the case of Bayes quadratic, accurately estimating the covariance matrix becomes challenging when dealing with high dimensionality and insufficient training sample patterns	Hidden Markov Model HMM [52, 103], Bayes quadratic [130]	0.1 0.3
Artificial Neural Networks (ANN)	NNs are discriminant algorithms specifically designed to identify root causes within a dataset, copying the operations of the human brain	Neural networks typically achieve high classification accuracy	EEG signals are noisy and constantly changing, making it difficult for models to learn without overfitting	Convolutional Neural Network (CNN) [67, 69, 129–131, 140–142], Complex-Valued Convolutional Neural Network (CVCNN) [125], are among the models employed	1.6 1.1 0.2 46.16 0.4 0.4
Hybrid classifiers (ensemble learners)	Hybrid classifiers employ techniques like boosting, voting, and stacking. Boosting involves a series of cascading classifiers. In voting, classifi ers assign scores to each class, resulting in a combined score and final class label	Reduction in variance resulting in improved classification accuracy	The quality measures vary depending on the context of the application	Ensemble methods have the capability to amalgamate various types of classifiers [135] Examples include random forest [10, 134], bagging tree [131, 135], XGBoost [114], and AdaBoost [136]	2.1 1.1 0.2 0.4 3.9

However, GANs tackle a crucial issue in EEG applications. By employing generative classifiers, GANs can mimic EEG signals, thereby expanding the training dataset, potentially improving the functioning of various DL classifiers. In a comparative study between Principal Component Analysis (PCA) and Recurrent Neural Network (RNN), RNN attained a control accuracy of 94.5% within a time expenditure of 0.61 ms, whereas PCA achieved a control accuracy of 93.1% with a time overhead of 0.48 ms [143]. Table 7 provides a comparative analysis of different classifiers utilized in EEG systems.

Table 7 Comparative analysis on various EEG-based classification methods

Citations	Classification algorithm applied	Advantages	Disadvantages
[144]	Linear Discriminant Analysis (LDA)	It requires minimal computational resources Easy to operate Yields effective outcomes	It struggles when the discriminatory function is based on variance rather than mean of the features. Additionally, it may not adequately capture complex structures in non Gaussian distributions
[145]	Support Vector Machine (SVM)	It demonstrates robust generalization capabilities Outperforms other linear classifiers in terms of performance	It exhibits significant computational complexity
[146]	Artificial Neural Network (ANN)	Simplicity in application Robust in nature Involves straight forward computations Demands minimal training data	Challenging to construct Performance linked to hidden layer neuron count
[147]	Naive Based Classifier (NBC)	Only a small training dataset is needed for parameter estimation Only the variance of class variables needs computation, without the necessity of computing the entire covariance matrix	Struggles to provide accurate estimates for the probabilities of correct classes
[30]	k-Neural Network (KNN)	Highly understandable Straightforward to implement and troubleshoot	Runtime performance suffers with large training sets Prone to the influence of irrelevant and redundant features Under performs when compared to alternative classification methods on challenging tasks

In [114], it was found that Artificial Neural Networks (ANNs) outperform SVM classifiers when dealing with limited data. A deep learning framework based on Convolutional Neural Networks (CNNs) is employed for continuous decoding of EEG signals related to motor imagery (MI) across various individuals. Results from the publicly accessible BCI competition IV-2b dataset reveal that employing stochastic gradient descent and adaptive moment estimation as training methods produce average decoding accuracies of 71.49% (with variance of 0.42) and 70.84% (with a variance of 0.42), respectively [148, 149].

4 Conclusion

Over the past four decades, a multitude of Brain-Computer Interface (BCI) applications have emerged across diverse fields, spanning entertainment to medicine and even data security frameworks. While BCI systems have primarily prioritized improving their precision, reliability, and user-friendliness, insufficient attention has been devoted to safeguarding these devices and the sensitive data they gather. Several studies have highlighted regarding rapid interest in BCI applications across various fields including medicine, organizational management, transportation, gaming, entertainment, security and authentication. Additionally, it showcases the diverse range of devices employed for capturing brain signals. Till now, BCI systems have primarily concentrated on improving their accuracy, dependability, and user accessibility, with insufficient emphasis placed on securing these devices and the confidential data they gather. This article presents a comprehensive review analysis focusing on EEG-based Brain-Computer Interface (BCI), with particular emphasis on examining its methodological strengths and weaknesses. The field of EEG-based BCI in smart home environment has gained considerable attention within the computing domain. Substantial progress in creating affordable BCI devices with improved usability has spurred numerous research endeavors. This work performs an extensive analysis on different algorithms and methods that can be part of EEG-based BCI in Smart Home Environment. These include: (1) EEG signal acquisition and segmentation, (2) EEG signal pre-processing, (3) EEG feature extraction, (4) EEG feature selection and dimensionality reduction, and (5) EEG signal classification methods. As observed in this review, computational methods lack standardized approaches across various applications, prompting researchers to persist in their quest for new solutions in an ongoing endeavor. Exploring the relationship between brain signals and emotions proves to be a complex challenge, with continuous presentation of innovative methods and implementations. We anticipate that many existing challenges will soon find resolution, thereby opening up a wide array of potential applications using EEG based BCI systems.

References

1. Weiss, K., Khoshgoftaar, T.M., Wang, D.: A survey of transfer learning. J. Big Data **3**(1), 9 (2016). https://doi.org/10.1186/s40537-016-0043-6
2. Ali, S.M.M., Augusto, J.C., Windridge, D.: A survey of user-centred approaches for smart home transfer learning and new user home automation adaptation, Appl. Artif. Intell. **33**(8), 747–774 (2019)
3. Rashidi, P., Cook, D.J.: Keeping the resident in the loop: Adaptingthe smart home to the user. IEEE Trans. Syst. Man Cybern. Part Syst. Hum. **39**(5). 949–59 (2009a). https://doi.org/10.1109/TSMCA.2009.2025137
4. Weiser, M.: The computer for the twenty-first century. Sci. Amer. **265**(3), 94–104 (1991)
5. Wu, C.-L., Fu, L.-C.: Design and realization of a framework for human–system interaction in smart homes. Trans. Syst. Man Cybern. Part Syst. Hum. **42**(1), 15–31 (2012). https://doi.org/10.1109/TSMCA.2011.2159584. Keywords: {Smart homes;Privacy;TV;Pervasive computing;Quality of service;Humans;Security;Human centric;human–computer interaction;multiagent systems;smart homes;user-centered design}
6. Chang, E., Dillon, T., Calder, D.: Human system interaction with confidentcomputing. In: Proceedings of Conference on Human System Interactions, pp. 1–11. Krakow, Poland (2008)
7. Bu, Y., Guo, P.: Voice orientation recognition: New paradigm of speech-based human-computer interaction. Int. J. Hum. Comput. Interact. (2023). https://doi.org/10.1080/10447318.2023.2233128
8. Turow, J.: Journalism and the voice intelligence industry. Digit. J. **9**(7), 1000–1006 (2021)
9. Pham Thi, T.D., Duong, N.T.: Do people intend to use AI voice assistants? An empirical study in Vietnam. J. Hum. Behav. Soc. Environ., 1–20 (2022)
10. Bazan-Sulzberger, A., Arribas-Sanchez, V., Sanz-Fernandez, J., de Castro-Gonzalez, J., Perez-Pomares, J.F.: Multipath effects characterization on air-to-air analogue voice AM communications. IEEE Trans. Aerosp. Electron. Syst. **53**(3), 1109–1118 (2017)
11. Park, S., Ha, J., Cha, H.-S., Lee, K.-G., Im, C.-H.: Development of an online home appliance control system for the elderly based on SSVEP-based brain-computer interface: A feasibility study. In: 2021 9th International Winter Conference on Brain-Computer Interface (BCI), pp. 1–2. Gangwon, Korea (South) (2021). https://doi.org/10.1109/BCI51272.2021.9385365
12. Si-Mohammed, H., et al.: Towards BCI-based interfaces for augmented reality: Feasibility, design and evaluation. IEEE Trans. Vis. Comput. Graph. (2019)
13. Volosyak, I., Gembler, F., Stawicki, P.: Age-related differences in ssvep-based BCI performance. Neurocomputing **250**, 57–64 (2017)
14. Cook, D.J.: Multi-agent smart environments. J. Ambient. Intell. Smart Environ. **1**(1), 51–55 (2009)
15. Mckeever, S., Juan, Y., Coyle, L., Bleakley, C., Dobson, S.: Activity recognition using temporal evidence theory. J. Ambient. Intell. Smart Environ. **2**(3), 253–69 (2010)
16. Rao, S.P., Cook, D.J.: Predicting inhabitant action using action and task models with application to smart homes. Int. J. Artif. Intell. Tools **13**, 81–100 (2004)
17. Bakar, U.A.B.U.A., Ghayvat, H., Hasanm, S.F., Mukhopadhyay, S.C.: Activity and anomaly detection in smart homes: A survey. In: Mukhopadhyay (ed.) Next Generation Sensors and Systems, Smart Sensors, Measurement and Instrumentation, vol 16, pp. 191–220. Cham: Springer (2016)
18. Ni, Q., Beln, A., Hern, G., De La Cruz, I.P.: The elderly's independent living in smart homes: A characterization of activities and sensing infrastructure survey to facilitate services development. Sensors **15**(5), 11312–62 (2015)
19. Azkune, G., Almeida, A., Ipiña, D.L.-D., Chen, L.: Combining users activity survey and simulators to evaluate human activity recognition systems (2015)
20. Jebara T (2004) Generative versus discriminative learning. In: Ismail, M. (ed.) Machine learning: discriminative and generative, pp. 17–60. Boston, MA: Springer US (2004)
21. Lei, Z., Yue, S., Chen, Y., Yuanchun, S.: SHSim: An OSGI-based smart home simulator. In: IEEE International Conference on Ubi-media Computing, pp. 87–90. Jinhua, China (2010)

22. Sarkar, A.M., Jehad, Y.-K.L., Lee, S.: A smoothed Naive Bayes-based classifier for activity recognition (2010)
23. Lance, B.J., Kerick, S.E., Ries, A.J., Oie, K.S., McDowell, K.: Brain-computer interface technologies in the coming decades. Proc. IEEE **100**, 1585–1599 (2012)
24. Yang, D., Huang, R., Yoo, S.-H., Shin, M.-J., Yoon, J.A., Shin, Y.-I., Hong, K.-S.: Detection of mild cognitive impairment using convolutional neural network: Temporal feature maps of functional near-infrared spectroscopy. Front. Aging Neurosci. **12** (2020)
25. Kosmyna, N., Tarpin-Bernard, F., Bonnefond, N., Rivet, B.: Feasibility of BCI control in a realistic smarthome environment. Front. Hum. Neurosci. **10**, 1–10 (2016)
26. Dhabale, R.: Literature survey on BCI conrolled robot **5**(7), 2051–2056 (2016)
27. Tanaka, M.: Effects of mental fatigue on brain activity and cognitive performance: A magnetoencephalography study. Anat. Physiol. **4**, 1–5 (2015)
28. Birn, R.M.: The role of physiological noise in resting-state functional connectivity. Neuroimage **62**, 864–870 (2012)
29. Naseer, N., Hong, K.-S.: fNIRS-based brain-computer interfaces: A review. Front. Hum. Neurosci. **9** (2015)
30. Abdulwahab, S.S., Khleaf, H.K., Jassim, M.H.: A survey in implementation and applications of electroencephalograph (EEG)-based brain-computer interface. Eng. Technol. J. (2021)
31. Zhang, X.: A survey on deep learning based brain computer interface: Recent advances and new frontiers. In: SIGIR 2019—Proceedings of 42nd International ACM SIGIR Conference on Research and Development in Information Retrieval, vol. 1, no. 1, pp. 1281–1284 (2019). https://doi.org/10.1145/nnnnnnn.nnnnnnn
32. Chaisaen, R., et al.: Decoding EEG rhythms during action observation, motor imagery, and execution for standing and sitting. IEEE Sens. J. 1–11 (2020). https://doi.org/10.1109/jsen.2020.3005968
33. Jebelli, H., Asce, A.M., Hwang, S., Lee, S., Asce, M.: EEG signal-processing framework to obtain high-quality brain waves from an off-the-shelf wearable EEG device **32**(1), 1–12. (2018). https://doi.org/10.1061/(ASCE)CP.1943-5487.0000719
34. Welcome to the OpenBCI Community OpenBCI Documentation. https://docs.openbci.com/docs/Welcome. Accessed 04 Aug 2020
35. MindWave. https://store.neurosky.com/pages/mindwave. Accessed 04 Aug 2020
36. Al-Saegh, A., Dawwd, S.A., Abdul-Jabbar, J.M.: (2021) Deep learning for motor imagery EEG-based classification: A review. Biomed. Signal Process. Control. **63**. Elsevier Ltd (2021). https://doi.org/10.1016/j.bspc.2020.102172
37. Lotte, F., et al.: A review of classification algorithms for EEG-based brain—computer interfaces to cite this version : A review of classification algorithms for EEG-based brain-computer interfaces. Hum. Brain Mapp. **38**(11), 270–278 (2018). https://doi.org/10.1002/hbm.23730
38. Dai, M., Zheng, D., Na, R., Wang, S., Zhang, S.: EEG classification of motor imagery using a novel deep learning framework. Sensors (Switzerland) **19**(3), 1–16 (2019). https://doi.org/10.3390/s19030551
39. Batres-Mendoza, P., et al.: Quaternion-based signal analysis for motor imagery classification from electroencephalographic signals. Sensors (Switzerland) **16**(3) (2016). https://doi.org/10.3390/s16030336
40. Vishwakarma, R., Khwaja, H., Samant, V., Gaude, P., Gambhir, M., Aswale, S.: EEG signals analysis and classification for bci systems: A review. Int. Conf. Emerg. Trends Inf. Technol. Eng. IC-ETITE 2020, 1–6 (2020). https://doi.org/10.1109/icETITE47903.2020.066
41. Schmalstieg, D., Bornik, A., Müller-Putz, G.R., Pfurtscheller, G.: Gaze-directed ubiquitous interaction using a brain-computer interface. In: Proceedings of the 1st Augmented Human International Conference (2010)
42. Vertegaal, R., Dickie, C., Sohn, C., Flickner, M.: Designing attentive cell phone using wearable eyecontact sensors. In CHI'02: CHI'02 extended abstracts on Human factors in computing systems, pp. 646–647. New York, NY, USA, ACM (2002)
43. Mullen, T.R., Huang, Q., Reich, J.: Towards ubiquitous brain-aware computing: A preliminary EEG study (2008)

44. Shalk, G., McFarland, D.J., Hinterberger, T., Birbaumer, N., Wolpaw, J.: BCI2000: A general-purpose brain-computer interface (BCI) system. IEEE Trans. BME **51**(6), 1034–1043 (2004)
45. Chandra Babu, K.M., Harsha Vardhini, P.A.: Brain computer interface based arduino home automation system for physically challenged. In: 2020 3rd International Conference on Intelligent Sustainable Systems (ICISS), pp. 125–130 (2020)
46. Ramesh, S., Krishna, M.G., Nakirekanti, M.: Brain computer interface system for mind controlled robot using bluetooth. Int. J. Comput. Appl. **104**, 20–23 (2014)
47. Shilaskar, S., Talwekar, S., Bhatlawande, S., Singh, S., Jalnekar, R.M.: An electroencephalogram based detection of hook and span hand gestures. In: 2022 IEEE Pune Section International Conference (PuneCon), pp. 1–6 (2022)
48. Zhang, W., Tan, C., Sun, F., Wu, H., Zhang, B.: A review of EEG-based brain-computer interface systems design. Brain Sci. Adv. **4**, 156–167 (2018)
49. Zakaria, M.H.F., Mansor, W., Lee, K.Y.: Time-frequency analysis of executed and imagined motor movement EEG signals for neuro-based home appliance system. In: Proceedings of the TENCON 2017—2017 IEEE Region 10 Conference, pp. 1657–1660. Penang, Malaysia (2017)
50. Rani, M.S.B.A., Mansor, W.B.: Detection of eye blinks from EEG signals for home lighting system activation. In: Proceedings of the 2009 6th International Symposium on Mechatronics and its Applications, pp. 1–4. Sharjah, UAE (2009)
51. Yang, D., Nguyen, T., Chung, W.: A bipolar-channel hybrid brain-computer interface system for home automation control utilizing steady-state visually evoked potential and eye-blink signals. Sensors (Basel, Switzerland) **20** (2020)
52. Shi, T., Wang, H., Cui, W., Ren, L.: Indoor space target searching based on EEG and EOG for UAV. Soft. Comput. **23**(21), 11199–11215 (2019)
53. Lee, D-H., Jeong, J.-H., Ahn, H.-J., Lee, S.-W.: Design of an EEG based drone swarm control system using endogenous BCI paradigms. In: Proc. IEEE 9th International Winter Conference on Brain–Computing Interface (BCI), pp. 1–5 (2021)
54. Kumar, R., Srinivas, K.K., Peddi, A., Vardhini, P.A.: Artificial intelligence based human attention detection through brain computer interface for health care monitoring. In: 2021 IEEE International Conference on Biomedical Engineering, Computer and Information Technology for Health (BECITHCON), pp 42–45 (2021)
55. Çigdem, I.A., Kayaa, M., Mishchenko, Y.: Distinguishing mental attention states of humans via an EEG-based passive BCI using machine learning methods. Expert Syst Appl **134**, 153–166 (2019)
56. Sisinni, E., Saifullah, A., Han, S., Jennehag, U., Gidlund, M.: Industrial internet of things: challenges, opportunities, and directions. IEEE Trans. Ind. Informat. **14**(11), 4724–4734 (2018)
57. Li, P., Meziane, R., Otis, M.J.-D., Ezzaidi, H., Cardou, P.: In: Proceedings of IEEE International Symposium on Robotic and Sensors Environments (ROSE), pp. 55–60 (2014)
58. Von Rosenberg, W., Chanwimalueang, T., Goverdovsky, V., Looney, D., Sharp, D., Mandic, D.P.: Smart helmet: Wearable multichannel ECG and EEG. IEEE J. Transl. Eng. Health Med. **4**(2700111) (2016)
59. Lotte, F., Congedo, M., Lécuyer, A., Lamarche, F., Arnaldi, B.: A review of classification algorithms for EEG-based brain-computer interfaces. J. Neural Eng. **4**(2), R1–R13 (2007)
60. Collins, N.: Hawking in the future brains could be separated from the body. Telegraph **20**, 2013 (2013)
61. Allison, B.Z., Wolpaw, E.W., Wolpaw, J.R.: Brain-computer interface systems: progress and prospects. Expert Rev. Med. Devices **4**, 463–474 (2007)
62. Jiang, L., Stocco, A., Losey, D.M., Abernethy, J.A., Prat, C.S., Rao, R.P.: Brainnet: A multi-person brain-to-brain interface for direct collaboration between brains. Sci. Rep. **9**, 1–11 (2019)
63. Coogan, C.G., He, B.: Brain-computer interface control in a virtual reality environment and applications for the internet of things. IEEE Access **6**, 10840–10849 (2018)

64. Jamil, N., Belkacem, A.N., Ouhbi, S., Lakas, A.: Non invasive electroencephalography equipment for assistive, adaptive, and rehabilitative brain-computer interfaces: A systematic literature review. Sensors **21**, 4754 (2021)
65. Mane, R., Chouhan, T., Guan, C.: BCI for stroke rehabilitation: Motor and beyond. J. Neural Eng. **17**, 041001 (2020)
66. Baniqued, P.D.E., Stanyer, E.C., Awais, M., Alazmani, A., Jackson, A.E., MonWilliams, M.A., Mushtaq, F., Holt, R.J.: Brain–computer interface robotics for hand rehabilitation after stroke: A systematic review. J. Neuroeng. Rehabil. **18**, 15 (2021)
67. Huang, Z., Wang, M.: A review of electroencephalogram signal processing methods for brain-controlled robots. Cogn. Robot. **1**(3), 111–124 (2021)
68. Lazarou, I., Nikolopoulos, S., Petrantonakis, P.C., Kompatsiaris, I., Tsolaki, M.: EEG based brain– computer interfaces for communication and rehabilitation of people with motor impairment: A novel approach of the 21st century. Front. Hum. Neurosci. **12**(January), 1–18 (2018). https://doi.org/10.3389/fnhum.2018.00014
69. Rojas, G.M., Alvarez, C., Montoya, C.E., de la Iglesia-Vayá, M., Cisternas, J.E., Gálvez, M.: Study of resting-state functional connectivity networks using EEG electrodes position as seed. Front. Neurosci. **12**, 235 (2018). https://doi.org/10.3389/fnins.2018.00235
70. Ajmeria, R., et al.: A critical survey of EEG-based BCI systems for applications in industrial internet of things. IEEE Commun. Surv. Tutor. **25**(1), 184–212 (2023). https://doi.org/10.1109/COMST.2022.3232576
71. Houssein, E.H., Hammad, A., Ali, A.A.: Human emotion recognition from EEG-based brain–computer interface using machine learning: a comprehensive review. Neural Comput. Applic. **34**, 12527–12557 (2022)
72. Jiang, W., Mei, W.: Review of the emotional feature extraction and classification using EEG signals. Cogn. Robot. (2021)
73. Akkar, P.H.A.R., Jasim, F.A.: Brain computer interface technology and applications, a survey review. Int. J. Comput. Appl. **7**(3), 101–111 (2017)
74. Khosla, A., Khandnor, P., Chand, T.: A comparative analysis of signal processing and classification methods for different applications based on EEG signals. Biocybern. Biomed. Eng. **40**(2), 649–690 (2020)
75. Ameera, A., Saidatul, A., Ibrahim, Z.: Analysis of EEG spectrum bands using power spectral density for pleasure and displeasure state. IOP Conf. Ser. Mater. Sci. Eng. **557**(1). https://doi.org/10.1088/1757-899X/557/1/012030
76. Girouard, A., et al.: Distinguishing difficulty levels with non-invasive brain activity measurements. In: Lecture Notes in Computer Science (including subseries Lecture Notes in Artificial Intelligence and Lecture Notes in Bioinformatics), vol. 5726 LNCS, no. PART 1, pp. 440–452 (2009). https://doi.org/10.1007/978-3-642-036552_50
77. Zhang, J., Yin, Z., Chen, P., Nichele, S.: Emotion recognition using multi-modal data and machine learning techniques: A tutorial and review. Inf. Fusion **59**, 103–126 (2020)
78. Torres, E.P., Torres, E.A., Hernández-Álvarez, M., Yoo, S.G.: EEG-based BCI emotion recognition: A survey. Sensors **20**(18), 5083 (2020)
79. LaRocco, J., Le, M.D., Paeng, D.-G.: A systemic review of available low-cost EEG headsets used for drowsiness detection. Front. Neuro. Inform. **14**(553352) (2020)
80. Bustamante, P.A., Lopez Celani, N.M., Perez, M.E., Quintero Montoya, O.L.: Recognition and regionalization of emotions in the arousal-valence plane. In: Proceedings of the 2015 Milano, Italy 37th Annual International Conference of the IEEE Engineering in Medicine and Biology Society (EMBC), vol. 2015, pp. 6042–6045. Milano, Italy (2015)
81. Sanei, S., Chambers, J.A.: EEG signal processing. Wiley & Sons: Hoboken, NJ, USA (2013). ISBN 9780470025819
82. Jiang, X., Bian, G.: Removal of artifacts from EEG signals: A review, pp. 1–18 (2019). https://doi.org/10.3390/s19050987
83. Tandle, A., Jog, N.: Classification of artefacts in EEG signal recordings and overview of removing techniques. Int. J. Comput. Appl. **975**(7), 8887 (2015)

84. Abhang, P.A., Suresh, C., Mehrotra, B.W.G.: Introduction to EEG-and speech-based emotion recognition. Elsevier: Amsterdam, The Netherlands (2016). ISBN 9780128044902
85. De Lathauwer, L., De Moor, B., Vandewalle, J.: An introduction to independent component analysis. J. Chemometr. Soc. **14**(3), 123–149 (2000)
86. Vigário, R., Sarela, J., Jousmiki, V., Hamalainen, M., Oja, E.: Independent component approach to the analysis of EEG and MEG recordings. IEEE Trans. Biomed. Eng. **47**(5), 589–593 (2000)
87. Sun, L., Liu, Y., Beadle, P.J.: Independent component analysis of EEG signals. In: Proceedings of IEEE International Workshop on VLSI Design and Video Technology, pp. 219–222 (2005)
88. Alhaddad, M.J.: Common average reference (CAR) improves P300 speller. Int. J. Eng. Technol. **2**(3), 21 (2012)
89. Syam, H.F., Lakany, H., Ahmad, R., Conway, B.A.: Comparing common average referencing to Laplacian referencing in detecting imagination and intention of movement for brain computer interface. In: Proceedings of MATEC Web Conference, vol. 140, p. 1028 (2017)
90. Alhaddad, M.J., Kamel, M., Malibary, H., Thabit, K., Dahlwi, F., Hadi, A.: P300 speller efficiency with common average reference. In: Lecture Notes in Computer Science, vol. 7326 LNAI, pp. 234–241 (2012)
91. Alhaddad, M.J., Kamel, M., Malibary, H., Thabit, K., Dahlwi, F., Hadi, A.: P300 speller efficiency with common average reference. In: Proceedings of the International Conference on Autonomous and Intelligent Systems, pp. 234–241. Aveiro, Portugal (2012)
92. Murugappan, M., Nagarajan, R., Yaacob, S.: Combining spatial filtering and wavelet transform for classifying human emotions using EEG Signals. J. Med. Biol. Eng. **31**, 45–51 (2011)
93. Burle, B., Spieser, L., Roger, C., Casini, L., Hasbroucq, T., Vidal, F.: Spatial and temporal resolutions of EEG: Is it really black and white? A scalp current density view. Int. J. Psychophysiol. **97**, 210–220 (2015)
94. Mazumder, I.: An analytical approach of EEG analysis for emotion recognition. In: Proceedings of the 2019 Devices for Integrated Circuit (DevIC), pp. 256–260. Kalyani, India (2019)
95. Subasi, A., Gursoy, M.I.: EEG signal classification using PCA, ICA, LDA and support vector machines. Expert Syst. Appl. **37**, 8659–8666 (2010)
96. Wold, S., Esbensen, K., Geladi, P.: Principal component analysis. Chemometr. Intell. Lab. Syst. **2**(1–3), 37–52 (1987)
97. Turnip, A., Junaidi, E.: Removal artifacts from EEG signal using independent component analysis and principal component analysis. In: Proceedings of IEEE 2nd International Conference on Technology, Informatics, Management, Engineering & Environment (2014)
98. Bugli, C., Lambert, P.: Comparison between principal component analysis and independent component analysis in electroencephalograms modelling. Biometrical J. Math. Methods Biosci. **49**(2), 312–327 (2007)
99. Pallavicini, F., Ferrari, A., Pepe, A., Garcea, G.: Effectiveness of virtual reality survival horror games for the emotional elicitation: Preliminary insights using resident evil 7: Biohazard. In: International Conference on Universal Access in Human-Computer Interaction. Springer: Cham, Switzerland (2018)
100. Yong, X., Ward, R.K., Birch, G.E.: Robust common spatial patterns for EEG signal preprocessing. In: Proceedings of the 30th Annual International Conference of the IEEE Engineering in Medicine and Biology Society, EMBS'08-"Personalized Healthcare through Technology", pp. 2087–2090. Boston, MA, USA (2011)
101. Li, X., Fan, H., Wang, H., Wang, L.: Common spatial patterns combined with phase synchronization information for classification of EEG signals. Biomed. Signal Process. Control **52**, 248–256 (2019)
102. Interfaces, B.: A tutorial on EEG signal processing techniques for mental state recognition in brain-computer interfaces. Guid. Brain Comput. Music. Interfacing (2014)
103. Haider, A.: A brief review of signal processing for EEG-based BCI: Approaches and opportunities. In: Proceedings of IEEE International Conference on Electro-Information Technology (EIT), pp. 389–394 (2021)

104. Jenke, R., Peer, A., Buss, M.: Feature extraction and selection for emotion recognition from EEG. IEEE Trans. Aect. Comput. **5**, 327–339 (2014)
105. Al-Fahoum, A.S., Al-Fraihat, A.A.: Methods of EEG signal features extraction using linear analysis in frequency and time-frequency domains. ISRN Neurosci. **2014**, 1–7 (2014)
106. Petrantonakis, P.C., Hadjileontiadis, L.J.: Emotion recognition from brain signals using hybrid adaptive filtering and higher order crossings analysis. IEEE Trans. Aect. Comput. **1**, 81–97 (2010)
107. Murugappan. M., Subbulakshmi, M.: Human emotion recognition through short time electroencephalogram (EEG) signals using fast fourier transform (FFT). In: 2013 IEEE 9th international colloquium on signal processing and its applications. IEEE, pp 289–294 (2013)
108. Yoon, H.J., Chung, S.Y.: EEG-based emotion estimation using Bayesian weighted-logposterior function and perceptron convergence algorithm. Comput. Biol. Med. **43**(12), 2230–2237 (2013)
109. Asadi Ghanbari, A., Nazari Kousarrizi, M.R., Teshnehlab, M., Aliyari, M.: An evolutionary artifact rejection method for brain computer interface using ICA. Int. J. Electr. Comput. Sci. **9**(9), 48–53 (2009)
110. Iacoviello, D., Petracca, A., Spezialetti, M., Placidi, G.: A real time classification algorithm for EEG-based BCI driven by self induced emotions. Comput. Methods Programs Biomed. **122**(3), 293–303 (2015)
111. Torse, R.M., Pujari, S.: Nonlinear blind source separation for EEG signal preprocessing in brain–computer interface system for epilepsy. Int. J. Comput. Appl. **50**(14), 12–19 (2012)
112. Mahajan, M.V., Rajput, S.M.: Classification of EEG using PCA, ICA and neural network. Int. J. Eng. Adv. Technol. **1**(1), 80–83 (2011)
113. Kumar, P.N., Kareemullah, H.: EEG signal with feature extraction using SVM and ICA classifiers. In: Proceedings of International Conference on Information Communication and Embedded Systems (ICICES), pp. 1–7 (2014)
114. Amin, H.U., et al.: Feature extraction and classification for EEG signals using wavelet transform and machine learning techniques. Aust. Phys. Eng. Sci. Med. **38**(1), 139–149 (2015)
115. Prochazka, A., Kukal, J., Vysata, O.: Wavelet transform use for feature extraction and EEG signal segments classification. In: Proceedings of IEEE 3rd International Symposium on Communications, Control and Signal Processing, pp. 719–722 (2008)
116. Alturki, F.A., AlSharabi, K., Abdurraqeeb, A.M., Aljalal, M.: EEG signal analysis for diagnosing neurological disorders using discrete wavelet transform and intelligent techniques. Sensors **20**(9), 2505 (2020)
117. Ting, W., Guo-Zheng, Y., Bang-Hua, Y., Hong, S.: EEG feature extraction based on wavelet packet decomposition for brain computer interface. Measurement **41**(6), 618–625 (2008)
118. Zhang, D.: Wavelet transform. In: Fundamentals of Image Data Mining, pp. 35–44. Heidelberg, Germany: Springer (2019)
119. Krishnaveni, V., Jayaraman, S., Aravind, S., Hariharasudhan, V., Ramadoss, K.: Automatic identification and removal of ocular artifacts from EEG using wavelet transform. Meas. Sci. Rev. **6**(4), 45–57 (2006)
120. Bălan, O., Moise, G., Moldoveanu, A., Leordeanu, M., Moldoveanu, F.: Fear level classification based on emotional dimensions and machine learning techniques. Sensors **19**, 1738 (2019)
121. Lee, Y.Y., Hsieh, S.: Classifying different emotional states by means of eegbased functional connectivity patterns. PLoS ONE **9**, e95415 (2014)
122. Li, Y., Yingle, F., Gu, L., Qinye, T.: Sleep stage classification based on EEG Hilbert–Huang transform. In: 2009 4th IEEE Conference on Industrial Electronics and Applications, pp 3676–3681. IEEE (2009)
123. Kubben, P., Dumontier, M., Dekker, A.: Fundamentals of clinical data science. Fundam. Clin. Data Sci. 1–219 (2018)
124. Lotte, F., Congedo, M., Lécuyer, A., Lamarche, F., Arnaldi, B., Anatole, L., Lotte, F., Congedo, M., Anatole, L., Abdulhay, E., et al.: A review of classification algorithms for EEG-based brain–computer interfaces to cite this version: A review of classification algorithms for EEG-based brain-computer interfaces. Hum. Brain Mapp. **38**, 270–278 (2018)

125. Elfenbein, H.A., Ambady, N.: Predicting workplace outcomes from the ability to eavesdrop on feelings. J. Appl. Psychol. **87**, 963–971 (2002)
126. Thejaswini, S., Ravi Kumar, K.M., Nataraj, A., K Students, A.K.: Analysis of EEG based emotion detection of DEAP and SEED-IV databases using SVM (2019)
127. Sander, D., Grandjean, D., Scherer, K.R.: A systems approach to appraisal mechanisms in emotion. Neural Netw. **18**, 317–352 (2005)
128. Pane, E.S., Wibawa, A.D., Pumomo, M.H.: Channel selection of EEG emotion recognition using stepwise discriminant analysis. In: Proceedings of the 2018 International Conference on Computer Engineering, Network and Intelligent Multimedia (CENIM), pp. 14–19. Surabaya, Indonesia (2018)
129. Hassanien, A.E., Azar, A.T. (eds.): Intelligent systems reference library 74 brain computer interfaces current trends and applications. Springer, Berlin/Heidelberg, Germany (2015)
130. Nicolas-Alonso, L.F., Gomez-Gil, J.: Brain computer interfaces, a review. Sensors **12**, 1211–1279 (2012)
131. Cao, Z.: A review of artificial intelligence for EEG-based brain computer interfaces and applications. Brain Sci. Adv. **6**(3), 162–170 (2020)
132. Aggarwal, S., Chugh, N.: Review of machine learning techniques for EEG based brain computer interface. Arch. Comput. Methods Eng. **29**, 3001–3020 (2022)
133. Hong, K.-S., Khan, M.J.: Hybrid brain–computer interface techniques for improved classification accuracy and increased number of commands: A review. Front. Neurorobot. **35** (2017)
134. Cecotti, H., Graser, A.: Convolutional neural networks for P300 detection with application to brain-computer interfaces. IEEE Trans. Pattern Anal. Mach. Intell. **33**, 433–445 (2010)
135. Thomas, J., Maszczyk, T., Sinha, N., Kluge, N., Dauwels, J.: Deep learning-based classification forbrain-computer interfaces. In: Proceedings of the 2017 IEEE International Conference on Systems, Man, and Cybernetics (SMC), Banff, AB, Canada, 5–8 October 2017. IEEE: Piscataway, NJ, USA (2017)
136. Tang, Z., Sun, S., Zhang, S., Chen, Y., Li, C., Chen, S.: A brain-machine interface based on ERD/ERS for an upper-limb exoskeleton control. Sensors **16**, 2050 (2016)
137. Amrani, G., Adadi, A., Berrada, M., Souirti, Z., Boujraf, S.: EEG signal analysis using deep learning: A systematic literature review. In: Proceedings of IEEE 5th International Conference on Intelligent Computing in Data Sciences (ICDS), pp. 1–8 (2021)
138. Li, H., et al.: Application of IOT deep learning in edge computing: A review. Acad. J. Comput. Inf. Sci. **4**(5), 13 (2021)
139. Längkvist, M., Karlsson, L., Loutfi, A.: A review of unsupervised feature learning and deep learning for time-series modeling. Pattern Recognit. Lett. **42**, 11–24 (2014)
140. Purkayastha, S.S., Jain, V.K., Sardana, H.K.: Topical review : A review of various techniques used for measuring brain activity in brain computer interfaces. Adv. Electron. Electr. Eng. **4**(5), 513–522 (2014)
141. Zhou, H.: Principles and applications in brain computer interfaces, vol. 20, pp. 1–13
142. Abibullaev, B., Keutayeva, A., Zollanvari, A.: Deep learning in EEG-based BCIs: A comprehensive review of transformer models, advantages, challenges, and applications. IEEE Access **11**, 127271–127301 (2023). https://doi.org/10.1109/ACCESS.2023.3329678
143. Lu, W., Wei, Y., Yuan, J., Deng, Y., Song, A.: Tractor assistant driving control method based on EEG combined with RNN-TL deep learning algorithm. IEEE Access **8**, 163269–163279 (2020)
144. Mahmood, N.T., Al-Muifraje, M.H., Salih, S.K., Saeed, T.R.: Pattern recognition of composite motions based on EMG signal via machine learning. Eng. Technol. J. **39**(02), 295–305 (2021). https://doi.org/10.30684/etj.v39i2A.1743
145. Abiri, R., Borhani, S., Sellers, E.W., Jiang, Y., Zhao, X.: A comprehensive review of EEG-based brain-computer interface paradigms. J. Neural Eng. **16**(1), 011001 (2019). https://doi.org/10.1088/17412552/aaf12e
146. Haider, A.: Electroencephalogram signal processing for hybrid brain computer interface systems (2018)

147. Osalusi, B., Abraham, A., Aborisade, D.: EEG classification in brain computer interface (BCI): A pragmatic appraisal **8**(1), 1–11 (2018). https://doi.org/10.5923/j.ajbe.20180801
148. Roy, S., Chowdhury, A., Mccreadie, K., Prasad, G.: Deep learning based inter-subject continuous decoding of motor imagery for practical brain-computer interfaces. Front. Neurosci. **14**, 918 (2020)
149. Choi, J., Kim, K., Lee, J., Lee, S.J., Kim, H.: Robust semi-synchronous BCI controller for brain-actuated exoskeleton system. In: Proceedings of the 2020 8th International Winter Conference on Brain-Computer Interface (BCI), Gangwon, Repulic of Korea, 26–28 February 2020. IEEE: Piscataway, NJ, USA (2020)

Smart Payment System-Using RFID Technology

Sarthak Johri, Perepi Rajarajeswari, Ishank Goel, and Aryan Mudgal

Abstract In order to improve transaction processes and strengthen security measures, this research study examines the integration of Radio Frequency Identification (RFID) technology into Smart Payment Systems. Examining the fundamentals of RFID technology, the study looks at how it might be used to make data sharing and contactless transactions easier. Important advantages like quicker transaction processing and more user ease are compared with drawbacks like infrastructure needs and security threats. The study provides an extensive analysis of payment options provided by RFID, based on case studies and current implementations. It also talks about potential future developments like improved authentication techniques and IoT integration. This study advances knowledge and implementation of RFID-based smart payment systems and offers industry stakeholders insightful information..

Keywords RFID · Security measures · Authentication · Contact-less transactions

S. Johri (✉) · P. Rajarajeswari · I. Goel · A. Mudgal
Vellore Institute of Technology, Vellore, Tamil Nadu, India
e-mail: sarthak.johri2021@vitstudent.ac.in

P. Rajarajeswari
e-mail: rajarajeswari.p@vit.ac.in

I. Goel
e-mail: ishank.goel2021@vitstudent.ac.in

A. Mudgal
e-mail: aryan.mudgal2021@vitstudent.ac.in

B. K. Mishra and A. Rocha (eds.), *Future of AI, IoT, and Sustainability*, Information Systems Engineering and Management 23,
https://doi.org/10.1007/978-3-031-76286-4_7

1 Introduction

Radio Frequency Identification (RFID) technology has emerged as a transformative force across various sectors, revolutionizing processes with its ability to wirelessly identify and track objects through radio waves. In the realm of transportation, particularly in college campuses, RFID holds immense potential for streamlining operations and enhancing user experience, notably in managing shuttle bus systems.

Implementation of RFID in college shuttle buses involves outfitting vehicles, passengers, and infrastructure with RFID tags and readers. Each bus is equipped with an RFID reader, while passengers possess RFID-enabled identification cards or mobile devices. As a passenger boards or exits the bus, their RFID tag is scanned automatically, facilitating seamless tracking of ridership data without manual intervention.

Linking RFID technology with transactions for college shuttle buses offers a myriad of benefits. It enables cashless payments, automated fare collection, and real-time monitoring of passenger flows. Moreover, RFID enhances security and accountability by accurately recording boarding and alighting instances. This integration not only modernizes shuttle bus operations but also fosters a more efficient and user-friendly transportation experience within college campuses.

The following are benefits of using RFID technology with Smart Payments:

Efficiency: Passengers may easily pay for shuttle services by tapping their RFID-enabled cards or devices, thanks to the rapid and smooth transactions made possible by RFID technology. This shortens wait times for passengers and expedites boarding by doing away with the necessity for cash handling or manual ticketing procedures.

Convenience: Passengers no longer have to worry about carrying cash or buying tickets in advance thanks to RFID-enabled smart payment systems. RFID transactions are contactless, which simplifies payment and improves passenger convenience overall.

Enhanced Security: When compared to conventional payment methods, RFID technology offers enhanced security features including encryption and authentication, which lowers the chance of fraud or theft. Since each RFID tag has a unique identification, counterfeiters find it challenging to copy or alter payment credentials.

2 Literature Survey

A Literature Survey on RFID-Based Smart Payment Shuttle Systems Utilizing RFID Tag, ESP8266, and Arduino.

Overview of RFID technology and its applications in transportation and payment systems. Introduction to ESP8266 and Arduino platforms for IoT applications. Purpose of the literature review: to investigate current research and advancements in RFID-based smart payment shuttle systems that incorporate RFID tags, ESP8266, and Arduino. Review of RFID Technology in Transportation Systems, including its use in ticketing, toll collection, and vehicle tracking. Evaluation of the benefits of RFID compared to traditional payment methods like cash and magnetic stripe cards. Discussion of obstacles and constraints in adopting RFID technology in transportation, such as security issues and compatibility challenges.

Arduino is widely used in IoT projects and has various capabilities. ESP8266 is crucial in providing Wi-Fi connectivity, while Arduino excels in sensor interfacing and data processing. IoT applications, like home automation and environmental monitoring, take advantage of ESP8266 and Arduino. Different industries, such as retail and healthcare, employ RFID-based payment systems. The systems' architecture, components, and functionalities are analyzed. Challenges and solutions in implementing RFID-based payment systems are identified. In this review of literature, the focus is on the integration of RFID technology, ESP8266, and Arduino in shuttle systems. The examination includes system architecture, such as RFID tag readers, ESP8266 modules for data transmission, and Arduino for data processing and control. The benefits and limitations of using this integrated approach for smart payment shuttle systems are evaluated. Case studies are presented to showcase real-world implementations of RFID-based smart payment shuttle systems using RFID tags, ESP8266, and Arduino. The analysis includes system design, deployment challenges, and performance metrics. Different implementation approaches are compared, and lessons learned from these case studies are discussed. Future trends and research directions in this field are also explored.

The conversation focuses on new developments and ideas in RFID-based smart payment shuttle systems. We will look into how these systems can be improved by connecting them to mobile payment platforms, using blockchain technology, and implementing advanced authentication methods. We will also discuss the main obstacles and possibilities for future research and advancement in this field. In conclusion, we will summarize the main discoveries from our research. We will emphasize the importance of RFID technology, ESP8266, and Arduino in creating smart payment shuttle systems. We will also urge for more research and creativity to tackle current issues and discover new chances in this area.

This literature review offers valuable insights for researchers, practitioners, and policymakers looking to develop and implement RFID-based smart payment shuttle systems by combining information from various sources on RFID technology, ESP8266, Arduino, and smart payment systems.

In large universities, there should also be a pre-planned schedule of shuttle arrival. During class times students need more shuttles near academic blocks and after classes students need more shuttles near relaxation and entertainment zones. The management should do this planning before only to make sure that students are comfortable with the schedule of shuttles.

In [1] the writers additionally employed IoT to produce an also used RFID tag. But he used that tag to book the ticket. The technique of RFID tagging he utilized in his research was extremely sophisticated and individuals require time to become used to it. Students generally select for a simple manner to commute between academic blocks. They won't choose a highly complex method to merely travel 1–2 km. The authors had decreased the procedure of cues but did not simplify the process or lower the time efficiency.

Authors in [2] RFID tags were also used. However, he purchased the ticket using that tag. It takes some getting acclimated to the somewhat complicated RFID tagging procedure he utilized in his research. In essence, students choose a convenient means of transportation between classrooms. They won't choose to go through a very difficult procedure to go 1–2 km. The writers simplified the procedure and decreased time efficiency, but they did not lessen the number of cues.

In [3] The respective writers employed RFID, but he did so in a different way by using an Android app. The driver's mobile device is connected to the ticket booking procedure. As passengers board the bus, the driver's mobile phone uses an interface to obtain the passenger ID from the RFID card reader. "SwipeNgo" maps the GPS coordinates and the stoppage name/number into a database in addition to the passenger ID. When a passenger exits the bus, the precise fare between the source and the destination is computed and subtracted from the remaining amount. Additionally, balance information is transmitted to the RFID setup, which shows the fare. However, the author of [3] claims that the driver must be trained to use the software.

Authors in [4] the recharging is done in specified spots for the RFID tag. But if students are busy, it is quite hard for them to travel to the recharge station and recharge the tag. Students will choose the simplest approach to recharge which is via mobile applications. Students like to easily recharge with one click utilizing smartphones instead of visiting recharge stations.

In [5] the authors utilized an RFID tag and also offered an Android application. In this application, customers may view which seats are free ahead to get aboard the bus. The sensor detects the incoming and leaving members from the bus and updates it in the app. But in our paper, we are also offering a feature whereby students can also see where the shuttle is. In a research [6], the authors have developed a GPS-based transportation that records the real-time location of the passenger and the bus and therefore estimates the distance between the two. The system also offers access to monitor the real-time position of the bus and employs an Artificial Neural Network [ANN] approach for producing the forecast on the bus arrival time. IoT-based solutions for smart bus transit have been offered in some of the recent publications like [7]. All IoT based systems are integrated with a controller and Wi-Fi module. Different controllers have been utilized in literature for smart bus transit systems. Though each controller has its own prominent qualities, most of the controllers utilized in the literature function with external GPRS/Wi-Fi modules for connectivity to the data store or application.

In [8] Research develops prototypes, presenting RFID implementation based on Raspberry Pi for electronic payment systems and user authentication at BRT. This study also has implications for the regulation of Makassar's BRT ticketing system, which charges a set rate at the start of each journey, regardless of the distance travelled. The authors have published a survey on the application of smartphone-based sensing for intelligent transportation systems in [7]. Smartphones with embedded sensors, like the gyroscope, accelerometer, and global positioning system (GPS), have been used to gather data about vehicles, traffic, and driver behaviour.

Authors in [9] present Urban Bus Navigator (UBN), a navigation system designed for passengers aboard urban buses. In order to prevent overcrowding on public transport, UBN tracks the wireless signals from mobile phones visible on buses in order to identify passenger crowds. It then recommends bus routes based on increased reality awareness. The fare in [10] will be calculated based on the passenger's actual distance travelled. A GPS tracking device is also included.

3 Methodology

The methodology for implementing the Smart Shuttle System involves several key steps, including assembling the hardware components, installing the necessary software, registering the RFID tags and student information in a database, detecting RFID tags and transmitting information via Wi-Fi, verifying student account balances, logging transaction details, and creating a web interface. The block diagram and schematic of the hardware setup illustrate the flow of information between the various components of the system. The detailed methodology for implementing the Smart Shuttle System is detailed below.

3.1 Hardware Setup

The Smart Shuttle System comprises several hardware components, including an RFID reader, Arduino microcontroller, buzzer, ESP8266 Wi-Fi module, and LCD screen. These components are assembled according to the system's design specifications (Figs. 1, 2, 3 and 4).

Fig. 1 Shows the Arduino used

Fig. 2 Shows the RFID tag

Fig. 3 (16 × 2) LCD module

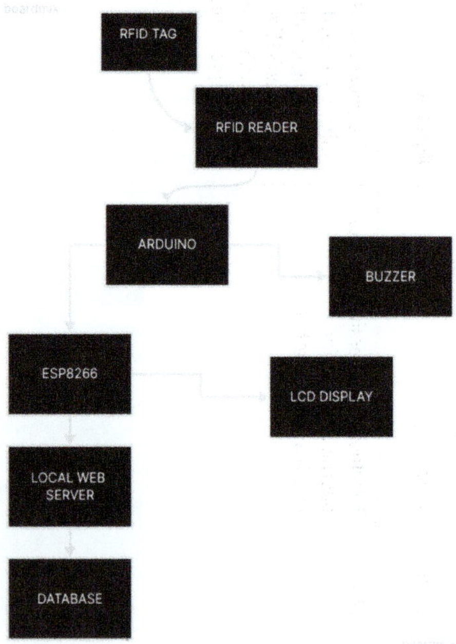

Fig. 4 Shows the different type of hardware components used in this project

3.2 *Software Installation and RFID Tag Registration:*

To implement the Smart Shuttle System, we first installed the necessary software components on our laptop and microcontroller. We installed the Arduino Integrated Development Environment (IDE) to program the Arduino microcontroller and ESP8266 Wi-Fi module. We also installed a PHP server and Apache server on the laptop to handle communication with the ESP8266 module.

We created a database of student information using MySQL and PHP, including unique ID numbers, names, registration numbers, passwords, and balance amounts. We assigned an RFID tag to each student and registered the tag ID with the corresponding student information in the database.

The ESP8266 RFID scanner and tags play integral roles in a smart shuttle payment system, enhancing efficiency, security, and convenience for passengers and operators alike. Here's a breakdown of their roles:

1. ESP8266 RFID Scanner:

   - The ESP8266 is a low-cost, Wi-Fi-enabled microcontroller widely used for IoT applications.
   - In the context of a smart shuttle payment system, the ESP8266 serves as the main hub for RFID scanning.
   - It is responsible for wirelessly communicating with RFID tags, processing the data received from them, and interfacing with the central payment system server.
   - The scanner is typically installed at the entry and exit points of the shuttle to register passengers as they board and disembark.
   - It ensures real-time tracking of passenger movements, facilitating accurate fare calculation and management.

2. RFID Tags:

   - RFID (Radio Frequency Identification) tags are small electronic devices that contain a unique identifier and communicate with RFID scanners via radio waves.
   - In a smart shuttle payment system, each passenger is provided with an RFID tag linked to their account.
   - These tags can be in the form of cards, key fobs, or even embedded in smartphones.
   - When a passenger boards the shuttle, they tap or pass their RFID tag near the scanner, initiating the payment process and registering their presence on the shuttle.
   - Similarly, when the passenger exits the shuttle, they again use their RFID tag to signal their departure, thereby completing the journey.

3. Role in Payment System:

- The ESP8266 RFID scanner acts as a bridge between the physical world (passengers and shuttles) and the digital payment system.
- Upon detecting a passenger's RFID tag, the scanner communicates with the central payment system server to verify the passenger's account details and process the fare transaction.
- The fare amount is automatically deducted from the passenger's account balance, ensuring a seamless and cashless payment experience.
- Real-time data from the scanners and RFID tags enable operators to monitor shuttle occupancy, track passenger flow, and analyze usage patterns, optimizing shuttle routes and schedules.

4. Enhanced Security and Convenience:

   - The use of RFID technology enhances security by reducing the reliance on cash transactions, minimizing the risk of theft and fraud.
   - Passengers benefit from the convenience of a contactless payment system, eliminating the need for physical tickets or cash exchanges.
   - Additionally, the automated fare collection process reduces boarding and disembarking times, improving overall shuttle efficiency and passenger satisfaction.

In summary, the ESP8266 RFID scanner and tags are essential components of a smart shuttle payment system, enabling seamless fare collection, accurate passenger tracking, and enhanced operational efficiency.

4 Results

The Smart Shuttle System was successfully implemented and tested. The RFID reader and buzzer were able to detect the RFID tag and indicate a successful tap. The ESP8266 was able to communicate with the laptop and display the transaction status on the LCD screen. The figure displays the output of the LCD screen of both Successful and failing transactions (Figs. 5, 6 and 7).

The Smart Shuttle System is capable of handling up to 23 transactions in 90 s, providing a fast and efficient payment method for students. The Tapping of the RFID tag is done simultaneously by two students to get the efficiency of the device. Around 23 tappings have taken place in less than 2 Mins.

To evaluate the performance of the smart shuttle system, we conducted several experiments and collected data on the time taken for transactions, the accuracy of the balance verification process, and the user satisfaction level. The data is as follows:

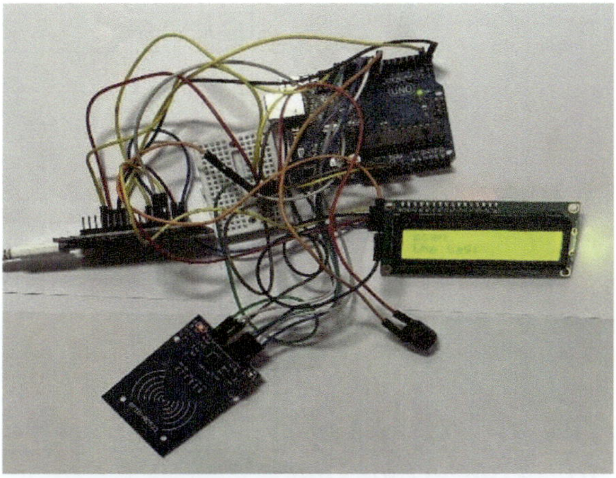

Fig. 5 Hardware setup of the project

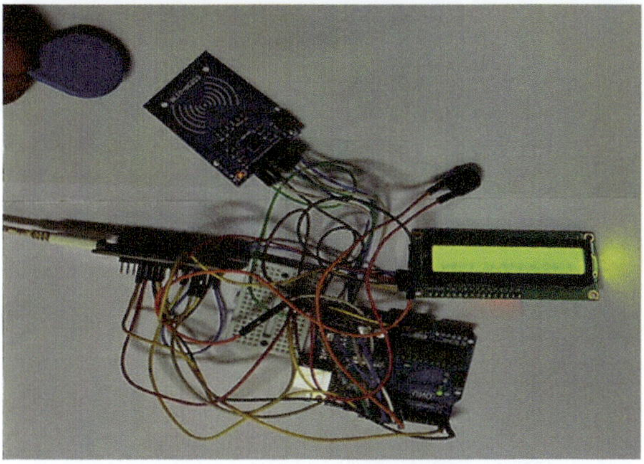

Fig. 6 When the RFID payment has a successful transaction

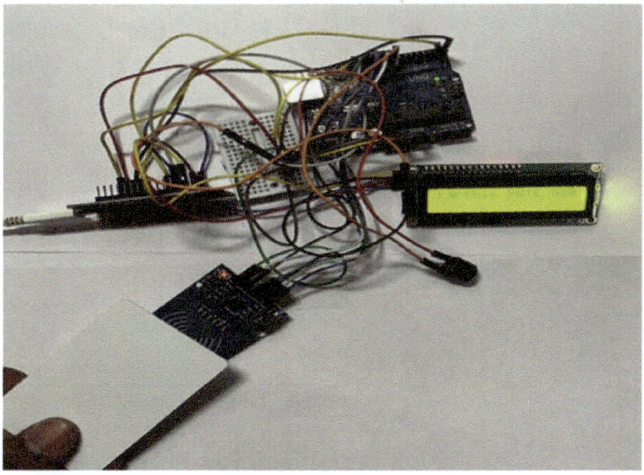

Fig. 7 When the RFID payment has a failed transaction

Table 1 Transaction time comparison

Payment method	Average transaction time (Seconds)
Cash	15
UPI	10
RFID	3.5

Table 2 Balance verification accuracy

Test case	Expected result	Actual result
Sufficient balance	Success	Success
Insufficient balance	Failure	Failure
Incorrect balance	Failure	Failure

4.1 Transaction Time

We timed three distinct payment methods: cash, UPI, and RFID, to see how long transactions took. Table 1 presents the findings. As anticipated, the fastest payment method was RFID, requiring an average of 3.5 s to complete each transaction. UPI and cash payments, on the other hand, required an average of 10 and 15 s, respectively.

4.2 Balance Verification

See Table 2.

4.3 User Happiness

We polled the students who utilised the smart shuttle system in order to gauge their degree of happiness. There were five survey questions, and Table 3 displays the findings. The majority of students, it turned out, were happy with the system and would rather use it than cash or UPI payments.

Table 3 User satisfaction survey results

Question	Yes (%)	No (%)
RFID payment feedback	80	20
Reduced waiting time	95	05
Hassle-free payments	90	10
Convenience	75	25
Reaching class on time	85	15%

Table 4 Transaction summary

Total transactions	Successful transactions	Failed transactions
500	480	20

4.4 *Data Analysis*

To learn more about the shuttle service's utilization trends, an analysis was done on the data that the smart shuttle system had acquired. The total number of transactions executed by the system throughout the testing phase is displayed in Table 4, together with the number of successful and unsuccessful transactions. Twenty transactions were unsuccessful because there was not enough money in the user's account, out of a total of 500 transactions.

Website Interface

Users can locate the shuttle, recharge their account, and check their balance among other things via the Smart Shuttle System's internet interface. The outcomes of the website interface are as follows:

The following is the Smart Shuttle System's home page or index page (Figs. 8).

Checking balance: By entering their username, users can check their current balance. The balance linked to the ID is retrieved by the website via a database query. Users can immediately ascertain their account balance by viewing the outcome on the website. An example of a user checking their balance is shown in the figure below (Figs. 9 and 10).

All things considered, the user interface of the website makes it easy for users to do a variety of tasks quickly and effectively (Fig. 11).

Fig. 8 The index page of the smart shuttle system project website

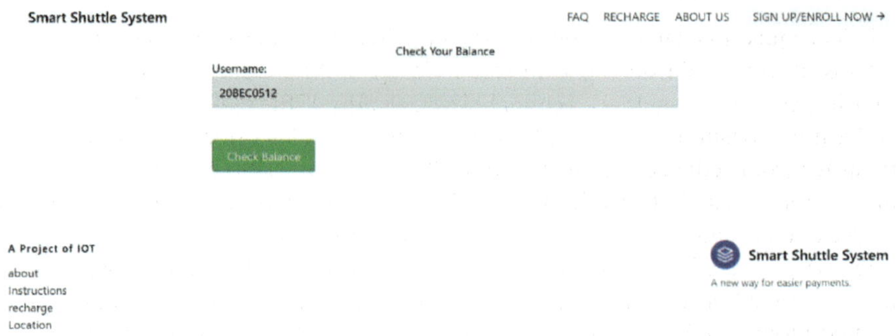

Fig. 9 Balance checking page

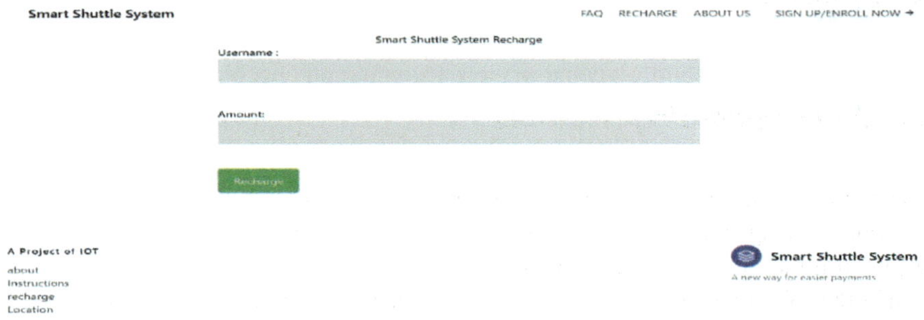

Fig. 10 Recharge option page to recharge the tag

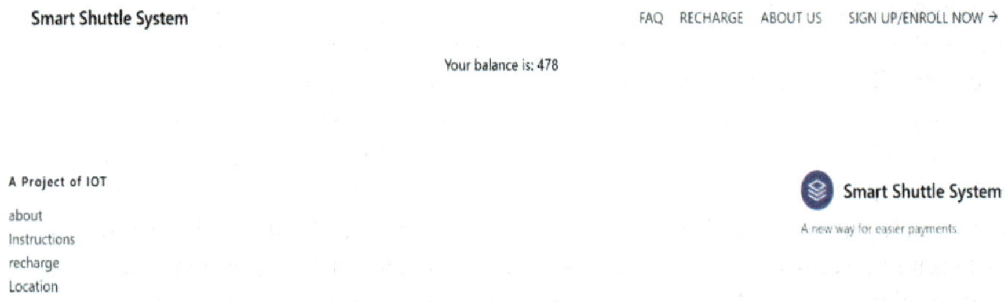

Fig. 11 Result of the balance checking

Recharge feature: By choosing their usernames and the desired amount to recharge, users may also utilize the online interface to recharge their accounts. Real-time payments using the payments gateway are among the upcoming enhancements.

5 Discussions

The Smart Shuttle System greatly enhances the efficiency of managing shuttle services, making it a fast and effortless process. The system is highly suitable for high-traffic areas due to its ability to efficiently handle up to 23 transactions during a span of 90 s.

The RFID reader is a very precise means of identifying users, and the buzzer offers feedback on the success or failure of a transaction.

User-Friendly Interface: The online interface offers a straightforward method for customers to top up their balance and track the shuttle's whereabouts. Implementing this can enhance the user experience and heighten user happiness.

Ensuring the security and protection of user information is of utmost importance due to the sensitive nature of the data stored in the system, including name, registration number, and password. Unauthorized access must be prevented. To enhance the system in the future, it may be upgraded by incorporating supplementary security features like encryption and user authentication.

An opportunity for enhancement lies in the creation of a mobile application for the Smart Shuttle System. The program would provide a user-friendly interface for users to check their account balance, recharge their balance, and follow the shuttle vehicles' real-time whereabouts. Furthermore, the system's security might be strengthened by integrating multi-factor authentication techniques such as biometric authentication. This would prohibit unwanted access to the system and secure users' sensitive data.

Finally, the system's hardware might be enhanced to boost the number of users that can be handled each second. This would assist in cutting waiting times and boost the overall efficiency of the system. Overall, these modifications would enhance the user experience, improve the system's security, and raise the system's efficiency.

6 Conclusion

The Smart Shuttle System delivers a hassle-free, cashless payment experience for students using the shuttle service. The solution employs IoT technology to provide a fast and effective payment mechanism and machine learning algorithms to estimate income for a certain date. The system has been successfully deployed and tested, indicating its potential to disrupt the old payment methods for shuttle services.

The method may also be readily applied to other locations, such as airports, bus terminals, and other transit hubs.

Acknowledgements We would like to express our sincere gratitude to everyone who has contributed to the success of this project. First and foremost, we would like to thank our project guide Perepi Rajarajeswari, ma'am, who provided us with invaluable guidance and support throughout the project.

We extend our thanks to VIT University for providing us with the necessary resources and support to complete this project.

Finally, We would also like to acknowledge our team members who worked tirelessly to bring this project to fruition. Their dedication, hard work, and enthusiasm were instrumental in the success of this project.

References

1. Hasan, M.F.M., Tangim, G., Islam, M.K., Khandokar, M.R.H., Alam, A.U.: RFID-based ticketing for public transport system: perspective megacity Dhaka. In: 2010 3rd International Conference on Computer Science and Information Technology, Chengdu, China, pp. 459–462 (2010). https://doi.org/10.1109/ICCSIT.2010.5564067
2. Geetha, S., Cecilia, D.: IoT enabled intelligent bus transportation system. In: 2017 2nd International Conference on Communication and Electronics Systems (ICCES), Coimbatore, India, pp. 7–11 (2017). https://doi.org/10.1109/CESYS.2017.832123
3. Chowdhury, P., Bala, P., Addy, D., Giri, S., Chaudhuri, A.R.: RFID and Android-based smart ticketing and destination announcement system. In: 2016 International Conference on Advances in Computing, Communications and Informatics (ICACCI), Jaipur, India, pp. 2587–2591 (2016). https://doi.org/10.1109/ICACCI.2016.7732447
4. Telluri, P., Manam, S., Oli, J.M.: Automated bus ticketing system using RFID. In: 2019 2nd International Conference on Intelligent Computing, Instrumentation and Control Technologies (ICICICT), Kannur, India, pp. 445–450 (2019). https://doi.org/10.1109/ICICICT46008.2019.8993159
5. Akter, R., Khandaker, M.J.H., Ahmed, S., Mugdho, M.M., Haque, A.K.M.B.: RFID based smart transportation system with android application. In: 2020 2nd International Conference on Innovative Mechanisms for Industry Applications (ICIMIA), Bangalore, India, pp. 614–619 (2020). https://doi.org/10.1109/ICIMIA48430.2020.9074869
6. Sungur, C., Babaoglu, I., Sungur, A.: Smart bus station-passenger information system. In: 2015 2nd International Conference on Information Science and Control Engineering, pp. 921–925 (24–26 April 2015)
7. Engelbrecht, J., Booysen, M.J., van Rooyen, G.J., Bruwer, F.J.: Survey of smartphone-based sensing in vehicles for intelligent transportation system applications. IET Intell. Transp. Syst. **9**(10), 924–935 (2015)
8. Noer, A., Hasanuddin, Z.B., Djamaluddin, D.: Implementation of RFID based raspberry Pi for user authentication and offline intelligent payment system. In: 2017 15th International Conference on Quality in Research (QiR): International Symposium on Electrical and Computer Engineering, Nusa Dua, Bali, Indonesia, pp. 251–255 (2017). https://doi.org/10.1109/QIR.2017.8168491
9. Handte, M., Foell, S., Wagner, S., Kortuem, G., Marrón, P.J.: An Internet-of-Things enabled connected navigation system for urban bus riders. IEEE Internet Things J. **3**(5), 735–744 (2016)
10. Oudah, A.: RFID-based automatic bus ticketing: features and trends. IOP Conf. Ser. Mater. Sci. Eng. **114**, 012146 (2016). https://doi.org/10.1088/1757-899X/114/1/012146

AI and IoT for Environmental Sustainability: A Comprehensive Analysis in Line with Sustainable Development Goals

Mahak Bhatia and Aled Williams

Abstract The world is vulnerable to climate variations which are illustrated in the terms of increase in temperature, intense weather conditions and so on. However, there is a substantial disparity in the understanding of the context of varying climatic conditions and how it is impacting the natural ecosystems. The AI and IoT outlook in this area support in understanding the impact on sustainability under dynamic climate elements. Over the past 5 years there is a rapid variation in global climate due to human activities. This trend seems to be continuous depending on the amount of heat trapped in the environment. The application of AI for analysing the constraints in attaining sustainability has drawn the focus of researchers in recent era. The chapter provides an insight on how technological advancement impacts environmental sustainability alongside how it promotes the United Nations Sustainable Development Goals (SDGs). However, so far no study is conducted that focuses on how AI impacts the dimensions of SDGs, i.e. social, environment and economic.

1 Global Impact of Technology on Environmental Sustainability

The term environmental sustainability [1] refers to the system that maintains global life-support, particularly, the environment, and maintains the environmental sources to replenish the natural resources. A mathematical equation is proposed that provides; environmental sustainability is a product of population, technology and affluence.

"Affluence" defines per capita consumption of resources while " technology" refers to the per unit of human well-being generated against per unit of environmental

M. Bhatia (✉)
Poornima Institute of Engineering & Technology, Jaipur, India
e-mail: drmahakbhatia@gmail.com

A. Williams
London School of Economics and Political Science, London, UK
e-mail: A.E.Williams1@lse.ac.uk

B. K. Mishra and A. Rocha (eds.), *Future of AI, IoT, and Sustainability*, Information Systems Engineering and Management 23,
https://doi.org/10.1007/978-3-031-76286-4_8

cost. Consequently, it is remarked that when the stage of equilibrium is achieved in terms of both population and energy per unit of production then we may approach towards attaining sustainability. In the current era, society is radically changing; driven by the digital revolution. Côrte-Real et al. [2] defines digital transformation as the set of innovations compiled together to transform existing rules.

The use of Information Computers Technology (ICT) and this digital transformation has influenced not only the environment but optimizes the efficiency of resources too. The evolution in digitalization has a systematic and wide- ranging impact on the optimization process that affects [3] sustainability.

Over the years the world is involved in technological innovations that contribute to preserve biodiversity [4] and to attain sustainability. Omri [5] elaborates how technology promotes economic growth and environmental sustainability. Researchers have explored the impact of the digital economy [6] on various aspects of sustainable development. Enhanced vitality is required for continuous economic growth coupled with population growth [7] resulting increase in the ability to maintain the sustainability through technological advancement. The contemporary requirement of sustainability can be evaluated by using the indicators of sustainability. Henceforth, to forecast the future change, the management policies are being evaluated. Sensing and communication technologies, IoT to achieve sustainability. In developing countries, IoT will bring paradigm changes to attain sustainability.

The world is at the environmental risks, alarmed by melting of glaciers, rise in temperature and intense climatic variations in terms of frosts, droughts and floods. However, there is a gap in in understanding that how the climatic variations and regional climate change is impacting the humanity and ecosystem making it difficult to adapt and improve the upcoming changes.

The revolution in the field of artificial Intelligence (AI) and deep learning (DL) has marked a paradigm shift in various sectors particularly to achieve sustainability. These transformative tools and techniques evolved as potential drivers in promoting sustainable practices. These technologies proliferated and made a significant impact in the domains of decision making, education, health and so on. The emergence of artificial intelligence (AI) has increased the productivity manifolds. AI has both positive as well as negative impacts to attain environmental sustainability.

Since IoT is still developing there is a need for a comprehension model. The applicability of IoT to attain sustainability can be embodied during the design phase. The concept of IoT using it for sustainability focuses on different environmental and sustainable objectives. Hence the application of AI and IoT supports in addressing the developmental challenges particularly pertaining to sustainability (Fig. 1).

AI emerged as a transformative technology that considers the concerns of sustainability to achieve SDGs. In recent era SDGs are the area where AI technology shows a rapid growth Goralski and Tan [8], Vinuesa and Sirmacek [9].

In the era of global change, in terms of development the world is facing a challenge in terms of environmental degradation. However, several decisions are taken by the policy makers to improve the environmental sustainability, viz, Sustainable Development Goals (SDGs). The aim of the initiative is to maintain the equilibrium between development and environmental anxiety. Thus, in this digital era artificial

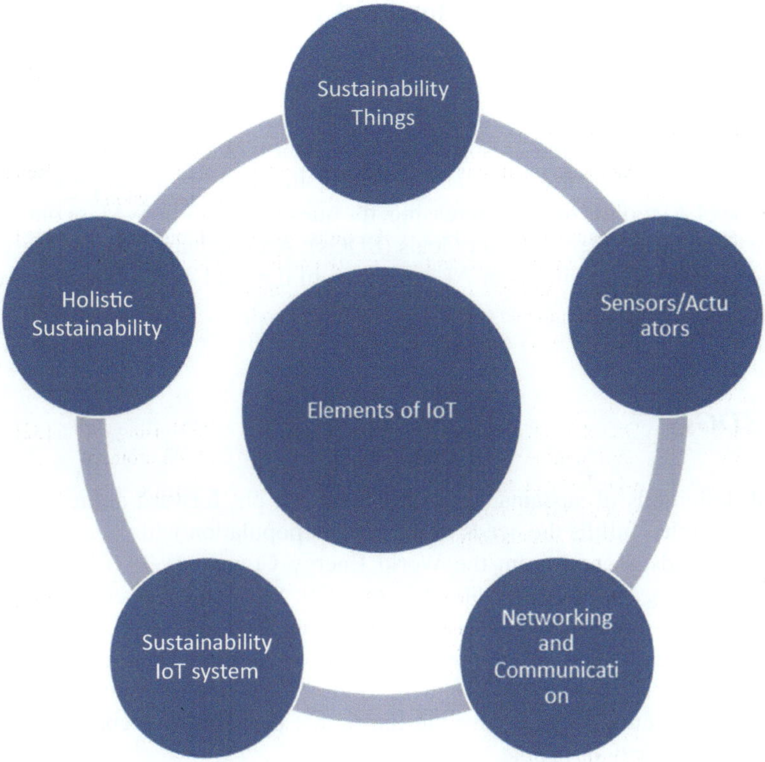

Fig. 1 Elements of IOT

Intelligence (AI) and technological advancement plays a significant role. However, the association of digitalization and environmental sustainability transcends [10] at industry and country level too. New products and platforms are being introduced by companies to promote environmental sustainability. Goralski and Tan [8] highlights the role of AI tools in Smart Water Management and optimizing operational costs. Yalina and Rozas [11] highlights that digital workplace can also support the achievement of environmental sustainability. The concept of environmental sustainability is so diverse that many times it includes trade-offs between the sufficient and self-interested solutions. This trade off can have both positive and negative impact on environmental sustainability. It is witnessed over the years that the increased pollution and deterioration of resources affects the sustainability of the environment. Henceforth, AI is crucial for handling the issues of environmental sustainability.

2 AI and Sustainable Development Goals

As we are stepping towards the Age of sustainability following the imprints of Sustainable Development Goals, defining the agenda of developing countries, AI is providing a platform to resolve cognitive issues. Incorporation of AI supports to achieve and promote the SDGs at experimentation and to support policy making. As the use of AI tools is rapidly increasing, the question arises how AI is connected to achieve Sustainable Development Goals (SDGs). Arumugham et al. [12] discussed that the prediction algorithms of AI contribute to the field of renewable sources of energy.

2.1 SDGs

Several definitions of sustainable development are put forward defining it as the development that fulfills the needs of the present population without compromising the future needs. Report from the World Energy Council (WEC) found that the demand of energy increased to 50–80% in 2020. Such an exponential increase in the demand put a pressure on energy infrastructure and causes an impact on environmental sustainability. Therefore, addressing environmental issues necessitates actions aimed at sustainable development. Hence, the UN [13] proposed SDG Goals with the aim to create a planet for the future that provides a blueprint for dignity, peace and prosperity of people.

Sustainable Development Goals coined by the United Nations are a follow up of Millennium Development Goals (MDGs) that focus on poverty and hunger whereas SDGs focus on 17 goals, covering 169 targets all across the globe (Fig. 2).

Economic and environmental sustainability are the two key dimensions of the goals. The aim of achieving SDGs by 2030 creates an obviate impact on sustainability. Goals 6, 7, 12, 13 and 14 concerned with environmental sustainability. These goals are related to [14] clean water and sanitation, affordable and clean energy, sustainable consumption of natural resources, climate change and life below water. Gupta [15] explain the trade-off and interdependencies among these goals and targets.

Optimizing the supply chain using tools of AI supports the optimization of logistics, reducing food waste enabling the farmers to adapt to the climate change by providing information about climate. Chien et al. [16] analyzed the impact of ICT on CO_2 emissions in BRICS countries. Increase in population, is creating higher demand for food and fiber, which in turn is exerting pressure on agroecosystems that [17] consequently causes environmental changes. However, by 2050 the diminishing soil carbon level, stressed supply chain will constrict to achieve SDG-2 (Zero Hunger). The emergence of digital technology is a hope of development to cater these challenges. 67 targets within the group can be achieved from AI based technologies by acting as an enabler for all the targets to attain. Moore [18] adheres that SDG

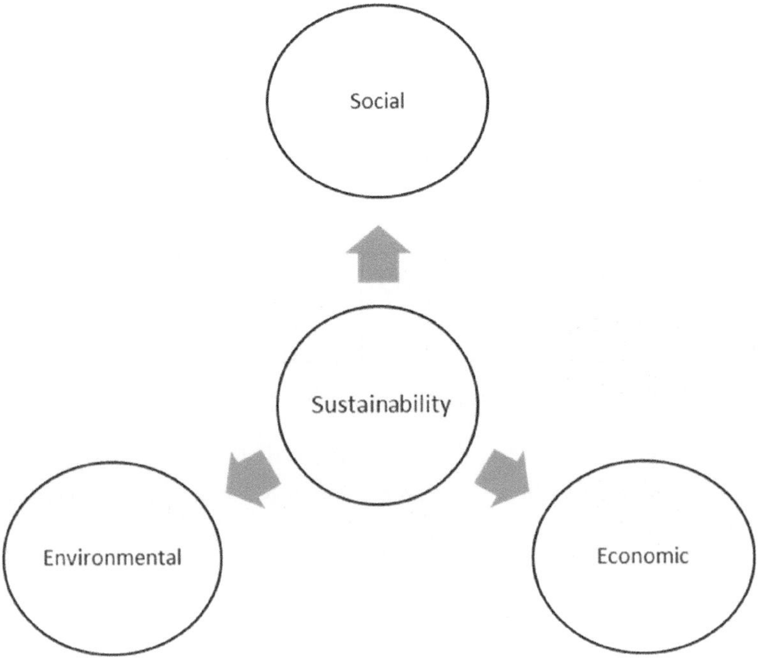

Fig. 2 Dimensions of SDG

goals must be taken as an opportunity to radically conceptualize the understanding of sustainability.

The process of developing and applying AI tools to assess environmental sustainability is a journey of achievements and setbacks. However the role of AI to attain sustainability is commendable.

2.1.1 SDG 6 and AI

SDG 6 aims to ensure universal access to clean water and sanitation, promoting hygiene for everyone while ensuring sustainable management and availability of adequate water resources in terms of both quality and quantity. Climatic variation, increasing population [19, 20] and agriculture demands have a significant impact on available water resources. According to the reports billions of people didn't get clean water even in places where plenty of water is available. To deal with this crisis a strategy is needed that focuses on improving the availability and accessibility of quality water. Assessing progress [21] towards achieving SDG 6 is challenging due to the high cost associated with evaluating the quality of water and hygiene services. Geographical Information System (GIS) is used to process the spatial information by which multi-criteria [19, 22] analysis can be done. A real time algorithm is proposed [23] to mitigate the performance of water reservoirs under climate change. The

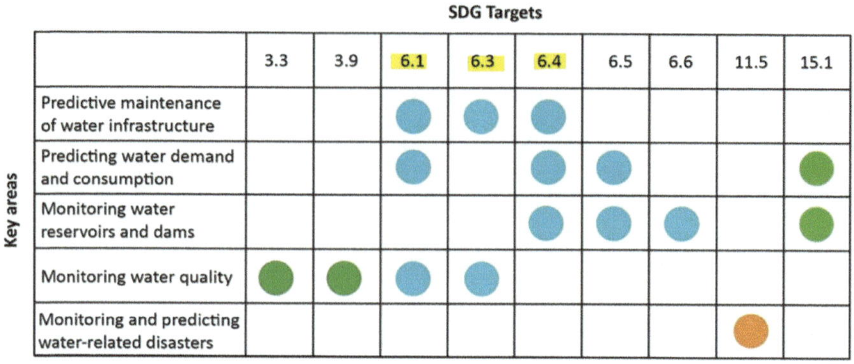

Fig. 3 Key areas and targets. *Source* [24]

algorithm was applied to discuss the case study of Morocco and results indicate the improved efficiency. Consequently, the results show how technological advancement contributes to improve the attainment of sustainability as a step to attain SDGs. AI is certainly involved in attaining and monitoring the water-related SDGs. the predicted AI-enabled water infrastructure contributes to achieve SDG targets 6.1, 6.3 and 6.4 (Fig. 3).

2.1.2 SDG 7 and AI

The SDG goal 7 ensures that all the people get access to affordable and sustainable energy by 2030. The goal has 3 targets that deals with [25]

1. Ensuring the accessibility and affordability of energy to all
2. Increase the percentage share of renewable sources of energy
3. Double the rate of improvement in energy efficiency

United Nations Department of Economic and Social Affairs [26] recorded that 660 million of the population still lack access to electricity and nearly 2 billion people are still using fuels as a source of energy. Hassani et al. [27] employed big data technology to predict energy poverty with the help of satellite images and concluded that Sub-African countries are adversely impacted by it. By combining different data sets, big data provides information about the current and future energy demand. SDG7 has to attain all the three core targets to achieve the aim of affordable energy to all. Ryan et al. [28] concludes how big data analysis facilitates environmental sustainability. SDG 7 is interlinked with SDG 11 as the sustainable cities are focusing on clean energy technologies.

To ensure access to energy for all by 2030, we must accelerate electrification, increase investments in renewable energy, improve energy efficiency and develop policies and regulatory frameworks

2.1.3 SDG 12 and AI

SDG 12 deals with responsible production and consumption. These two parameters are difficult to sustain for a sustainable world. Dubey et al. [29] shows the role of organizations to attain SDG 12. They concluded that predictive analysis and big data has a sustainable impact on production and consumption and has a positive moderate impact on the linking data.The high correlation is identified between CSR and SDG 12 [30]. SDG 12 indicators [31] shows the relation between AI, SDG and marketing. The market offers solutions to fulfill consumer needs, yet it also contributes to increased fuel consumption, thereby impeding progress towards sustainability. AI in marketing must support in making sustainable decisions. Reasonable production and consumption accelerates the SDGs and integrating it with technology will help to improve sustainability.

2.1.4 SDG 13 and AI

SDG 13 deals with climate change. Majorly, targeted goals linked with it are:

- Strengthening resilience and climate related adaptive capacity
- Integrating changes in climatic conditions into national strategies, planning and policies
- Creating awareness regarding climate change
- Implementing the commitment by UN framework on climate change
- Promoting climate change related planning and management

To reduce the damage caused due to climatic hazards there is a need for predictions and in-time warning. Data driven [32, 33] models are proposed for predicting flash floods and weather predictions. These approaches strengthen the achievement of SDG 13. Agriculture being one of the important sources of greenhouse emissions contributes significantly to climate change. Vice-versa the changing climatic conditions has an adverse impact on the agriculture sector, [34] thus, sustainable agriculture practices plays a crucial role in climate mitigation and adaptation. Hence, there is a need to address the issue of climate change in national policies.

2.1.5 SDG 14 and AI

SDG 14 deals with the resources below water for sustainable development. Digitalization fosters the achievement of SDG 14. To monitor the pollution near [35] the coastal region the indicators of water pollution are being implemented. However, some researchers have concluded that there exists a gap in research regarding the implementation of the digital paradigm to attain SDG 14. Digital and technological advancements used to maintain the sustainability [36] of the supply chain. IoT technologies provide an opportunity to collaborate between physical devices and applications [37]. This data driven system monitors the temperature to get a real time view. Kroeker et al. [38] apply meta analysis to study the impact of acidification on marine biodiversity. Furthermore, aquaculture and wastewater management practices contribute to coastal eutrophication and the proliferation of algal blooms leading to the reduction in oxygen. Constantly high rate of eutrophication was found in the Arabian Sea.

3 Conclusion

Human activities are the major cause of environmental change, SDGs provide a change to society to limit human activities back into safe operating spaces. These changes require a proper understanding and insight towards social, economic and technological progress. Hence, localizing the agenda, monitoring and participation will be the key to achieve environmental sustainability. Evolving technological tools like AI and IoT has played a vital role in achieving sustainability. The UN SDGs provide a framework to achieve better and sustainable generations. The technological innovations are necessary but to get a positive outcome a sustainable economic understanding is required. The study contributes how the advancement in technology impacts environmental sustainability and supports SDGs. The association of AI into sustainable development showcases a transformative and translative opportunity for policymakers, and industries. By comprehensive literature survey a holistic framework highlighting the landscape of AI-driven SDGs is showcased. To align AI with SDGs the involvement of stakeholders plays a vital role. However, to ensure a responsible governance robustness of the AI tools is required. Study explores the potential of AI to follow a roadmap of SDGs.

Though attaining SDGs are the growing concerns, there is a long way to go to achieve the 17 SDG goals particularly when comparing developed and developing countries. For the latter the goals need to be taken care of at the local level. All the SDGs are interlinked and hence the AI tools can support in achieving the goals simultaneously. Hence for future research it is recommended to find the interconnected goals and the role of AI in accelerating the innovations promoting the SDGs. some of the questions need to explored to address the future research viz., how the decision of implementing AI for achieving one SDG impacts the other 16 goals. To check the robustness of AI tools a data driven framework is required. In conclusion, it can be affirmed that data-driven and AI approaches have proven pivotal and are expected to remain so in advancing the SDGs, though their full potential has yet to be realized. While achieving all 17 of the UN's SDGs by 2030 may be an ambitious goal, it is not without challenges. Nevertheless, it is required to continue our dedication for working towards fulfilling the SDGs, as they represent crucial objectives for global sustainability and well-being.

References

1. Goodland, R.J.A., Daly, H.E., & El Serafy, S.: The urgent need for rapid transition to global environmental sustainability. Environ. Conserv. **20**(4), 297–309 (993). https://doi.org/10.1017/S0376892900023481
2. Côrte-Real, N., Ruivo, P., Oliveira, T., Popovič, A.: Unlocking the drivers of big data analytics value in firms. J. Bus. Res. **97**, 160–173 (2019). https://doi.org/10.1016/j.jbusres.2018.12.072
3. Liu, R., Koehler, A., Gailhofer, P., Gensch, C.-O., & Wolff, F.: Impacts of the Digital Transformation on the Environment and Sustainability. (2019)
4. Kwilinski, A., Lyulyov, O., Pimonenko, T.: Environmental sustainability within attaining sustainable development goals: the role of digitalization and the transport sector. Sustainability **15**(14), 11282 (2023). https://doi.org/10.3390/su151411282
5. Omri, A.: Technological innovation and sustainable development: does the stageof development matter? Environ. Impact Assess. Rev. **83**, 106398 (2020). https://doi.org/10.1016/j.eiar.2020.106398
6. Sun, X., Chen, Z., Shi, T., Yang, G., Yang, X.: Influence of digital economy onindustrial wastewater discharge: Evidence from 281 Chinese prefecture-level cities. J. Water Clim. Change **13**(2), 593–606 (2022). https://doi.org/10.2166/wcc.2021.447
7. Hess, P.N.: Economic Growth and Sustainable Development. Routledge (2016). https://doi.org/10.4324/9781315722467
8. Goralski, M.A., Tan, T.K.: Artificial intelligence and sustainable development. Int. J. Manag. Educ. **18**(1), 100330 (2020). https://doi.org/10.1016/j.ijme.2019.100330
9. Vinuesa, R., Sirmacek, B.: Interpretable deep-learning models to help achieve the Sustainable Development Goals. Nat. Mach. Intell. **3**(11), 926–926 (2021). https://doi.org/10.1038/s42256-021-00414-y
10. Feroz, A.K., Zo, H., Chiravuri, A.: Digital transformation and environmental sustainability: A review and research Agenda. Sustainability **13**(3), 1530 (2021). https://doi.org/10.3390/su13031530
11. Yalina, N., Rozas, I.S.: Digital workplace: Digital transformation for environmental sustainability. IOP Conf. Ser. Earth Environ. Sci. **456**(1), 012022 (2020). https://doi.org/10.1088/1755-1315/456/1/012022
12. Arumugham, V., Ghanimi, H.M.A., Pustokhin, D.A., Pustokhina, I.V., Ponnam, V.S., Alharbi, M., Krishnamoorthy, P., Sengan, S.: An artificial-intelligence-based renewable energy prediction program for demand-side management in smart grids. Sustainability **15**(6), 5453 (2023). https://doi.org/10.3390/su15065453
13. Martin. (n.d.). SDGs—Fast Facts. United Nations Sustainable Development. Retrieved April 24, 2024, from https://www.un.org/sustainabledevelopment/sdg-fast-facts/
14. Arora, N.K., Mishra, I.: United Nations sustainable development goals 2030 and environmental sustainability: race against time. Environ. Sustain. **2**(4), 339–342 (2019). https://doi.org/10.1007/s42398-019-00092-y
15. Gupta, A.D.: Water-energy-food (WEF) nexus and sustainable development. In: Salam, P.A., Shrestha, S., Pandey, V.P., Anal, A.K. (eds.) Geophysical Monograph Series, 1st ed., pp. 221–241. Wiley (2017). https://doi.org/10.1002/9781119243175.ch19
16. Chien, F., Anwar, A., Hsu, C.-C., Sharif, A., Razzaq, A., Sinha, A.: The role of information and communication technology in encountering environmental degradation: Proposing an SDG framework for the BRICS countries. Technol. Soc. **65**, 101587 (2021). https://doi.org/10.1016/j.techsoc.2021.101587
17. Okolo, C.C., Gebresamuel, G., Retta, A.N., Zenebe, A., Haile, M.: Advances inquantifying soil organic carbon under different land uses in Ethiopia: a review and synthesis. Bull. Natl. Res. Centre **43**(1), 99 (2019). https://doi.org/10.1186/s42269-019-0120-z
18. Moore, H.L.: Global prosperity and sustainable development goals. J. Int. Dev. **27**(6), 801–815 (2015). https://doi.org/10.1002/jid.3114

19. Giupponi, C., Gain, A.K.: Integrated spatial assessment of the water, energy andfood dimensions of the sustainable development goals. Reg. Environ. Change **17**(7), 1881–1893 (2017). https://doi.org/10.1007/s10113-016-0998-z
20. Monni, S., Iorio, M., Realini, A.: Water as freedom in the Brazilian Amazon. Entrepreneurship Sustain. Issues **5**(4), 812–826 (2018). https://doi.org/10.9770/jesi.2018.5.4(8)
21. Salam, A.: Internet of Things for environmental sustainability and climate change. In: Salam, A. (ed.) Internet of Things for Sustainable Community Development, pp. 33–69). Springer International Publishing (2020). https://doi.org/10.1007/978-3-030-35291-2_2
22. Gain, A.K., Giupponi, C., Wada, Y.: Measuring global water security towards sustainable development goals. Environ. Res. Lett. **11**(12), 124015 (2016). https://doi.org/10.1088/1748-9326/11/12/124015
23. Elhassnaoui, I., Moumen, Z., Tvaronavičienė, M., Ouarani, M., Ben-Daoud, M., Serrari, I., Lahmidi, I., Wahba, M.A.S., Bouziane, A., Ouazar, D., Hasnaoui, M.D.: Management of water scarcity in arid areas: A case study (Ziz Watershed). Insights Reg. Dev. **3**(1), 80–103 (2021). https://doi.org/10.9770/ird.2021.3.1(5)
24. Mehmood, H., Liao, D., & Mahadeo, K.: A review of artificial intelligence applications to achieve water-related sustainable development goals. In: 2020 IEEE/ITU International Conference on Artificial Intelligence for Good (AI4G), pp. 135–141. (2020). https://doi.org/10.1109/AI4G50087.2020.9311018
25. United Nations Department of Economic and Social Affairs. (2023a). Synergy Solutions for a World in Crisis: Tackling Climate and SDG Action Together: Report on Strengthening the Evidence Base (1st ed.) United Nations (2023). https://doi.org/10.18356/9789213585238
26. United Nations Department of Economic and Social Affairs.: The Sustainable Development Goals Report 2023: Special Edition. United Nations (2023b). https://doi.org/10.18356/9789210024914
27. Hassani, H., Yeganegi, M.R., Beneki, C., Unger, S., Moradghaffari, M.: BigData and energy poverty alleviation. Big Data Cogn. Comp. **3**(4), 50 (2019). https://doi.org/10.3390/bdcc3040050
28. Ryan, M., Antoniou, J., Brooks, L., Jiya, T., Macnish, K., Stahl, B.: The ethical balance of using smart information systems for promoting the United Nations' sustainable development goals. Sustainability **12**(12), 4826 (2020). https://doi.org/10.3390/su12124826
29. Dubey, R., Gunasekaran, A., Childe, S.J., Luo, Z., Wamba, S.F., Roubaud, D., Foropon, C.: Examining the role of big data and predictive analytics on collaborative performance in context to sustainable consumption and production behaviour. J. Clean. Prod. **196**, 1508–1521 (2018). https://doi.org/10.1016/j.jclepro.2018.06.097
30. Gunawan, J., Permatasari, P., Tilt, C.: Sustainable development goal disclosures: Do they support responsible consumption and production? J. Clean. Prod. **246**, 118989 (2020). https://doi.org/10.1016/j.jclepro.2019.118989
31. Hermann, E.: Artificial intelligence in marketing: Friend or foe of sustainable consumption? AI & Soc. **38**(5), 1975–1976 (2023). https://doi.org/10.1007/s00146-02101227-8
32. Poolman, E., Rautenbach, H., Vogel, C.: Application of probabilistic precipitation forecasts from a deterministic model towards increasing the lead-time of flash flood forecasts in South Africa. Water SA **40**(4), 729 (2014). https://doi.org/10.4314/wsa.v40i4.18
33. Weyn, J.A., Durran, D.R., Caruana, R.: Improving data-driven global weather prediction using deep convolutional neural networks on a cubed sphere. J. Adv. Model. Earth Syst. **12**(9), e2020MS002109 (2020). https://doi.org/10.1029/2020MS002109
34. Griggs, D.J., Nilsson, M., Stevance, A., McCollum, D.: A guide to SDG interactions: From science to implementation. International Council for Science (ICSU) (2017). https://doi.org/10.24948/2017.01
35. Mao, Z., Xue, X., Tian, H., Michael, A.U.: How will China realize SDG 14 by2030?—a case study of an institutional approach to achieve proper control of coastal water pollution. J. Environ. Manage. **230**, 53–62 (2019). https://doi.org/10.1016/j.jenvman.2018.09.028
36. Del Río Castro, G., González Fernández, M. C., Uruburu Colsa, Á.: Unleashing the convergence amid digitalization and sustainability towards pursuing the Sustainable Development Goals (SDGs): a holistic review. J. Clean. Prod. **280**(1) (2021)

37. Awan, I., Younas, M., Ünal, P., Aleksy, M. (eds.).: Mobile Web and Intelligent Information Systems: 16th International Conference, MobiWIS 2019, Istanbul, Turkey, August 26–28, 2019, Proceedings (vol. 11673). Springer International Publishing. https://doi.org/10.1007/978-3-030-27192-3
38. Kroeker, K.J., Kordas, R.L., Crim, R., Hendriks, I.E., Ramajo, L., Singh, G.S., Duarte, C.M., Gattuso, J.: Impacts of ocean acidification on marine organisms: quantifying sensitivities and interaction with warming. Glob. Change Biol. **19**(6), 1884–1896 (2013). https://doi.org/10.1111/gcb.12179

Machine Learning Approaches for Anomaly Detection in IoT Systems

Amitava Podder, Shivnath Ghosh, Piyal Roy, Subrata Paul, Saptarshi Kumar Sarkar, and Anuska Dutta

Abstract The Internet of Things (IoT) has emerged as a revolutionary technology that enables the exchange of data between physical objects in a seamless manner via networked sensors and devices. This article examines the far-reaching consequences of Internet of Things (IoT) systems in various sectors, such as transportation, healthcare, industry, Smart cities, and industrial IoT (IIoT). IoT facilitates the oversight and administration of vital infrastructure within Smart Cities, thereby advancing the goals of sustainability and efficiency. By optimising production processes and reducing expenses via predictive maintenance and supply chain management, the IIoT transforms industrial environments. IoT enables individualised treatment, proactive health management, and remote patient surveillance in the healthcare industry, thereby improving the accuracy of diagnoses and the timeliness of treatments. Smart farming practices propelled by the Internet of Things benefit agriculture by facilitating real-time monitoring of soil and crop conditions to optimise yield and irrigation. Moreover, IoT intelligent systems reshape the transportation industry in terms of sustainability, efficiency, and safety. The importance of the Internet of Things (IoT) resides in its ability to produce extensive quantities of data and extract valuable insights, which guide decision-making and improve the overall standard of living in various industries. Information analytics, predictive maintenance, Internet of Things, IoT systems, Smart Cities, Industrial IoT, Healthcare, Agriculture, and Transportation.

Keywords Healthcare · Agriculture · Transportation · Data analytics · Real-time monitoring · Predictive maintenance · Smart farming · Remote patient monitoring · Intelligent transportation systems · Sustainability · Efficiency · Precision irrigation · Yield maximization · Decision-making · Quality of life · Connected devices

A. Podder · S. Ghosh · P. Roy · S. Paul (✉) · S. K. Sarkar · A. Dutta
Brainware University, Barasat, West Bengal, India
e-mail: sp.cs@brainwareuniversity.ac.in

B. K. Mishra and A. Rocha (eds.), *Future of AI, IoT, and Sustainability*, Information Systems Engineering and Management 23,
https://doi.org/10.1007/978-3-031-76286-4_9

1 Introduction

The Internet of Things, or IoT, has become a paradigm-shifting technology that simplifies the linking and sharing of data across a vast array of physical objects. Networked sensors, actuators, and other devices that gather and transmit data across network infrastructures make up Internet of Things systems. This makes it possible to automate, regulate, and monitor processes in real time.

IoT systems hold tremendous significance across numerous domains, including but not limited to

Smart Cities

In large part because of IoT technology, urban landscapes are becoming more sustainable, efficient, and livable. Smart city initiatives employ Internet of Things (IoT) sensors and devices to monitor and manage critical infrastructure, such as electricity grids, transportation networks, and waste management systems [1].

Industrial IoT (IIoT)

Thanks to the Internet of Things, industrial settings may become "smart factories" by utilizing linked gear and equipment that interact with one another to optimize production processes, enhance efficiency, and decrease downtime. IIoT reduces costs and improves operations by enabling supply chain management, asset tracking, and predictive maintenance [2].

Healthcare

Internet of Things technologies are changing the way healthcare is delivered by offering personalized treatment plans, proactive health management, and remote patient monitoring. Thanks to wearable technologies, integrated sensors, and Internet of Things-enabled medical equipment, healthcare professionals are better able to collect health data in real time, increase diagnosis accuracy, and deliver timely treatments [3].

Agriculture

IoT solutions are driving the digital transformation of agriculture, often referred to as "smart farming." By deploying sensors and actuators in fields, farmers can monitor soil moisture levels, temperature, and crop health in real time, enabling precision irrigation, crop optimization, and yield maximization [4]. It is indeed the future of communication that has transformed Things (Objects) of the real world into smarter devices. The functional aspect of IoT is to unite every object of the world in such a manner that humans have the ability to control them via Internet.

Transportation

IoT technologies are reshaping the transportation sector by enabling intelligent transportation systems (ITS) that enhance safety, efficiency, and sustainability. Connected vehicles, traffic management systems, and smart infrastructure enable real-time traffic monitoring, congestion mitigation, and route optimization [5].

The significance of IoT systems lies in their ability to generate vast amounts of data from diverse sources and leverage this data to drive insights, inform decision making, and create value across various sectors. Through the utilisation of sophisticated analytics and networked devices, the Internet of Things (IoT) empowers enterprises to seize novel prospects, enhance operational efficiency, and elevate the standard of living for both people and communities.

Importance of Anomaly Detection in IoT

Anomaly detection is crucial for ensuring the reliability, security, and efficacy of IoT systems. As IoT networks continue to grow in size and complexity, the ability to identify potential threats and spot abnormal behaviour becomes more and more crucial. The following points emphasise the importance of identifying anomalies in the larger picture of IoT:

Early Threat Detection

Anomaly detection helps organisations spot potential security lapses, cyberattacks, or malicious activity before it becomes worse via early identification of deviations from normal behaviour in Internet of Things networks [6].

Enhanced System Reliability

IoT devices and sensors are always on the lookout for anomalous behaviour, and anomaly detection contributes to the dependability and availability of critical systems. Early anomaly detection lowers the possibility of service disruptions and system breakdowns by facilitating prompt corrective action [7].

Data Integrity and Trustworthiness

Anomalous detection methods assist in guaranteeing the accuracy and dependability of Internet of Things data by identifying anomalies, inconsistencies, or mistakes in the data. This is particularly important for industries like healthcare, where reliable data is required to make judgements that can be defended [8].

Optimized Resource Utilization

Anomaly detection makes it easier to allocate and use assets in the Internet of Things networks efficiently by spotting unusual patterns in resource utilisation or network traffic. Through the identification of inefficiencies or irregularities, organisations may optimise resource utilisation and improve overall system performance [9].

Proactive Maintenance and Fault Prediction

Predictive maintenance strategies benefit from the early signals of equipment malfunctions or breakdowns found via IoT anomaly detection. By proactively resolving any issues before they develop, organisations may reduce downtime, save maintenance costs, and extend the life of their equipment [10].

Compliance and Regulatory Requirements

Anomaly detection helps businesses comply with regulations by identifying deviations from widely recognised norms or standards. This is particularly important in sectors like health care, banking, and transportation where abiding by the law is mandatory [11].

To sum up, anomaly detection lowers risks, ensures the reliability and validity of IoT systems, & helps businesses recognise and manage unusual activities. For these reasons, it is crucial to IoT safety and administration strategies.

Role of Machine Learning in Anomaly Detection

In Internet of Things (IoT) systems, machine learning (ML) algorithms are essential for anomaly detection because they can automatically find trends, deviations, and outliers in huge, heterogeneous datasets. The use of machine learning algorithms enables Internet of Things devices to efficiently identify anomalous activity and possible security risks. The significance of machine learning in anomaly detection is highlighted by the following point:

Complex Pattern Recognition

Machine learning algorithms are excellent at discovering intricate links and patterns in data, which makes them useful for identifying abnormalities in a variety of IoT datasets. Without requiring explicit programming, supervised, unsupervised, and semi-supervised machine learning techniques may recognize anomalous behavior based on learnt patterns [12].

Adaptability to Dynamic Environments

Machine learning algorithms are able to identify new or undetected abnormalities and adjust to changing IoT settings. ML-based anomaly detection systems may update their models in real-time, guaranteeing resilience against new threats and shifting operational circumstances, by continually learning from incoming data streams [13].

Scalability and Efficiency

Scalable anomaly detection solutions are made possible by the massive amounts of data generated by IoT devices, which ML algorithms can process and analyze with efficiency. Deep learning and ensemble learning are two methods that are wellsuited for real-time anomaly detection in Internet of Things systems because they can handle high-dimensional data streams with little processing cost [14].

Feature Extraction and Representation Learning

Finding pertinent aspects that aid in anomaly detection is made possible by machine learning, which makes it easier to automatically extract features and learn representations from unprocessed IoT data. With the ability to learn hierarchical data representations, machine learning models can better detect anomalies by collecting both high- and low-level characteristics [15].

Anomaly Detection in Multimodal Data

Multimodal data streams are frequently produced by IoT systems from a variety of sources, such as sensors, photos, text, and time-series data. The freedom to evaluate and integrate many data modalities provided by machine learning techniques allows for thorough anomaly identification across a variety of IoT data sources [16].

Continuous Improvement and Adaptation

Through feedback loops and model retraining, machine learning enables anomaly detection systems to gradually improve their performance. ML models may improve their anomaly detection skills and adjust to shifting IoT settings and threat landscapes by utilizing past data and user input [17]. In summary, the development of dependable, scalable, and flexible anomaly detection techniques for Internet of Things (IoT) systems depends heavily on machine learning. These solutions enable businesses to recognise and mitigate security threats, ensure data integrity, and maintain the reliability of Internet of Things deployments.

2 Literature Review

The detection of anomalies in IoT systems has garnered a lot of attention because of the extensive usage of Internet of Things, or IoT, sensors in a range of sectors, including smart towns and cities, medical care, manufacturing automation, and agriculture. The ability to identify unusual behaviour and potential security threats in IoT data streams is essential for guaranteeing the reliability, safety, and effectiveness of IoT installations. This review of the literature provides a thorough analysis of current research projects, strategies, and challenges in the subject of IoT anomaly detection.

Historical Perspectives and Traditional Approaches:

Typical statistical techniques, rule-based systems, and simple thresholding strategies were the cornerstones of early IoT anomaly detection research. These techniques often made use of manually constructed attributes and pre-established criteria to detect deviations from regular behaviour [12]. The moving average & HoltWinters seasonal decomposition methods, for instance, were widely used for time-series anomaly detection in IoT sensor data. However, these methods were less flexible in adjusting to changing IoT environments and lacked the scalability necessary for large-scale deployments.

Machine Learning-Based Approaches:

In recent years, machine learning (ML) techniques have emerged as the industry standard for anomaly detection in Internet of Things (IoT) systems. Machine learning techniques, including supervised, unsupervised, and semi-supervised learning methods, may automatically detect patterns and anomalies in IoT data streams [10]. Supervised learning techniques such as Random Forests and Support Vector

Machines (SVM) train models using labelled data in order to perform classification tasks. Gaussian Mixture Models (GMM) and K-means clustering are examples of unsupervised methods that find anomalies based just on data distributions. Semi-supervised approaches such Self-Training Support Vector Data De-scription (ST-SVDD) are used for training models using both labelled and unlabeled data.

Deep Learning and Neural Network-Based Approaches:

Deep learning models, particularly Convolutional Neural Networks (CNNs) and Recurrent Neural Networks (RNNs), have gained popularity for anomaly detection in the Internet of Things due to their ability to automatically extract intricate features from unprocessed sensor data [18]. While CNNs are most suitable for image-based anomaly detection tasks in IoT surveillance systems, RNNs excel at processing sequential data, including time-series sensor readings. While these deep learning methods provide state-of-the-art outcomes, they may need a large amount of computer power and labelled training data.

Challenges and Future Directions:

Even with major progress, there are still a number of issues with IoT anomaly detection. The development of reliable and scalable anomaly detection systems is hampered by the dynamic nature of IoT settings and the variety of IoT data sources [8]. The healthcare industry has grown more quickly as a result of the quick advancements in automation and digitization. These developments have produced innovative models that open up new avenues for providing treatment at a lower cost. Furthermore, for anomaly detection models to be adopted in vital IoT applications like industrial automation and healthcare, they must be reliable and easy to understand. Prospective research avenues encompass tackling issues related to scalability, refining the interpretability of models, and fortifying anomaly detection systems' resistance against dynamic security risks in Internet of Things contexts. In summary, the literature on IoT anomaly detection reflects a shift towards data-driven approaches, with an increasing emphasis on machine learning and deep learning techniques. Future research efforts are needed to address the challenges of scalability, interpretability, and security in IoT anomaly detection. One of the most important factors in guaranteeing the overall, physical, social, and mental health of all people on the planet is access to healthcare.

Anomaly detection using machine learning algorithms:

It is essential in many different fields, such as manufacturing, fraud detection, cybersecurity, and healthcare monitoring. These methods seek to find patterns in data that substantially depart from typical behavior. Let's examine some well-known machine learning techniques for anomaly identification, as well as their uses, benefits, and drawbacks.

Supervised Learning Techniques:

Using labeled data, supervised learning entails training a model with each instance being assigned a class label denoting either normal or abnormal behavior. However, it might be difficult and costly to obtain tagged anomaly data.

3 Classification-Based Approaches

3.1 Support Vector Machines (SVM)

Using a hyperplane that it builds in a high-dimensional space, SVM divides data points into many classes. Since one-class SVM develops a judgment border around the normal data points, it is especially helpful for anomaly identification [19].

3.2 Decision Trees and Random Forests

The feature space is iteratively divided into smaller sections by decision trees. By combining the predictions from several decision trees, random forests increase the model's resistance to outliers [20].

4 Neural Networks

4.1 Autoencoders

Neural networks trained to reproduce their input at the output layer are called autoencoders. Measuring the reconstruction error, or the difference between the input and the result, allows anomalies to be found. By introducing probabilistic modeling, variational autoencoders (VAEs) make it possible to handle uncertainty more effectively [21].

Unsupervised Learning Techniques:

Since unsupervised learning does not require annotated examples of anomalies, it is especially helpful in situations where labeled data is insufficient or unavailable.

5 Density-Based Approaches

5.1 K-Means Clustering

K-Means partitions data into clusters based on their distance to centroids. Anomalies are points that are distant from any cluster centroid [22].

5.2 DBSCAN (Density-Based Spatial Clustering of Applications with Noise)

DBSCAN classifies outliers as points that are located in low-density areas and clusters together densely packed points (Ester et al. [23].

6 Probabilistic Approaches

6.1 Gaussian Mixture Models (GMM)

The GMM presupposes that a combination of multiple Gaussian distributions produced the data. Cases having low likelihood under the fitted model are called anomalies [24].

6.2 Isolation Forest

Separation By dividing the feature space and choosing features at random, forest separates anomalies. Anomalies stand out because they require fewer partitioning stages to isolate [25].

Semi-Supervised Learning Techniques:

As is frequently the case in anomaly detection scenarios, semi-supervised algorithms make use of a limited amount of labeled data and a big number of unlabeled data.

7 Self-Training

7.1 Self-Training with SVM

On the labeled data, a model is first trained. Next, the model predicts labels for data that isn't tagged, and iteratively adds high-confidence guesses to the training set.

8 Generative Models

8.1 Generative Adversarial Networks (GANs)

A discriminator network plus a generator make up a GAN. The discriminator separates genuine data from synthetic data, whereas the generator creates synthetic data. When the discriminator is unable to discern between actual and fake data, this is known as an anomaly.

Applications and Advantages:

Cybersecurity:

Detecting unusual network activities that might indicate a cyberattack.

Fraud Detection:

Identifying financial transactions or activities that are dishonest.

Healthcare Monitoring:

Identifying anomalies in patient data to help in early disease identification.

Manufacturing:

Monitoring sensor data from apparatus to identify mistakes or anomalies.

Limitations:

Scalability:

Some techniques may have difficulties with large datasets because to their computational complexity.

Labeling:

Supervised techniques need labelled data, which may be costly or difficult to get.

Data Quality:

Anomalies are often singular events leading to skewed models and imbalanced datasets.

To sum up, machine learning techniques for anomaly detection give a multitude of instruments for identifying deviations from the standard in a number of domains. Although each approach has advantages and disadvantages, the decision is dependent on a number of factors, such as the availability of labelled data, the features of the data distribution, and processing resources. The accuracy and scalability of anomaly detection systems are constantly being enhanced by developments in machine learning methods.

Feature engineering and data preprocessing are crucial phases in the machine learning process that transform raw data into a format appropriate for modelling. These processes ensure that the data is ready for analysis, assist to improve input feature quality, and enhance model performance and interpretability. In this discussion, we will delve into the intricacies of features engineering and data preparation, elucidating its significance, techniques, and impact on machine learning systems.

Feature Engineering:

The practice of implementing new features or altering existing ones in order to enhance the effectiveness of machine learning models is known as feature engineering. The difference in perspectives between developers and consumers makes it challenging to design software systems that function well. It entails selecting, blending, and refining characteristics to increase their discriminative, informative, and job-specific value. One of the main reasons for challenges in developing and delivering effective software is the difference in perspective between the developer and the user of the system.

Importance of Feature Engineering:

Augmented Model Performance:

Well-designed features significantly boost the prediction accuracy of machine learning models by capturing pertinent correlations and trends in the data.

Dimensionality Reduction:

Through the creation of condensed representations that retain significant information while removing irrelevant and noisy characteristics, feature engineering lowers the dimensionality of data.

Enhanced Interpretability:

Developing significant features makes it easier to understand and evaluate the behavior of the model, which provides deeper insights into the underlying data.

Techniques of Feature Engineering:

Modification of Features:

Skewed feature distributions can be normalized using methods such the Box-Cox transformation, square root transformation, and logarithmic transformation.

Feature Scaling:

Techniques like standardization and Min-Max scaling guarantee that features are scaled to a consistent range, preventing the dominance of certain characteristics during the training of the model.

Feature Deletion:

Essential data patterns are reduced to linear combinations of the original characteristics using Principal Component Analysis (PCA) and other approaches.

Effective processing of categorical data:

It is made possible by converting categorical variables into binary vectors.

Feature Selection:

The identification and selection of salient features is facilitated by recursive feature elimination (RFE) and feature significance ranking based on ensemble approaches.

Example of Feature Engineering:

To gather pertinent information on house characteristics and how they impact price, new features in a housing dataset, such as "Age of the House," "Total Square Footage," and interaction terms between features, can be created.

Data Preprocessing:

Preparing raw data for analysis and modeling involves cleaning, converting, and organizing it. We call this procedure "data preparation." It involves tasks including handling missing data, scaling features, eliminating outliers, and encoding categorical variables.

Importance of Data Preprocessing:

Enhanced Data Quality:

Preprocessing techniques that improve data cleanliness, consistency, and accuracy are what enable machine learning models to be trustworthy and accurate.

Amplified Model Performance:

Machine learning algorithms operate noticeably better when the right preprocessing is used to regulate missing data, reduce noise, and level feature distributions. Algorithm Compatibility:

Customised data preparation is necessary for optimal performance and interoperability with many machine learning techniques.

Techniques of Data Preprocessing:

Handling Missing Values:

Strategies like imputation, deletion, or treating missing values as a separate category may be used to fill in data gaps.

Outlier Detection and Removal:

Outliers may be prevented from having an undue influence on the outcomes of model training by identifying and removing them.

Normalization and Standardization:

Scaling numerical features improves algorithm performance and convergence.

Encoding Categorical Variables:

Techniques like encoding labels and one-hot encoding are used to transform categorical data into numerical representations.

Feature Discretization:

Boulle asserts that binning continuous features lowers model complexity and enhances interpretability.

Example of Data Preprocessing:

In a consumer transaction dataset, preprocessing consists of imputed missing income values, scaling numerical parameters like aged and transaction amount, and a one-hot encode of categorical variables like payment method. Performance assessment is crucial to ensuring the effectiveness, reliability, and scalability of Internet of Things (IoT) systems. A multitude of sensors, devices, networks, and other devices interact to collect and process data, and many aspects of the Internet of Things (IoT) must be evaluated using evaluation metrics and performance methods of analysis. Data quality, system responsiveness, resource use, and overall system efficiency are some of these factors. In this session, we will discuss common metrics and methodologies as well as the importance of performance analysis and evaluation metrics in the Internet of Things. Additionally, we will provide relevant references.

Importance of Evaluation Metrics and Performance Analysis in IoT:

Evaluation metrics and results analysis are critical to the design and deployment of Internet of Things systems for many reasons.

Quality Assurance:

Evaluation metrics help to guarantee the Internet of Things systems meet necessary quality requirements, including reliability, accuracy, and robustness.

Optimization:

Performance analysis approaches, which identify inefficiencies, bottlenecks, and feasible improvement areas, enable IoT system optimization.

Resource Management:

Metrics related to resource consumption and utilisation enable the efficient allocation of resources, such as electricity, bandwidth, and computing power.

User Experience:

User-perspective performance assessment improves the user experience by ensuring accurate and timely responses to customer orders and requests.

Evaluation Metrics in IoT:

A broad variety of subjects are covered by IoT assessment metrics, such as system performance, data quality, and user satisfaction. These measurements may be divided into many groups:

Data Quality Metrics:

9 Accuracy

Assesses how accurate sensor data is in comparison to ground truth or reference data [1].

10 Precision and Recall

Determine what proportion of relevant data IoT systems were able to locate and extract with success.

11 Completeness

Determines the extent to which the dataset is devoid of missing values and contains all relevant data components.

12 Latency

Measures the latency between data from sensors and the Internet of Things system or end users, which has an impact on real-time responsiveness.

13 Throughput

Evaluates the Internet of Things system's efficacy and capacity by measuring the speed at which data is transferred, obtained, processed, or managed [1].

14 Reliability

Demonstrates how Internet of Things systems' processing of information, communication, and transmission are dependable and consistent [8].

15 Energy Efficiency

Evaluates the energy consumption of Internet of Things networks and devices, which is critical for prolonging battery life and reducing environmental impact (Zanella et al. 2014).

16 Bandwidth Utilization

Measures how efficiently Internet of Things devices and applications utilise the available network bandwidth to optimise transmission of data and communication.

17 Computational Overhead

Determines how much processing power is required for data processing, analysis, and decision-making in Internet of Things systems.

18 Response Time

Reflects how long it takes for IoT devices to respond to user actions or requests, which affects their experience and satisfaction.

19 User Perception

Assesses consumers' subjective perceptions of the usability, reliability, and overall value of Internet of Things technologies.

20 User Engagement

Determines how effectively the system satisfies users' needs by analysing the degree to which users actively engage and communicate using Internet of Things applications and services.

IoT Performance Analysis Techniques:

IoT performance analysis employs a variety of techniques to assess system behaviour, identify performance bottlenecks, and optimise system parameters. A few techniques that are often used are listed below:

Modelling and Simulation:

Using simulation tools like NS-3, OMNeT++, and MATLAB/Simulink, IoT systems are modelled, numerous scenarios are generated, and system performance is examined under different conditions.

Experimental Testing: Real-world experiments and field trials are conducted to analyse IoT system performance in actual deployment scenarios, with the goal of validating simulation results and measuring real-time performance.

Benchmarking: Benchmarking allows for the objective comparison and evaluation of different systems or implementations by measuring IoT system efficiency against established standards or benchmarks.

Performance Monitoring: By continually observing the sensors, equipment, networks, as well as applications that comprise an Internet of Things system, anomalies, and deviations from expected behaviour may be found.

Optimisation Techniques:

A few of the IoT system optimisations that are carried out by using optimisation algorithms such genetic algorithms, ant colony optimisation, and particle swarm optimisation include distributing resources, routing, and scheduling.

In conclusion, performance analysis and evaluation metrics are critical to the creation, creation, and operation of Internet of Things systems in order to guarantee their effectiveness, reliability, and efficiency. By using appropriate measurements and analytical techniques, stakeholders may assess various aspects of IoT system performance, address performance bottlenecks, and modify system configurations to meet user requirements and intended objectives.

References

1. Gubbi, J., et al.: Internet of Things (IoT): A vision, architectural elements, and future directions. Futur Gener Comput Syst **29**(7), 1645–1660 (2013)
2. Tao, F., Wang, Y.: The role of big data and IoT in smart manufacturing. J. Ind. Inf. Integr. **1**, 6–13 (2016)

3. Shae, Z., Ali, S. R.: Internet of Things in healthcare: applications, benefits, and challenges. In: 2015 6th International Conference on Information and Communication Systems (ICICS), pp. 1–6. IEEE, (2015).
4. Mukhopadhyay, S.C.: Internet of Things: Challenges and opportunities. In: 2014 IEEE Emerging Technology and Factory Automation (ETFA), pp. 1–4. IEEE (2014)
5. Abdel-Aty, M., Ahmed, M.: Internet of things (IoT) applications for transportation. In: Transportation Research Board 96th Annual Meeting, No. 17–05378. (2017)
6. Kumar, P., Paul, A.: Security threats in IoT based applications and its prevention mechanisms: a review. In: 2017 International Conference on Computing Methodologies and Communication (ICCMC), pp. 1192–1195. IEEE (2017)
7. Ren et al.: Survey on IoT security technology and challenges. In: 2020 IEEE 4th Information Technology, Networking, Electronic and Automation Control Conference (ITNEC), pp. 242–247. IEEE (2020)
8. Al-Fuqaha, A., et al.: Internet of Things: A survey on enabling technologies, protocols, and applications. IEEE Commun. Surv. Tutor. **17**(4), 2347–2376 (2015)
9. Zhang, L., et al.: Internet of things (IoT) security: current status, challenges and prospective solutions. In: 2014 International Conference on Computing, Networking and Communications (ICNC), pp. 765–771. IEEE (2014)
10. Baccour, N., et al.: Anomaly detection in IoT systems: A survey and roadmap. Futur. Gener. Comp. Syst. **99**, 650–659 (2019)
11. Alaba, F. O., et al.: Internet of things security: A survey. J. Netw. Comp. Appl. **88**, 10–28 (2017)
12. Chandola, V., Banerjee, A., Kumar, V.: Anomaly detection: A survey. ACM Comput. Surv. (CSUR) **41**(3), 1–58 (2009)
13. Hodge, V.J., Austin, J.: A survey of outlier detection methodologies. Artif. Intell. Rev. **22**(2), 85–126 (2004)
14. Goodfellow, I., et al.: Deep learning. MIT press (2016)
15. Bengio, Y., Courville, A., Vincent, P.: Representation learning: A review and new perspectives. IEEE Trans. Pattern Anal. Mach. Intell. **35**(8), 1798–1828 (2013)
16. Hasan, M.M., et al.: A survey on multimodal deep learning. Neurocomputing **261**, 211–233 (2017)
17. Aggarwal, Charu, C.: Outlier analysis. Springer Science & Business Media (2013)
18. LeCun, Y., Bengio, Y., Hinton, G.: Deep learning. Nature **521**(7553), 436–444 (2015)
19. Cortes, C., Vapnik, V.: Support-vector networks. Mach. Learn. **20**(3), 273–297 (1995)
20. Breiman, L.: Random forests. Mach. Learn. **45**(1), 5–32 (2001)
21. Kingma, D.P., Welling, M. (2013). Auto-encoding variational Bayes. arXiv preprint arXiv: 1312.6114
22. MacQueen, J.: Some methods for classification and analysis of multivariate observations. Proc. Fifth Berkeley Symp. Math. Statist. Probabil. **1**(14), 281–297 (1967)
23. Ester, M., Kriegel, H.P., Sander, J., Xu, X.: A density-based algorithm for discovering clusters in large spatial databases with noise. In: Proceedings of the Second International Conference on Knowledge Discovery and Data Mining (KDD-96), pp. 226–231. (1996)
24. Fraley, C., Raftery, A.E.: Model-based clustering, discriminant analysis, and density estimation. J. Am. Stat. Assoc. **97**(458), 611–631 (2002)
25. Liu, F.T., Ting, K.M., Zhou, Z.H.: Isolation forest. In: 2008 Eighth IEEE International Conference on Data Mining, pp. 413–422 (2008)

Tech-Enabled Sustainability: Leveraging Technology for a Greener Future Through Smart Climate Agriculture

Jyotirmay Mahapatra, Jagjeet Singh, Debasis Gountia, and Ramesh K. Sahni

Abstract The global climate is undergoing a remarkable change with rises in temperature, increases in global warming effects, melting glaciers, ocean acidification, acid rains, extreme weather events, changing rainfall patterns, and changing ecosystems. These factors significantly reduce productivity and crop production, leading to global food scarcity, severe hunger, poverty, and health issues. The adoption of smart climate agricultural practices, which are essential for maximizing resources and lowering agriculture-related emissions, can address these kinds of problems. Advanced technology allows for the prediction of crop loss due to disease infestation, insects, pests, fog, and other factors, enabling proactive remedial actions to minimize losses. Artificial intelligence (AI) utilizes sensor-based, collected data relating to crop health, environment, and soil properties for better prediction and environmentally conscious decision making for effective handling of agricultural crops. IoT based irrigation, fertigation, agricultural vehicles, animal management, and green house microclimate regulations are some of the advanced systems using sensors, communications, artificial intelligence, and actuators together to be required to maintain agricultural activities as and when required. While satellites and Unmanned aerial vehicle (UAVs) help in the collection of a pool of raw data, useful findings, and critical information, they are essentially needed for the algorithm of AI and its effective working to be embedded in various agricultural machinery. A variable rate of input application based on collected critical information is possible through precisely regulating microcontrollers and actuators. In addition, computer vision systems analyze critical information precisely and use it to actuate agricultural robots to perform weeding or harvesting activities. Finally, tech enabled smart climate agriculture offers

J. Mahapatra (✉) · J. Singh · R. K. Sahni
ICAR-Central Institute of Agricultural Engineering, Bhopal, MP, India
e-mail: jyotirmayfmpe@gmail.com

J. Mahapatra
College of Agricultural Engineering and Technology, Odisha University of Agriculture and Technology, Bhubaneswar, Odisha, India

D. Gountia
School of Computer Sciences, Odisha University of Technology and Research, Bhubaneshwar, Odisha, India

B. K. Mishra and A. Rocha (eds.), *Future of AI, IoT, and Sustainability*, Information Systems Engineering and Management 23,
https://doi.org/10.1007/978-3-031-76286-4_10

a holistic approach to sustainable food production that balances environmental stewardship, economic viability, and social responsibility, ultimately benefiting farmers, consumers, and the planet.

Keywords Agricultural technology · Climate · Sustainability · Precision agriculture · Smart agriculture

1 Introduction

Smart climate agriculture refers to the integration of advanced technologies and data-driven approaches in agricultural practices to mitigate the effects of climate change and enhance agricultural sustainability and productivity. Smart climate agriculture aims to optimize resource use, reduce environmental impact, and adapt to changing climate conditions to enhance sustainability and productivity in agriculture. Some key components of smart climate agriculture include precision farming, climate resilient crop varieties, water management, carbon sequestration, data analytics and decision support system (DSS), renewable energy integration, and climate smart livestock management.

Precision farming utilizes technologies such as GPS, sensors, drones, and satellite imagery to collect data on soil properties, weather conditions, and crop health [1, 2]. This data is then used to optimize inputs such as water, fertilizers, and pesticides, leading to more efficient resource utilization and higher yields. Developing and deploying crop varieties that are adapted to specific climate conditions, including drought-resistant varieties, heat-tolerant crops, and varieties with enhanced pest and disease resistance enables farmers to get good return regardless of adverse climatic conditions. Implementing smart irrigation systems that monitor soil moisture levels and weather forecasts to deliver water precisely when and where it is needed. This reduces water waste, minimizes the risk of waterlogging or drought stress, and conserves valuable water resources. Practices such as conservation tillage, cover cropping, and agroforestry to enhance carbon sequestration in agricultural soils and vegetation helps not only in mitigating climate change by sequestering carbon dioxide from the atmosphere but also enhances soil health and fertility. In the case of animal husbandry rotational grazing, improved feed management, and breeding for climate resilience reduces greenhouse gas emission from livestock production and increases efficiency. Incorporation of renewable energy sources such as solar panels and wind turbines into agricultural operations can reduce reliance on fossil fuels and decrease greenhouse gas emissions. Data analytics and DSS applies, advanced analytics and modeling techniques are applied to analyze large datasets and provide farmers with actionable insights and recommendations. This includes tools for predicting weather patterns, optimizing planting schedules, and managing pest and disease outbreaks.

Through the implementation of intelligent climate agricultural techniques, farmers have found ways to enhance the durability, sustainability, and efficiency of their operations, while also supporting worldwide endeavors to alleviate and adjust to climate change. Supportive policies and incentives that encourage the adoption of climate-smart agricultural practices are required to be developed and implemented. This may include subsidies for sustainable practices, research funding for climate-resilient crop varieties, and carbon pricing mechanisms. Collaboration between farmers, researchers, policymakers, and agricultural extension services is crucial for the successful implementation of smart climate agriculture practices.

1.1 Advantages of Smart Climate Agriculture

Smart climate agriculture offers several advantages over traditional agricultural practices. It aids farmers in adjusting to climate change events, such as extreme weather events, droughts, and temperature fluctuations. By using datadriven approaches and climate-resilient crop varieties, farmers can maintain productivity even in challenging environments. Optimizing inputs such as water, fertilizers, and pesticides based on real-time data and precision farming techniques reduces resource wastage and enhances utilization efficiency. Advanced technologies and data analytics empower farmers to make informed decisions about crop management, resulting in higher yields and improved crop quality. By continuously monitoring soil health, weather conditions, and crop development, farmers can implement timely interventions to maximize productivity. Despite the initial investments in technology and infrastructure, smart climate agriculture can lead to long-term improvements in farm profitability. Through reducing input costs, increasing yields, and gaining access to premium markets for sustainably produced crops, smart climate agriculture demonstrates its tangible benefits. Consumers are increasingly demanding sustainably produced, climate-friendly agricultural products. It enables farmers to align with market demands by adopting environmentally friendly practices and showcasing a strong commitment to sustainability. Smart climate agriculture practices promote sustainable farming methods that minimize negative environmental externalities, such as greenhouse gas emissions, soil erosion, and habitat destruction.

1.2 Disadvantage of Smart Climate Agriculture

Although smart climate agriculture offers numerous benefits, it also presents certain disadvantages and challenges. The adoption of advanced technologies and infrastructure required for smart climate agriculture can involve significant upfront investments. This includes the cost of purchasing precision farming equipment, sensors, drones, and data analytics software, as well as investments in renewable energy systems and climate-resilient crop varieties. Implementing smart climate agriculture practices often requires specialized knowledge and technical skills in areas such as data analytics, remote sensing, and precision farming techniques. Farmers may need to undergo training and education to effectively use these technologies and interpret the data generated.

Disparities in access to technology and digital infrastructure can exacerbate inequalities within the agricultural sector. Farmers in underserved areas may lack access to reliable internet connectivity, hindering their ability to adopt climate agriculture practices. The collection, storage, and analysis of large volumes of agricultural data raise concerns regarding data privacy and security. Farmers may be hesitant to share sensitive information about their operations with technology providers or third-party platforms, fearing potential data breaches or misuse. While technical failures, such as equipment malfunctions or software glitches, could disrupt farm operations and lead to financial losses. Lastly, the adoption of smart climate agriculture practices may lead to changes in traditional farming practices and livelihoods, potentially disrupting local economies and social structures.

2 Precision Agriculture

It is the practice of administering the appropriate quantity of input (seed, fertilizer, pesticide, water, etc.) at the appropriate time and place in order to increase output, reduce input, improve product quality, and safeguard the environment. Climate change, population growth, limited resources, unavailability of manpower, hunger and poverty like major challenges are addressed by AI powered precision agriculture.

Climate change has a huge impact on agriculture production and productivity. Since it exacerbates frequent extreme weather events such as drought, flood, cyclone, forest fire, heat wave, etc. However, technology has potential to minimize the impact of climate change to a certain extent. Technology adoption like robotics, AI, and digital mapping can improve yields and production through efficient resource usage, climate resilient agricultural varieties, and timely operations in agriculture. Agricultural production plays a crucial role in reducing hunger and poverty. Precision technologies allow the judicious use of agriculture inputs and help to reduce farming costs and minimize human labor efforts [3]. Novel technologies such as robotics, AI, and computer vision being used in agriculture can succeed in bringing and retaining the youth population to agriculture.

2.1 Steps Followed in Precision Agriculture

Generally, there are four steps followed for the application of precision agriculture, as shown in Fig. 1. However, each step of precision agriculture has further sub-steps to take the decision based on the collected data, as depicted in Table 1.

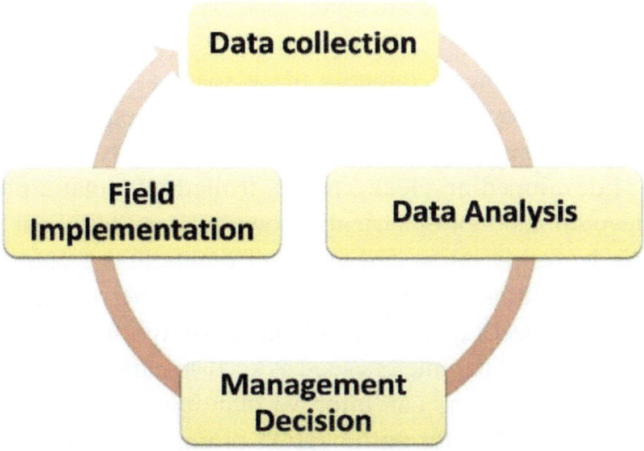

Fig. 1 Steps involve in precision agriculture

Table 1 Steps for precision agriculture with following sub-steps

1. Data collection	2. Analysis
• Weather related data • Soil properties data • Crop characteristics data • Yield monitoring and mapping date • Insect pest and disease related data	• Measuring variability and potential causes of variability? • To what extent do measured crop and soil properties differ? • Determine crop productivity and quality variations?
3. Decisions made by management	4. Farming
• Can the variability mitigated? • Will there be a decrease in inputs, an increase in yield, or a gain? • Is the modification profitable? • How put this change into practice?	• For applying the decisions • Variable rate irrigation • Variable rate seeding/planting • Variable rate pesticide application • Variable rate fertilizer application

2.2 *Precision Agricultural Machineries Coupled with AI for Achieving Smart Climate Agriculture*

Precision agricultural machinery combined with AI technology offers significant potential to optimize farming operations and improve productivity, sustainability, and profitability. By fine-tuning parameters and decision-making processes based on real-time data and predictive analytics, AI-driven precision agriculture systems enable farmers to make more informed and efficient use of resources, ultimately leading to better outcomes for both growers and the environment. Here are some examples of precision agricultural machinery and how AI can be utilized for parameter optimization.

2.2.1 Precision Planters

These machines accurately place seeds at the desired spacing and depth, ensuring uniform crop emergence. The plants get equal amounts of sunlight, water, nutrients, and improved aeration, which ultimately maximizes yield [4]. Mahapatra et al. [5] developed an electronic controlled pneumatic precision planter and suggested developments of electrically powered easy to adjust agricultural equipment. These can be easily integrated with IoT technology to get real time seeding information. AI algorithms can analyze soil data, weather conditions, and historical yield data to optimize seeding rates and planting configurations based on specific field conditions. By adjusting parameters such as seed spacing, depth, and variety selection, AI-driven precision planters can improve planting accuracy and efficiency. Researchers have also used ANN for optimization of design and operational parameters of seed drill and planters for precise seeding. ANN based optimization is observed to be more efficient than conventional statistical procedure.

2.2.2 Precision Sprayers

These machines precisely apply pesticides and herbicides on intended target areas with weed or pest infestations while minimizing spray drift and off-target damage [6, 7]. There are many precision spraying systems, available in market such as sensor based sprayer, electro-static sprayer, tunnel sprayer. These sprayers, especially tunnel sprayer might be worked very effectively with proper design and development as per crop morphology as recommended by researchers [8, 9]. These sprayers can be further improved by applying AI. AI algorithms can analyze data from remote sensors, drones, and machine vision systems to identify weed species, assess pest pressure, and optimize spray nozzle settings for maximum efficacy. By adjusting parameters such as spray pressure, droplet size, and nozzle orientation, AI-driven precision sprayers can improve spray coverage and reduce chemical usage, resulting in cost savings and environmental benefits [10]. Therefore, there is requirement of wide scale adoption of these AI based precision spraying systems for sustainable development and ecological footprint [11]. The efforts of various researchers in the development of precision sprayers and their outcomes are shown in Table 2.

2.2.3 Variable Rate Fertilizer Applicators

These machines apply fertilizers at variable rates across a field based on soil nutrient levels, crop requirements, and yield goals. AI algorithms can process data from soil sensors, satellite imagery, and crop health monitoring systems to create prescription maps that optimize fertilizer application rates in real-time. By adjusting parameters such as fertilizer type, application rate, and timing, AI-powered variable rate fertilizer applicators can minimize nutrient waste, reduce environmental impact, and enhance crop nutrition. A GPS-based variable rate fertilizer applicator was developed

Table 2 Studies on precision sprayers along with their key findings

Researchers	Methodology	Arrangement/outcomes
[12]	• Image processing using Borland C/C + + 4.5 compiler with image resolution 480 × 512: (R × C) pixel	• Herbicides saved from 48 to 58.4% corresponding to 0.5 and 1.5% weed coverage, respectively • Overall sprayer accuracy: 100, 75 and 47.8% in bare soil, weed, and crop plant, respectively
[13]	• In precision sprayer, CNN model used for detection of soybean and weed with SSD-MobileNetV1 model	• Spraying system saved up to 48.89% spray volume at 3.0 km/h
[14]	• Real-time sensors used in precision spraying system	• Mean reduction of pesticides applied: 68.28% (Soybean) and 52.02% (Maize) • Crop cost reduced by 56.16% of the cost of applied pesticide

by Chandel et al. [15], as shown in Fig. 2. It included a DC motor actuator, microprocessor, MCU (microcontroller), fluted roller type metering mechanism, differential global positioning system, and other components.

Numerous efforts have been made by the researchers for the development of variable rate fertilizer applicators with different sensors and tested with different operational parameters or arrangements, as given in Table 3.

These sensor and map based variable rate fertilizer applicators have been successfully tested in various crops with different fertilizers (granular form) and recognized

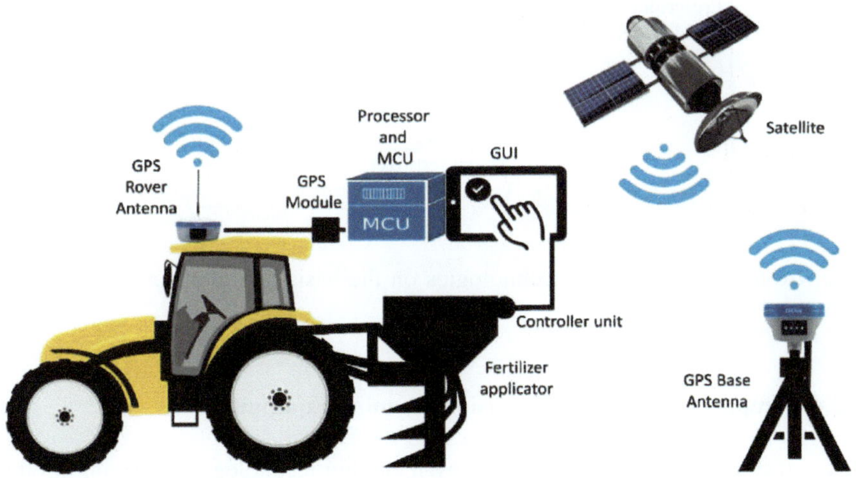

Fig. 2 Conceptual illustration of variable rate fertilizer applicator

Table 3 Studies on map and sensor based variable rate fertilizer applicators

Researchers	Sensors or map technology	Parameters used in study	Arrangement/outcomes
[16]	Phosphorous with VIS–NIR spectrophotometer	• Travelling speed • Sensor distance	• Soil sensor fixed0.91 m ahead of fertilizer outlets • Sensor position varied with forward speed of tractor (0.5 m ahead at 1 kmh^{-1} and in front of tractor at 7 and 8 kmh^{-1}
[17]	Prescription map	• Fertilizer type: urea and super phosphate • Operating speed: 0.83, 1.66 and 2.5 m/s • Application rate: 75, 125 and 175 kg/ha	• The accuracy of application rates declined as forward speed increased • Application rate error: 5.4%
[18]	Nitrogen sensor green seeker handheld sensor	• Nitrogen: 0, 90, 180, and 270 kg/hm^2 • Response time and applied N fertilizer	• Sensing width of sensor: 0.50 m • Response time: 3.49 to 4.90 s • CV (applied fertilizer): 2.34 to 5.1%
[19]	Prescription map (Grid size: 10 × 10 m)	• Operating speed: 0.33 and 0.66 m/s^2 • Discharge Rate, g/s	• Fertilizer application rate: 34 and 428 kg/ha at 0.2 and 0.8 m/s working speed respectively • Response times: 1.5 to 3.03 s
[15]	Prescription map (Grid size 8 × 8 m)	• Operating speed: three level (0.55, 0.83 and 1.11 m/s) • Fertilizer to be applied, kg • Actual fertilizer, kg (Error %)	• Fertilizer application rate: 5 and 300 kg/hm^2 at length of exposure 0 and 44 mm • Accuracy of the system: 89.3% and 98.1% CV of applied fertilizer: 11.7 to 15%

as environmentally friendly technologies on the basis of significant saved applied inputs, as shown in Table 4.

2.2.4 Impacts of AI Based Precision Farming Equipment

These techniques encourage farming by including technology, improving productivity, and reducing the input of resources and human efforts. This indirectly reduces the area of land that is left as arable waste or fallow. Smart climate agriculture prevents the degradation of soil fertility caused by fertilizer use, aligning agricultural practices

Table 4 Studies on variable rate fertilizer applicator with their economic and environmental impact

References	Crops and fertilizer	Methodology	Outcomes
[20]	Maize with NPK	• Variable-rate fertilization using GIS	• Improve maize yield and saved cost of applied fertilizer
[21]	Citrus with N	• Categorized tree size based on volume (0–240 m^3)	• Saved 38 to 40% urea and contributes to sustainable environment
[22]	Wheat with N	• VRA with a pendulum meter	• Saved 5 kg/hm^2 of applied fertilizer with improved yield up to 4.24.4%
[23]	Wheat with N	• Assessed top dressing of N fertilizer. requirements on-the-go on the basis of crop canopy reflectance	• Nitrogen rate decreased by 44% • Wheat yield improved by 15%
[24]	Maize with N	• Use of EPIC simulation model in VRA	• Leaching of nutrients was decreased up to 2.24–4.48 kg/ha

N: Nitrogen, VRA: variable rate applicator

with sustainable soil management. Smart climate agriculture enhances productivity and minimizes losses through effective pest management strategies, optimizing agricultural outcomes. Smart climate agriculture assists farmers in taking corrective actions based on both current and projected weather patterns, enhancing their ability to adapt to changing conditions. Enhanced productivity and reduced room for error in tasks like packing and sorting. Improved production standards lead to increased market accessibility.

3 Role of Smart Devices and IoT

Smart devices and the Internet of Things (IoT) play a significant role in advancing climate-smart agriculture. IoT sensors gather real-time data on soil moisture, soil nutrition, temperature, humidity, light intensity, sunshine hours, air composition, crop stress, and more, allowing farmers to make data-driven decisions. Weather stations are built with multiple important sensors and connected to the network for easy installation and access. Advancements in high-speed communication technology like 5G, and storage technology (edge or cloud) are taking IoT to new height. Data mining is a key process in IoT that helps in extracting meaningful relations, and insights from large amount of data collected using sensors. Algorithms like clustream, decision tree, auto regressive integrated moving average, exponential smoothing, isolation forest, one-class SVM, frequent pattern growth, regression analysis, Apriori algorithm, neural networks, gradient boosting machines etc. are used for data mining. Based on the analyzed data farmers can take actions by remotely regulating actuators connected to smart devices near them. Farmers can monitor their crops and livestock remotely, reducing the need for physical inspections and saving time. Smart devices analyze data to predict weather patterns, pest outbreaks, or disease occurrences, helping farmers plan better. IoT enables precise irrigation systems that water plants only when needed, reducing water wastage. Smart devices optimize energy use on farms, from machinery to storage facilities, cutting down on carbon emissions. IoT wearables on animals track their health, behavior, and location, allowing early detection of illnesses. Smart feeders dispense food based on individual animal needs, reducing waste and ensuring proper nutrition. IoT devices track temperature during transport and storage, ensuring food quality and reducing spoilage. Some important studies on IoT based animal raising and autonomous agricultural vehicles are summarized in Tables 5 and 6 respectively.

4 Autonomous Tractors

These machines perform field operations such as tillage, planting, and harvesting autonomously, guided by GPS and AI-based navigation systems. AI algorithms can analyze real-time data from onboard sensors and external sources to optimize tractor performance and path planning, taking into account factors such as soil compaction, terrain variability, and weather conditions. By adjusting parameters such as speed, steering angle, and implement depth, AI-enabled autonomous tractors can optimize fuel efficiency, reduce operator fatigue, and enhance overall productivity.

Autonomous soil cultivation systems use AI-driven autonomous tractors or implements to perform site-specific soil cultivation operations, such as tillage or soil preparation. AI algorithms analyze soil data and topographic information to optimize cultivation patterns and implement settings. By adjusting cultivation parameters based on field conditions, autonomous soil cultivation systems improve soil structure, water infiltration, and nutrient distribution, enhancing overall soil health and productivity.

Autonomous tractors depend on a range of sensors to detect their surroundings and make immediate judgements. The sensors aid in the tractor's navigation, obstacle avoidance, and precise task execution. Autonomous tractors commonly utilise a variety of sensors such as GPS (Global Positioning System), IMU (Inertial Measurement Unit), LiDAR (Light Detection and Ranging), Radar, camera, ultrasonic, wheel encoders, GNSS (Global Navigation Satellite System), temperature, humidity, and soil sensors. Sensed data is fed to control system for processing and decision making. Accordingly, it plans for navigation, speed, and operations. These autonomous tractors or agricultural vehicles are often integrated with IoT technology for remote monitoring and control, real time data collection, and precision agricultural application. The communication used can be wireless, such as through Wi-Fi, cellular networks, or dedicated short-range communication systems. Autonomous tractors are

Table 5 IoT based animal raising

Activity	Parameters and sensors	Microcontroller	Data transmission	References
Monitoring the health of each animal	Inferential sensing through ear mounted sensors for animal and ambient temperature, neck mounted collar sensors for feeding and watering proximity data, and a three-axis accelerometer		Zigbee radios mounted in the animal collars	[25]
Animal feed with its milk production monitoring	Temperature, humidity, CO_2 levels, animal activity level, etc. Through sensors at internal systems on farm and farm's external temperature and humidity through the API	e-Livestock Controller	Internet	[26]
Animal health monitoring	Rumination, heart rate and body temperature, with surrounding temperature and RH	PIC18F4550 microcontroller	Zigbee module	[27]
Animal health, animal missing monitoring	Core body temperature, heart rate, ambient temperature and humidity GPS location, and motion		Telemonitoring system	[28]
Animal health monitoring and tracking	Heart rate, temperature, pulse rate and respiratory, and GPS	Arduino UNO	ZigBee (wireless communication technology)	[29]
Animal healthcare system monitoring	Temperature, humidity, heart rate, weight through load cell	Node MCU	Wi-Fi router	[30]

typically powered by diesel engines, but electric or hybrid powertrains are becoming more common as the actuator for regulation are also electric motors. These actuators mostly regulate steering, throttle, brakes, and depth of operation of attached equipment.

Advanced software and algorithms govern the behavior of autonomous tractors, including path planning, obstacle avoidance, and task scheduling. Path planning aims to minimize travel time and energy consumption while avoiding collisions and optimizing coverage. Obstacle avoidance employ techniques such as path deviation, dynamic replanning, or local trajectory adjustments to safely maneuver around obstacles while staying on course. Task scheduling takes into account factors such as crop type, growth stage, weather conditions, equipment availability, and operational priorities to optimize productivity and resource utilization. It aims to minimize

Table 6 IoT based autonomous agricultural vehicles

Activity	Parameters and sensors	Microcontroller	Data transmission	References
Monitoring plowing, sowing, and spreading fertilize and irrigation operation in Agri-bot vehicle	Moisture sensor, temperature sensor, and humidity sensor along with length and width of the field as reference	Arduino UNOs and two node MCUs Arduino UNOs and node MCUs	Mobile and cloud server	[31]
Autonomous navigation	Diesel level, hydraulic oil level, engine temperature, wheel angles, wheel speeds, GPS data, remote control data and hitch height	PXI platform	CAN using ISO 11783 protocol	[32]
Lane control system	Vehicle speed, engine speed, battery SoC (State of Charge), battery SoH (State of Health), GPS navigator, tyre pressure, motor current consumption	STM microcontroller	ZigBee	[33]

idle time, reduce overlap between tasks, and maximize the overall efficiency of farm operations. Machine learning algorithms can analyze data collected from sensors and historical operational data to learn patterns, optimize decision-making, and enhance predictive capabilities. For example, AI algorithms can learn to recognize different types of crops or weeds, identify optimal planting patterns, or predict equipment maintenance needs based on usage patterns.

5 Harvesting Robots

These machines automate the harvesting process for crops such as fruits, vegetables, and nuts, using AI-driven robotic arms and computer vision systems to identify and pick ripe produce with precision and efficiency. AI algorithms can analyze images of crops and assess ripeness indicators such as color, size, and texture to determine optimal harvesting times and techniques. By adjusting parameters such as grip strength, harvesting force, and motion trajectories, AI-powered harvesting robots can minimize damage to crops, increase harvesting speed, and improve overall product quality. Multiple efforts have been made for the development of crop specific harvesting robots. Here, a list of research projects and their methodologies, along with their outcomes, is shown in Table 7.

Table 7 Studies on Robot harvesters for different crops with outcomes

Researchers	Crops	Methodology	Outcome
[34]	Apple	• 5-DoF manipulator structure • Pneumatically actuated spoon-shaped end effector • Vision-enabled module for harvesting operations	• Harvesting rate: 77% • Mean harvesting time: 15 s per apple
[35]	Sweet peppers	• Use vision algorithms with a novel end-effector design	• Harvesting rate: 58% • Grasping rate: 81% • Detachment rate: 90%
[36]	Litchi	• 6-DoF manipulator • DeepLab V3 semantic image segmentation • Vision-guided harvesting robot utilizing a Kinect V2 RGB-D camera • Position the camera 500–800 mm from trees	• Experimental accuracy: 83.33%
[37]	Sweet peppers	• Five steps, like segmentation of sweet pepper, peduncle segmentation, grasp identification, attachment, and detachment actions for harvesting sweet pepper • RGB-D camera employed to capture a three dimensional color image	• Harvesting accuracy: 76%
[38]	Kiwifruit	• The system comprised an end-effector, a cartesian coordinate manipulator, a programmable logic controller, an electric vehicle, and a machine vision device • A Kinect sensor mounted on the harvesting robot rack enables machine vision	• Harvesting accuracy: 94.2%
[39]	Pear and apple	• To detect fruit used sensors with computer vision • To harvest the fruit used an inverse kinematics and a fast path planning method with random sampling	• Accuracy for fruit detection: >95% • Maximum time taken for harvesting:10 s per fruit

6 Remote Sensing and Imaging Systems

These systems utilize AI driven remote sensing technologies, such as satellite imagery, drones, and aerial sensors, to monitor crop health, detect stress factors, and assess field variability. AI algorithms analyze imaging data to generate maps and recommendations for targeted interventions, such as variable rate application of inputs or early detection of pest outbreaks. By providing real-time insights into field conditions, remote sensing and imaging systems enable proactive decision making and precision management of agricultural field applications. The conceptual diagram of a remote sensing system for agricultural field applications is illustrated in Fig. 3.

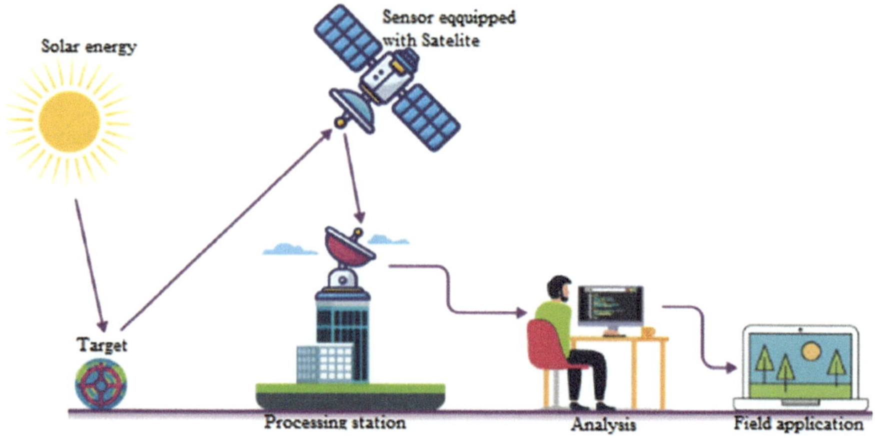

Fig. 3 Conceptual diagram for remote sensing system

7 Smart Decision Support System for Precision Agriculture Using Integrated Soil, Crop, Climate, and Machine Data

DSS is a computer-based information system designed to support decision-making processes within organizations by providing relevant data, analysis, and models. However, AI-powered DSS integrate data from multiple sources, including crop images, sensor data, weather forecasts, soil analyses, and historical records, to provide comprehensive insights and recommendations for farmers. These systems help farmers make informed decisions about crop management practices, disease prevention measures, pesticide application, and resource allocation. DSS includes data management, model management, user interface, knowledge base, communication and collaboration tools, data analysis, and external sources together.

The foundation of a DSS is its data management component, which involves collecting, storing, and organizing relevant data from various sources such as market reports, sensor networks, or government databases. DSS often includes mathematical models or analytical tools that help analyze data and simulate different scenarios to support decision-making. These models can range from simple statistical algorithms to complex optimization models or machine learning algorithms.

The user interface component of a DSS provides a platform for users to interact with the system, access data, and perform analyses. This interface may include dashboards, visualization tools, and reporting features that present information in a clear and intuitive manner. A DSS may incorporate a knowledge base or expert system that captures and codifies domain-specific knowledge and rules. This knowledge base can help guide decision-making by providing context, best practices, and recommendations based on past experiences or expert judgment. A DSS may integrate with external systems and data sources, allowing the DSS to leverage data from external sources such as market data providers, weather services, or supply chain partners, enhancing the depth and accuracy of its decision support capabilities. Some DSS include features for communication and collaboration among users, allowing them to share information, discuss options, and collaborate on decision-making processes. These tools may include messaging systems, discussion forums, or collaborative workspaces. DSS provides tools and functionalities to analyze data and evaluate alternative courses of action. This may involve performing "what-if" analyses to explore different scenarios, conducting sensitivity analyses to assess the impact of changes in key variables, or generating forecasts and predictions based on historical data and trends.

7.1 Benefits of Implementation of DSS

The implementation of DSS offers numerous benefits across various domains and industries. Here are some key advantages of implementing a DSS:

DSS provides decision-makers with timely, relevant, and accurate information to support their decision-making processes by integrating data from multiple sources and conducting analyses. DSS empower decision-makers to efficiently assess options, evaluate alternatives, and determine optimal courses of action, enhancing decision-making processes. It automates repetitive tasks, reducing manual data processing and thereby saving precious human time. It helps to optimize resource allocation by providing insights into resource utilization, demand forecasting, and performance metrics by analyzing data and simulating different scenarios. Future risks can be identified and mitigated through early warning signals, predictive analytics, and the scenario planning capabilities of DSS. Faster, more informed, and more intelligent decisions can be made by applying DSS in agriculture.

7.2 Challenges or Issues with Decision Support Systems

While DSS offer numerous benefits, they are not without their challenges and issues. DSS relies heavily on accurate and reliable data. Issues such as incomplete, inconsistent, or outdated data can undermine the effectiveness of the system and lead to poor decision-making. Integrating data from multiple sources and systems can be complex and challenging. Different data formats, incompatible systems, and varying levels of data granularity can hinder the seamless integration of data within the DSS.

The accuracy and validity of models used within a DSS are critical for generating reliable insights and recommendations. However, the model assumptions, parameter estimation errors, and model uncertainty can introduce biases and inaccuracies into the decision-making process. The success of a DSS depends on user adoption and acceptance. Resistance to change, a lack of user training, and usability issues can impede user acceptance of the system. Overreliance on technology without human judgment and intervention can be problematic in certain cases. Hence, the applicant must have basic skills in DSS as well as agriculture. So that the DSS is not blindly followed but passes through human judgment.

DSS may raise ethical and legal concerns related to data privacy, security, and bias. Issues such as unauthorized access to sensitive data, unintended consequences of automated decision-making, and algorithmic bias can undermine trust in the system and lead to legal liabilities. Implementing and maintaining a DSS can be costly and resource-intensive. Investments in technology infrastructure, data acquisition, model development, and user training may require significant financial and human resources.

7.3 Types of Decision Support Systems

DSS are categorized into five types based on their functionality, underlying technologies, and the nature of decision-making they support, as shown in Fig. 4 [40].

Model driven DSS is a computer-based system that assists decision-makers in solving complex problems by using mathematical models and analysis techniques. It utilizes various models to simulate different scenarios, predict outcomes, and provide recommendations or insights to aid in decision-making processes. It finds applications in various domains, including finance, operations management, marketing, healthcare, and environmental management. Data driven DSS leverages data analytics and insights to aid decision-making processes within organizations. Unlike Model-Driven DSS, which primarily rely on mathematical models, Data-Driven DSS focus on analyzing large volumes of data to derive patterns, trends, and actionable insights. It has applications across industries such as retail, healthcare, finance, marketing, supply chain management, and manufacturing. Communication driven DSS are designed to facilitate communication, collaboration, and information sharing among decision-makers to support the decision-making process. These systems focus on improving the exchange of ideas, opinions, and knowledge among stakeholders involved in making decisions. Knowledge driven DSS relies on expert knowledge, rules, and reasoning to support decision-making processes. In addition, it prioritizes the incorporation of human expertise and domain knowledge into the

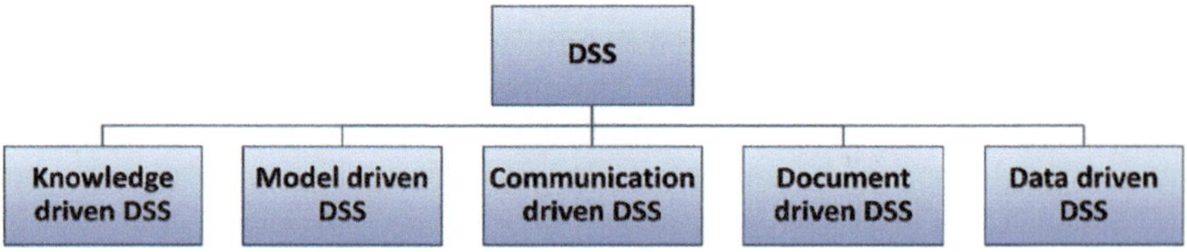

Fig. 4 Types of decision support systems

decision-making process. Document driven DSS assists decision-makers by organizing, managing, and analyzing documents and other unstructured information. However, it primarily handles unstructured textual data in the form of documents, reports, emails, and other written communication.

8 Crop Diagnosis and Disease Detection by AI Technologies

Artificial Intelligence (AI) plays a significant role in crop diagnosis and disease detection by leveraging various techniques such as machine learning, computer vision, and data analytics. AI algorithms can analyze data from sensors embedded in the field, such as temperature, humidity, soil moisture, and nutrient levels, to monitor the environmental conditions and identify potential stressors or risk factors that may contribute to crop diseases. By correlating sensor data with disease outbreaks, AI systems can provide early warnings and predictive insights to farmers. AI techniques enable the analysis of remote sensing data and satellite imagery to monitor large-scale crop health and detect anomalies across vast agricultural landscapes. AI algorithms can detect changes in vegetation indices, canopy cover, and spectral signatures indicative of crop diseases, allowing for timely intervention and targeted management strategies.

AI algorithms can analyze images of crops taken by drones, satellites, or smartphones to identify visual patterns associated with healthy plants or signs of diseases, pests, nutrient deficiencies, or other abnormalities. Computer vision techniques enable AI systems to detect subtle differences in color, texture, shape, and size that may indicate potential issues with crop health. Machine learning models can be trained on large datasets of images, sensor data, and historical records of crop diseases to learn patterns and correlations between various factors and the presence of specific diseases. These models can then be used to classify new images and accurately diagnose crop diseases in real-time.

AI-driven precision agriculture technologies enable targeted interventions at the individual plant or sub-field level, optimizing resource use and minimizing the spread of diseases. By combining AI-based crop diagnosis with precision agriculture techniques such as variable rate application of inputs, farmers can address crop health issues more effectively while reducing environmental impacts and improving productivity.

9 AI-Powered Yield Prediction System for Agriculture Crops

Yield prediction systems play a crucial role in precision agriculture techniques and consist of two separate techniques, such as yield mapping and yield monitoring.

A yield monitoring system comprises various components such as sensors, GPS, and specialized software. The GPS component determines the precise harvest location, while sensors measure the amount of crop harvested. The software integrates data from both the sensors and the GPS to generate a yield map of the field. This system is compatible with most harvesting equipment, including combines, grain carts, and forage harvesters.

The process of putting a field's crop yield into a graphic format is called yield mapping. GPS data and the information gathered by the yield monitoring system are combined to create yield maps. The agricultural yield in various field sections is displayed on the generated map. Yield maps are instrumental in identifying areas of a field that exhibit exceptionally high or low performance. They facilitate the creation of variable rate treatment maps, optimizing the use of fertilizers and other agricultural inputs based on the specific needs of different field zones.

9.1 Components of Yield Prediction Systems

The system typically comprises multiple components that collaborate to deliver valuable data, aiding growers in making informed decisions. A yield monitor, installed on the combine harvester, collects crop yield data through various sensors. A GPS receiver tracks the location of the combine harvester as it traverses the field, recording the corresponding yield data. By integrating GPS positional data with yield data, farmers can generate yield maps that illustrate variations in crop yields across their fields. They can use this to determine which regions might benefit from more irrigation, fertilizer, or soil amendments. The display monitor, typically situated in the cab of the combine harvester, provides real-time data on yields, moisture content, and other relevant metrics. Yield monitoring systems are equipped with software that enables farmers to analyze and interpret the collected data. This software facilitates the creation of yield maps, tracks crop yields over time, and identifies trends in crop growth and development.

9.2 Stepwise Procedure for Yield Prediction

The detailed description of the various steps to be involved in yield prediction is given below.

The first step in AI-based yield prediction involves collecting diverse data sources relevant to crop growth and yield using different sensors. After data collection, preprocessing is conducted to clean and prepare the data for analysis. This process may involve managing missing values, normalizing the data, and transforming it into a format compatible with AI algorithms. AI models like regression, random forests, decision trees, neural networks, support vector machines, and ensemble methods are used for yield prediction, selected based on factors such as data characteristics, relationship complexity, and desired accuracy levels.

The chosen AI model undergoes training on pre-processed data to discern patterns and correlations among input variables (such as weather conditions, soil characteristics, and agricultural practices) and crop yield. Following training, the model undergoes validation and testing using an independent dataset to assess its performance and ability to generalize. After validation, the trained AI model can be used to predict crop yield for future growing seasons based on input data such as weather forecasts, soil conditions, and management plans. AI-based yield prediction models can be continually refined and improved over time as more data becomes available and as the model learns from its predictions and feedback from users.

9.3 Benefits of Yield Prediction in Precision Agriculture

The yield prediction system enables farmers to monitor crop yields at the field level through a yield monitoring tool, offering critical insights into the variability of yields within different areas of the field. This system enhances crop management by providing precise and timely data on crop yields, allowing farmers to make well informed decisions regarding future planting and harvesting practices. By utilizing this information, farmers can optimize their resource usage, including fertilizer, water, and energy, thereby reducing waste and increasing efficiency. Additionally, yield prediction aids farmers in maintaining detailed records of crop yields, which can be invaluable for future planning and decision-making.

10 Conclusions

The global demand for food is increasing while the resources are limited. The qualitative aspects of food like non-GMO food, organic food are gradually gaining demand. At the same time the unpredictable climate variation, emissions due to intensive mechanization creates a higher challenge. Proper management of resources, timely decision making, and applying inputs as required are important steps that can be taken to achieve a sustainable solution to the above problem. These steps essentially require high level of knowledge, skill, and experience. Human memory, knowledge, and experience are limited and often leads to error. The discussed technology like sensors, communication technologies, data storage, AI, algorithms, smart devices, actuators do the job for human reliably. These technologies have proven quite efficient in managing agricultural activities. Data collection is easier with drones, satellite, and weather stations. Robots, and autonomous tractor have simplified labor intensive farm operations. Continuous monitoring and field application is possible through IoT and smart devices. All these technologies together can reduce the equivalent total energy requirements in crop production by maximizing production and minimizing loss.

References

1. Mahapatra, J., Sahoo, R.: Application of drones in farming. Agric. Magaz. **3**(3), 332–335. ISSN: 2583–1755. https://theagricultureonline.com/volume-03-issue-03-december-2023. (2023)
2. Tiwari, P.S., Sahni, R.K., Kumar, S.P., Kumar, V., Chandel, N.S.: Precision agriculture applications in horticulture. Pantnagar J. Res. **17**(1), 1–10 (2019a). https://www.gbpuat.res.in/uploads/archive/Vol.17(1).1.pdf
3. Tiwari, P.S., Singh, K.K., Sahni, R.K., Kumar, V.: Farm mechaniza on–trends and policy for its promo on in India. Indian J. Agric. Sci. **89**(10), 1555–1562 (2019b). https://doi.org/10.56093/ijas.v89i10.94575
4. Kumar, M., Sahni, R.K., Agarwal, K.N.: Development of a mechanically assisted pneumatic metering mechanism for high-speed planting of soybean. J. Sci. Indus. Res. **83**(1), 84–91 (2024a). https://doi.org/10.56042/jsir.v83i1.2863
5. Mahapatra, J., Tiwari, P.S., Singh, K.P., Nandede, B.M., Sahni, R.K., Pagare, V., Singh, J., Shrinivasa, D.J., Mandal, S.: An adjustable pneumatic planter with reduced source vibration for better precision in field seeding. Sensors **24**(11), 3399 (2024)
6. Rathnayake, A.P., Sahni, R.K., Khot, L.R., Hoheisel, G.A., Zhu, H.: Intelligent sprayer spray rates optimization to efficiently apply chemicals in modern apple orchards. J. ASABE **65**(6), 1411–1420 (2022). https://doi.org/10.13031/ja.14654
7. Sahni, R.K., Schrader, M.J., Rathnayake, A.P., Khot, L.R., Hoheisel, G.A., Zhu, H.: Evaluation of suitable base spray rate estimation methods for precision chemical applications in vineyards with different training systems. Am. J. Enol. Vitic. **75**(1), 0750009 (2024). https://doi.org/10.5344/ajev.2024.22064
8. Singh, J., Din, M., Agrawal, K.N., Jyoti, B., Roul, A.K., Kumar, M., Singh, K.: Estimation of leaf area and leaf area density for design optimization of a recycling tunnel sprayer. JSIR **81**(2), 173–179 (2022)
9. Singh, J., Din, M., Agrawal, K.N., Jyoti, B., Roul, A.K., Pagare, V.: Assessment of morphological attributes of Guava tree for design of a novel tunnel spraying system. Environ. Conserv. J. **25**(1), 1–9 (2024)
10. Kumar, S.P., Jat, D., Sahni, R.K., Jyoti, B., Kumar, M., Subeesh, A., Parmar, B.S., Mehta, C.R.: Measurement of droplets characteristics of UAV based spraying system using imaging techniques and prediction by GWO-ANN model. Measurement 114759 (2024b). https://doi.org/10.1016/j.measurement.2024.114759
11. Sahni, R.K., Khot, L.R.: Fixed spray systems for perennial specialty crops. In: Zhang, Q. (ed.) Encyclopedia of Smart Agriculture Technologies. Springer, Cham (2023). https://doi.org/10.1007/978-3-030-89123-7_195-1
12. Tian, L., Reid, J.F., Hummel, J.W.: Development of a precision sprayer for site-specific weed management. Trans. ASAE **42**(4), 893–900 (1999). https://doi.org/10.13031/2013.13269
13. Sanchez, P.R., Zhang, H.: Evaluation of a CNN-based modular precision sprayer in broadcast-seeded field. Sensors **22**(24), 9723 (2022). https://doi.org/10.3390/s22249723
14. Zanin, A.R.A., Neves, D.C., Teodoro, L.P.R., da Silva Júnior, C.A., da Silva, S.P., Teodoro, P.E., Baio, F.H.R.: Reduction of pesticide application via real-time precision spraying. Sci. Rep. **12**(1), 5638 (2022). https://doi.org/10.1038/s41598-022-09607-w
15. Chandel, N.S., Mehta, C.R., Tewari, V.K., Nare, B.: Digital map-based site-specific granular fertilizer application system. Curr. Sci. 1208–1213 (2016). https://doi.org/10.18520/cs/v111/i7/1208-1213
16. Maleki, M.R., Mouazen, A.M., De Ketelaere, B., Ramon, H., De Baerdemaeker, J.: On-the-go variable-rate phosphorus fertilisation based on a visible and near-infrared soil sensor. Biosys. Eng. **99**(1), 35–46 (2008). https://doi.org/10.1016/j.biosystemseng.2007.09.007

17. Loghavi, M., Forouzanmehr, E.: Design, development and field evaluation of a map-based variable rate granular application control system. In 2010 Pittsburgh, Pennsylvania, June 20-June 23, 2010, p. 1. American Society of Agricultural and Biological Engineers (2010). https://doi.org/10.13031/2013.29807
18. Mirzakhaninafchi, H., Singh, M., Bector, V., Gupta, O.P., Singh, R.: Design and development of a variable rate applicator for real-time application of fertilizer. Sustainability **2021**(13), 8694 (2021). https://doi.org/10.3390/su13168694
19. Kim, Y.J., Kim, H.J., Ryu, K.H., Rhee, J.Y.: Fertiliser application performance of a variable-rate pneumatic granular applicator for rice production. Biosys. Eng. **100**(4), 498–510 (2008). https://doi.org/10.1016/j.biosystemseng.2008.05.007
20. Wang, G., Hao, Y., Chen, Y.: Research on the model of variable-rate fertilization in maize based on geographic information system. In: 2015 6th International Conference on Manufacturing Science and Engineering, pp. 1756–1763. Atlantis Press (2015). https://doi.org/10.2991/icmse-15.2015.318
21. Zaman, Q.U., Schumann, A.W., Miller, W.M.: (2005). Variable rate nitrogen application in Florida citrus based on ultrasonically-sensed tree size. Appl. Eng. Agric. **21**(3), 331–335 (2005). https://doi.org/10.13031/2013.18448
22. Ehlert, D., Schmerler, J., Voelker, U.: Variable rate nitrogen fertilisation of winter wheat based on a crop density sensor. Precision Agric. **5**, 263–273 (2004). https://doi.org/10.1023/B:PRAG.0000032765.29172.ec
23. Solie, J.B., Stone, M.L., Needham, D.E., Washmon, C.N., Raun, W.R., Johnson, G.V., Thomason, W.E.: In-season N fertilization using an inseason estimate of potential yield. In: Proceedings of the 5th International Conference on Precision Agriculture, Bloomington, Minnesota, USA, 16–19 July, 2000, pp. 1–8. American Society of Agronomy (2000). https://doi.org/10.2134/agronj2002.8150
24. Roberts, R., English, B., Mahajanashetti, S.: (2001). Environmental and economic effects of spatial variability and weather. In: Blackmore, S., Grenier, G. (eds.) Proceedings of the 3rd European Conference on Precision Agriculture, pp. 545–550. AGRO, Montpellier, France (2001)
25. Yazdanbakhsh, O., Zhou, Y., Dick, S.: An intelligent system for livestock disease surveillance. Inf. Sci. **378**, 26–47 (2017). https://doi.org/10.1016/j.ins.2016.10.026
26. Gomes, J., Esteves, I., Neto, V.V.G., David, J.M.N., Braga, R., Arbex, W., Kassab, M., de Oliveira, R.F.: A scientific software ecosystem architecture for the livestock domain. Inf. Softw. Technol. **160**, 107240 (2023). https://doi.org/10.1016/j.infsof.2023.107240
27. Kumar, A., Hancke, G.P.: A zigbee-based animal health monitoring system. IEEE Sens. J. **15**, 610–617 (2015). https://doi.org/10.1109/jsen.2014.2349073
28. Smith, K., Martinez, A., Craddolph, R., Erickson, H., Andresen, D., Warren, S.: An integrated cattle health monitoring system. In: 2006 International Conference of the IEEE Engineering in Medicine and Biology Society (2006). https://doi.org/10.1109/iembs.2006.259693
29. Keertana, P., Vanathi, B.: IoT based animal health monitoring and tracking system using Zigbee. Int. J. Res. Trends Innov. **2**(4), 234–8 (2017)
30. Shinde, T.A., Prasad, J.R.: IoT based animal health monitoring with naive Bayes classification. Int. J. Emerg. Trends Technol. **1**(2), 252–7 (2017)
31. Rai, H.M., Gupta, D., Mishra, S., Sharma, H.: Agri-Bot: IoT based unmanned smart vehicle for multiple agriculture operation. In: 2021 International conference on simulation, automation & smart manufacturing (SASM) (2021). https://doi.org/10.1109/sasm51857.2021.9841182
32. Bakker, T., van Asselt, K., Bontsema, J., Müller, J., van Straten, G.: Autonomous navigation using a robot platform in a sugar beet field. Biosys. Eng. **109**, 357–368 (2011). https://doi.org/10.1016/j.biosystemseng.2011.05.001
33. Palaniappan, A., Muthiah, R., Sundaram, M.T.: ZigBee enabled IoT based intelligent lane control system for autonomous agricultural electric vehicle application. SoftwareX **23**, 101512 (2023). https://doi.org/10.1016/j.softx.2023.101512
34. De-An, Z., Jidong, L., Wei, J., Ying, Z., Yu, C.: Design and control of an apple harvesting robot. Biosys. Eng. **110**(2), 112–122 (2011). https://doi.org/10.1016/j.mechatronics.2021.102644

35. Lehnert, C., English, A., McCool, C., Tow, A.W., Perez, T.: Autonomous sweet pepper harvesting for protected cropping systems. IEEE Robot. Automation Lett. **2**(2), 872–879 (2017). https://doi.org/10.1109/LRA.2017.2655622
36. Li, J., Tang, Y., Zou, X., Lin, G., Wang, H.: Detection of fruit-bearing branches and localization of litchi clusters for vision-based harvesting robots. IEEE Access **8**, 117746–117758 (2020). https://doi.org/10.1109/ACCESS.2020.3005386
37. Lehnert, C., McCool, C., Sa, I., Perez, T.: Performance improvements of a sweet pepper harvesting robot in protected cropping environments. J. Field Robot. **37**(7), 1197–1223 (2020). https://doi.org/10.1002/rob.21973
38. Mu, L., Cui, G., Liu, Y., Cui, Y., Fu, L., Gejima, Y.: Design and simulation of an integrated end-effector for picking kiwifruit by robot. Inform. Process. Agric. **7**(1), 58–71 (2020). https://doi.org/10.1016/j.inpa.2019.05.004
39. Yoshida, T., Onishi, Y., Kawahara, T., Fukao, T.: Automated harvesting by a dual-arm fruit harvesting robot. Robomech J. **9**(1), 19 (2022). https://doi.org/10.1186/s40648-022-00233-9
40. Anonymous.: BotPenguin. Decision Support System: Types, Advantages and Use Cases. https://botpenguin.com/glossary/decision-support-system. (2024). Last accessed 29 Oct 2023

Potato Disease Classification in an Uncontrolled Environment Using Deep Learning Approaches

Anupam Anubhav, Tanisha Samantaray, and Chittaranjan Pradhan

Abstract Solanum tuberosum also known as Potato is one of the most important crops worldwide, contributing significantly to global food security. However, potato production is slowed down by a variety of diseases that can cause crucial yield losses if not effectively managed. Here, we explore the application of deep learning techniques for the classification of potato diseases based on leaf symptoms, aiming to assist farmers in timely disease diagnosis and management decisions. In this article, we use the convolution neural network (CNN) method; also, examine seven classes of potato diseases with the names: Bacteria, Fungi, Nematode, Phytophthora, Pest, Virus, and Healthy. We used a database of 3076 potato images. This proposed plan achieves an accuracy of 99% with respect to the EfficientNetV2B3 model, MobileNetV3-Large model, VGG-16 model, ResNet50 model, and DenseNet121 model.

Keywords Convolutional neural network · Deep learning · Inception V3 · Potato leaf disease

1 Introduction

Potato disease represents a significant threat to potato farming worldwide, with some of the most notorious diseases being late blight, early blight, black scurf, and bacterial wilt. In India, where potato cultivation is crucial for agricultural sustainability and food security, these diseases pose extreme challenges to farmers.

However, the contemporary agricultural research is based on machine learning techniques, particularly deep learning techniques. Within this context, the endeavor to apply such technologies to the domain of potato disease classification emerges as a critical pursuit. Implementing CNN-based disease detection systems can enable early diagnosis, allowing farmers to take timely preventive measures such as targeted

A. Anubhav · T. Samantaray · C. Pradhan (✉)
Kalinga Institute of Industrial Technology, Bhubaneswar, India
e-mail: chitaprakash@gmail.com

B. K. Mishra and A. Rocha (eds.), *Future of AI, IoT, and Sustainability*, Information Systems Engineering and Management 23,
https://doi.org/10.1007/978-3-031-76286-4_11

pesticide application or crop rotation to minimize losses. Furthermore, by providing insights into disease prevalence and distribution, these systems can support informed decision-making at both the individual farm level and on a broader scale, thereby contributing to more effective disease management strategies.

This paper seeks to address the pressing need for accurate and timely disease detection in potato farming, where the identification of leaf symptoms plays a pivotal role in mitigating yield losses and ensuring crop health. This article endeavors to bridge this gap by harnessing CNN-based models for the classification of potato diseases, predicated on leaf symptom images. By capitalizing on the capabilities of deep learning, the aim is to achieve heightened accuracy and robustness in disease detection, thereby furnishing farmers with a dependable tool for preemptive disease management. The ensuing sections of this report delineate the methodology employed, the experimental findings, and the ensuing discussions, with a view to advancing the discourse on effective potato disease management strategies.

2 Literature Survey

This study focuses on using a method to recognize patterns in images to figure out what category they belong to. We're not just looking at potato diseases here; this technique could be applied to sorting images into different categories for any purpose. When farmers check for diseased potatoes, they visually inspect the plants for any odd coloring or leaf abnormalities that could indicate a problem. Similarly, if we want a computer to do this job, we need to break down the image into different sections based on the patterns we're interested in. This process is called image segmentation. Image segmentation is like dividing a picture into different parts, each representing a distinct object or region [1]. It's like creating a map that outlines all the boundaries within an image, allowing computers to understand and analyze each part separately. This process is crucial for tasks like object recognition, scene understanding, and image editing, as it provides a detailed breakdown of what's in the image. Just as in real life, where we might categorize things into groups, image segmentation does the same thing computationally, making it easier for machines to comprehend visual data. Similar analysis is performed by the following authors:

Liakos et al. [2] explained the applications of machine learning in agriculture, including disease classification in crops. It discussed various algorithms and their potential for automating disease diagnosis and classification tasks in potatoes and other crops. Singh et al. [3] synthesized recent research efforts in potato disease detection and classification. It discussed the challenges and opportunities in automating disease classification tasks, such as data acquisition, feature selection, and model validation.

Oppenheim et al. [4] employed convolutional neural networks to classify potato diseases, demonstrating the model's effectiveness in automated disease identification, a significant step towards improving potato crop management. Arshabhi et al.

[5] focused on utilizing deep learning methods for the detection and classification of potato diseases. Their study likely involved collecting potato plant dataset, training convolutional neural networks (CNNs), on this dataset, and evaluating their performance in accurately identifying and categorizing various potato diseases. Their findings helped in development of automated tools for early disease detection and management in potato cultivation, potentially improving crop yield and reducing losses.

3 Background Techniques

A deep learning technique, in particular tensorflow methodologies, in agricultural domain names has garnered large attention in recent years. Within this context, the usage of Convolutional Neural Networks (CNNs) has became a prominent methodology for addressing diverse demanding situations in crop disorder type and control. This phase gives an overview of the essential standards underlying CNNs and opinions present literature applicable to the use of deep learning in potato disease type.

3.1 Convolutional Neural Networks (CNN)

CNNs are neural network structures tailored to learn hierarchical patterns from input images. They comprise convolutional layers, where filters are employed to detect local features like edges and textures, pooling layers that down sample feature maps spatially, and fully connected layers, which amalgamate the extracted features for classification purposes as shown in Fig. 1. In the realm of potato disease classification, CNNs scrutinize pixel patterns in potato leaf images to discern disease-related characteristics such as lesions or discoloration.

3.2 Image Processing

Image preprocessing involves several techniques applied to raw input images to prepare them for CNN training. Resizing ensures all images have the same dimensions, which is essential for efficient batch processing. Normalization scales pixel values to a standard range (e.g., 0 to 1), helping the model converge faster during training. Augmentation introduces variations to the training data, such as rotating or flipping images, to increase dataset diversity and improve model generalization.

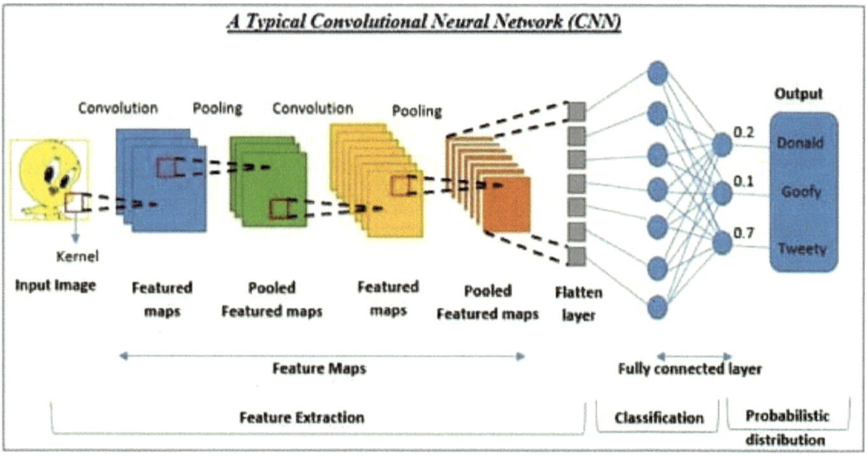

Fig. 1 Sample convolutional neural network architecture [6]

3.3 *Dataset Collection and Annotation*

Building a dataset for potato disease classification requires collecting a diverse range of potato leaf images, including both healthy and diseased samples [7]. Each image is meticulously labeled to indicate the presence of specific diseases or symptoms, such as virus, fungi or nematode. These annotations serve as ground truth labels during model training, enabling the CNN to learn to differentiate between healthy and diseased potato leaves.

3.4 *Transfer Learning*

Transfer learning means using already-trained CNN models, like ones trained on big image datasets such as ImageNet, and tweaking them for the particular job of classifying potato diseases. In this fine-tuning process, the model's existing weights are modified using the potato leaf dataset, so it becomes better at spotting disease-related characteristics. This method speeds up training and enhances model accuracy, which is particularly useful when you don't have a lot of labeled data to work with as shown in Fig. 2.

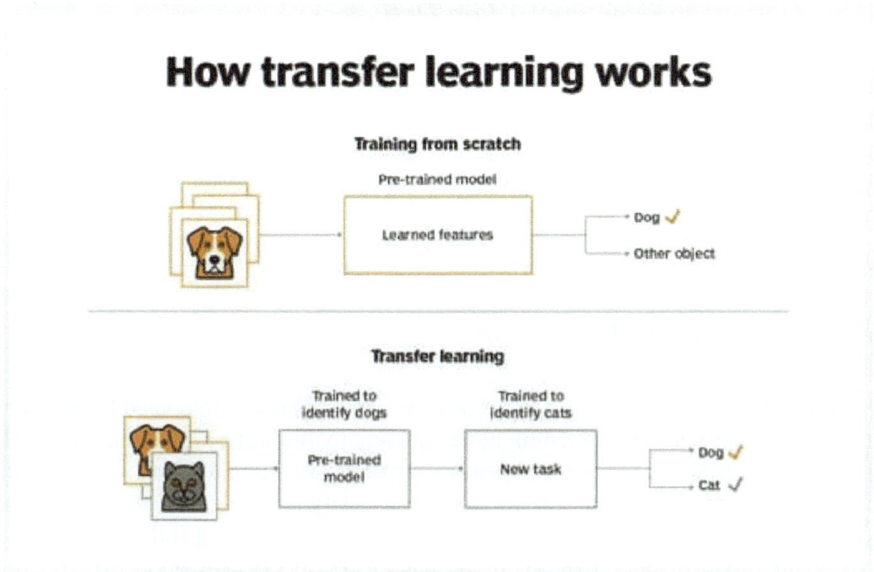

Fig. 2 Working of transfer learning [8]

3.5 *Model Architecture Selection*

Choosing the right CNN architecture is critical for achieving optimal performance in potato disease classification. Inception, for example, employs inception modules that capture features at different scales, making it suitable for detecting fine-grained details in potato leaf images while minimizing computational overhead.

3.6 *Hyperparameter Tuning*

Hyperparameters are settings that control the learning process of the CNN model, such as varying learning rate, different batch size, and dropout rate. It involves systematically exploring different variations of these settings to find specific configuration that maximizes model performance. Techniques like grid search or random search are mostly used to efficiently search the hyperparameter space and identify the optimal settings for the given task [9].

3.7 Optimizer Selection (Adam Optimizer)

The choice of optimizer plays a significant role in training CNN models. Adam can effectively navigate the model's parameter space and accelerate convergence during training. Its robust performance makes it well-suited for optimizing the parameters of CNN models, including those fine-tuned from pre-trained architectures like Inception v3.

3.8 Training and Evaluation

Training the CNN model involves iteratively feeding batches of labeled training data through the network and adjusting its parameters to minimize prediction errors. The model's performance is evaluated using a separate validation dataset, where metrics like accuracy measure its ability to correctly classify potato leaves into healthy and diseased categories. Additionally, a test dataset, unseen during training, is used to assess the model's generalization ability and ensure its effectiveness.

4 Proposed Methodology

In the presented article, various stages of research are depicted, as illustrated in Fig. 3, forming a research framework. The outlined stages in this proposed research framework are as follows:

4.1 Data Collection

The dataset utilized in this study comprises images of potato leaves categorized into seven categories:- ['Fungi', 'Bacteria', 'Healthy', 'Pest' 'Virus', 'Nematode', 'Phytophthora']. The dataset was taken from The National Center for Biotechnology Information website under the name "Potato Leaf Disease Dataset in an uncontrolled environment" [10]. The amount of data used is shown in Table 1.

4.2 Data Preprocessing

Data pre-processing involves eliminating non-relevant portions of the image to isolate the region of interest and reducing image noise. Images with excessive noise are

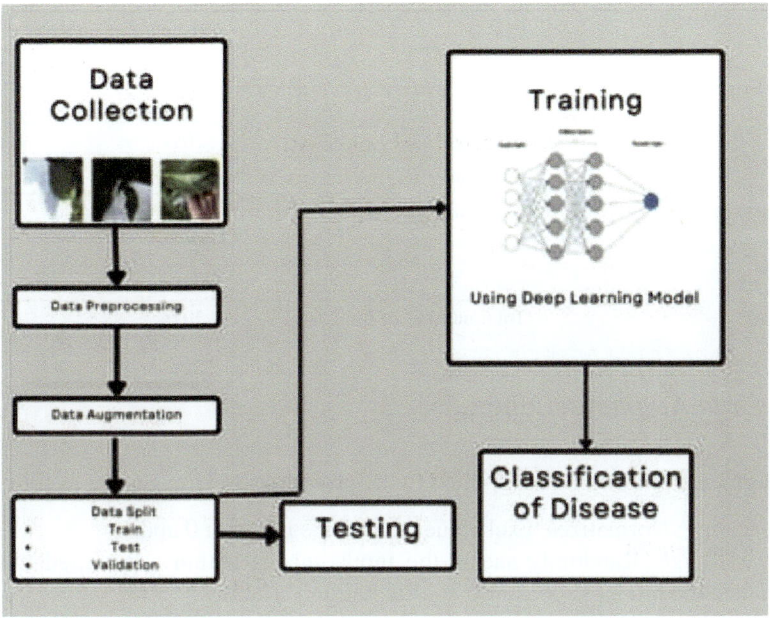

Fig. 3 Methodology followed

```python
from tensorflow.keras.optimizers import Adam
from tensorflow.keras.callbacks import LearningRateScheduler
import math

# Define a function to adjust the learning rate
def lr_schedule(epoch, lr):
    # Learning rate schedule can be adjusted based on the epoch
    if epoch < 10:
        return lr  # Keep the initial learning rate for the first 10 epochs
    else:
        return lr * math.exp(-0.1)  # Exponential decay schedule

# Define the initial learning rate
initial_lr = 0.0001  # Best performing learning rate

# Create an instance of the LearningRateScheduler callback
lr_scheduler = LearningRateScheduler(lr_schedule)
```

Fig. 4 lr_schedule function code.

excluded from utilization. Regarding the datasets, input photos have been standardized to 224 × 224 pixels post-collection from diverse sources and varying dimensions.

Table 1 Dataset used

Class	Number of images
Virus	532
Phytophthora	347
Healthy	201
Bacteria	569
Nematode	68
Pest	611
Fungi	748
Total number of images	3076

4.3 Data Augmentation

The augmentation techniques applied to the pre-processed images are as follows:

- Re-scaling: Normalizes pixel values to a range between 0 and 1.
- Shear Range: Randomly shears the input images within a specified range to enhance the model's robustness to orientation variations.
- Zoom Range: Ambiguously zooms the input images within a specified range to help the model learn to recognize objects at different scales.
- Horizontal Flip: Randomly flips the input images horizontally, effectively doubling the training data and improving the model's ability to generalize.

These augmentation techniques aim to improve the range of the training dataset, thereby improving the model's performance and generalization capabilities.

4.4 Data Splitting

At this juncture, 3076 potato leaf images are employed. These images are partitioned into training and testing datasets using two different ratios: 80:20 and 70:30 for data division. The outcomes obtained from each data split will be utilized in subsequent evaluations to determine the optimal approach.

4.5 Training

For this article, we utilized transfer learning with the InceptionV3 pre-trained convolutional neural network architecture. The steps involved in training the model are outlined below:

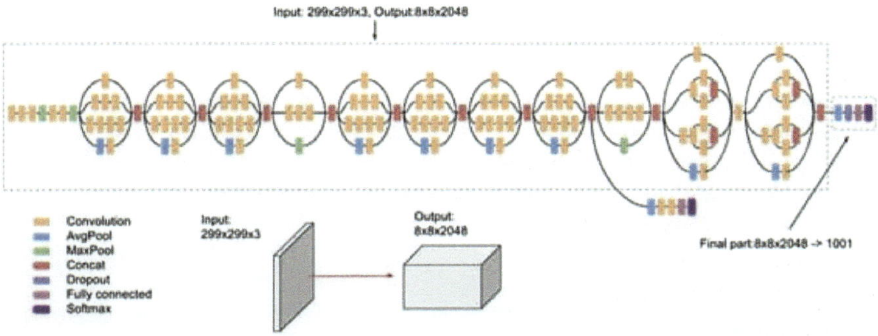

Fig. 5 A high-level diagram of the model [12]

Model Initialization

We instantiated the InceptionV3 model with input shape IMAGE_SIZE + [4] and loaded pre-trained weights from the 'imagenet' dataset [11]. To prevent retraining of existing weights, we set all layers of the InceptionV3 model to non-trainable.

Model Modification

Added custom layers on top of the InceptionV3 base model to adapt it to our specific task.

Learning Rate Schedule

Defined a function 'lr_schedule(epoch, lr)' to adjust the learning rate based on the epoch as shown in Fig. 4.

We initialized the learning rate to 0.0001, which was determined as the best-performing learning rate for our task. Utilized an exponential decay schedule for learning rate adjustment after the initial epochs.

Layer Training

We selectively froze the layers of the InceptionV3 model for fine-tuning as shown in Fig. 5 and Fig. 6.

By leveraging transfer learning and fine-tuning techniques, our model benefits from pre-trained feature extraction while adapting to the nuances of our dataset through training.

4.6 Testing

In our endeavor to develop an effective model for the task at hand, we conducted a series of experiments exploring different configurations [13]. These experiments

```
# view the structure of the model
model.summary()
```

```
                                                                          ']

 mixed10 (Concatenate)        (None, 5, 5, 2048)      0         ['activation_85[0][0]',
                                                                 'mixed9_1[0][0]',
                                                                 'concatenate_1[0][0]',
                                                                 'activation_93[0][0]']

 flatten (Flatten)            (None, 51200)           0         ['mixed10[0][0]']

 dense (Dense)                (None, 512)             2621491   ['flatten[0][0]']
                                                      2

 dense_1 (Dense)              (None, 7)               3591      ['dense[0][0]']

==================================================================================================
Total params: 48021287 (183.19 MB)
Trainable params: 26218503 (100.02 MB)
Non-trainable params: 21802784 (83.17 MB)
__________________________________________________________________________________________________
```

Fig. 6 Summary of proposed model

encompassed variations in the number of epochs, data-splitting strategies, and evaluation metrics.

Number of Epochs

We conducted experiments with varying numbers of epochs during model training to assess the impact on convergence and generalization performance [14, 15]. This exploration allowed us to identify the optimal balance between model complexity and training time.

Data Splitting

Data splitting is a significant step in evaluating machine learning models, as it directly influences their performance. The quality of the training, testing and validation datasets significantly impact the model's effectiveness. Ensuring a balanced distribution of data across these sets is essential for achieving reliable results and accurate assessments of the model's generalization capabilities.

Evaluation Metrics

These are pivotal in assessing the performance of our model. We explored a range of metrics, including accuracy, precision, and recall to gain insights into different aspects of model performance. This comprehensive evaluation enabled us to make informed decisions regarding the model's efficacy and suitability for real-world applications.

By systematically varying these parameters and meticulously analyzing the results, we gained valuable insights into the behavior of our model under different conditions. These experiments contributed to the cleansing of our model and enhanced our understanding of the underlying problem domain.

4.7 *Classification*

Deep learning has revolutionized various domains with its ability to learn intricate patterns directly from data. CNNs) a prominent class of deep learning models, have particularly excelled in image-related tasks. Leveraging the power of CNNs, notably the InceptionV3 architecture, we tailored the model to classify potato leaf images into seven distinct classes. This fine-tuning process involved adjusting the pre-trained CNN's parameters to better recognize the unique features of potato leaves. Through this specialized approach, we achieved accurate classification results, underscoring the efficacy of CNNs in addressing complex image classification tasks and advancing agricultural research and crop management practices.

5 Result Analysis

Our study on predicting potato leaf diseases using CNNs yielded encouraging results. When we tested our CNN model on a set of potato leaf images, it performed well, showcasing the accuracy and effectiveness of our method in classifying these diseases.

The model was trained for 10, 20 and 30 epochs achieving an accuracy of 95.60, 98.86, and 99.51% respectively. We also trained our model to split data into 70% for training, 15% for testing, and 15% for validation and achieved an accuracy of 98.38%.

Accuracy: Achieved an accuracy of 0.9951, indicating the proportion of correct classifications out of total predictions.

Precision: Attained a precision score of 0.9955, reflecting the proportion of correctly predicted positive cases out of all predicted positive cases.

Recall: Achieved a recall score of 0.9943, highlighting the proportion of correctly predicted positive cases out of all actual positive cases.

In addition to the numerical results presented above, visual representations of the performance metrics further elucidate the efficacy of our CNN model in predicting potato leaf diseases. The following graphs provide a extensive overview of the model's accuracy, precision, and recall across varying epochs and data partitioning strategies.

The following Graphs and Table are for Data split into 80% training 10% testing and 10% validation:

Here is a detailed version of our best-performing model:

Table 2 summarizes the accuracy, precision, recall, and loss values for each epoch:

This table provides a extensive overview of the model's performance metrics for each epoch during training. Figures 7, 8, 9 and 10 summarizes the accuracy, loss, recall and precision respectively for 30 Epochs.

Figures 11, 12, 13 and 14 summarizes the precision, recall, accuracy and loss respectively for 20 Epochs.

Table 2 Table for each epoch

Epoch	Loss	Accuracy	Precision	Recall
1	1.5486	0.4575	0.5867	0.2947
2	0.8983	0.6846	0.7655	0.5979
3	0.6546	0.7554	0.8241	0.7094
4	0.4452	0.8445	0.8808	0.8152
5	0.3615	0.8783	0.8967	0.8519
6	0.267	0.9039	0.9203	0.8885
7	0.2408	0.9198	0.9302	0.906
8	0.1738	0.9398	0.9463	0.9333
9	0.1593	0.943	0.9505	0.9385
10	0.1554	0.9471	0.9523	0.9422
11	0.1073	0.9634	0.9676	0.9601
12	0.0808	0.9703	0.9738	0.9678
13	0.0624	0.9809	0.9829	0.9801
14	0.0672	0.9788	0.9792	0.9764
15	0.0425	0.9845	0.9857	0.9829
16	0.0313	0.9894	0.9906	0.9886
17	0.049	0.9809	0.9812	0.9796
18	0.0327	0.9894	0.9906	0.9886
19	0.0412	0.9841	0.9853	0.9837
20	0.0208	0.9931	0.9931	0.9923
21	0.0213	0.9931	0.9935	0.9923
22	0.0264	0.9931	0.9935	0.9927
23	0.0223	0.9923	0.9935	0.991
24	0.0145	0.9951	0.9959	0.9951
25	0.0184	0.9947	0.9951	0.9939
26	0.0209	0.9927	0.9935	0.9927
27	0.0135	0.9967	0.9967	0.9963
28	0.0153	0.9959	0.9959	0.9959
29	0.0156	0.9939	0.9943	0.9935
30	0.0128	0.9951	0.9955	0.9943

We also performed by splitting the dataset into 70% training 15% validation and 15% testing and achieved the following results as shown in Fig. 15.

In addition to accuracy, precision, and recall, the F1 score gives a more comprehensive evaluation of the model's performance, especially in binary classification tasks like ours. It considers both precision and recall, offering a balanced assessment of how well the model identifies positive cases while reducing false positives and false negatives. Our model consistently achieved an F1 score of 0.9965 across

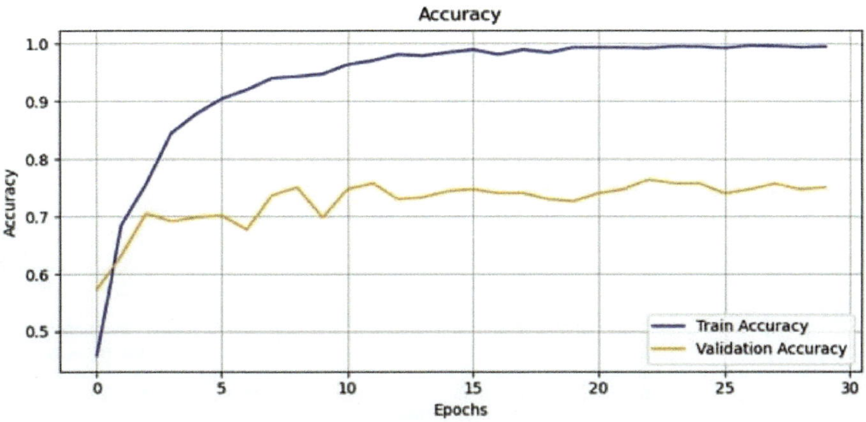

Fig. 7 Accuracy versus epochs graph for 30 epochs

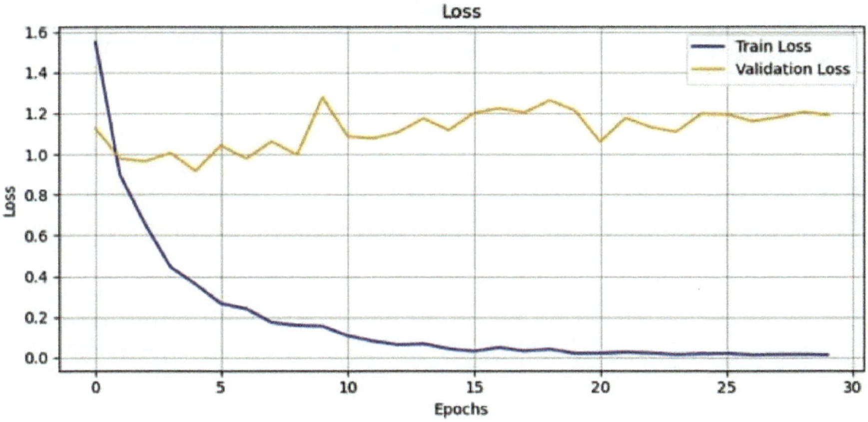

Fig. 8 Loss versus epochs graph for 30 epochs

various epochs, demonstrating its strong ability to distinguish between healthy and diseased potato leaves.

These graphical representations not only confirm the effectiveness of our CNN model in predicting potato leaf diseases but also provide valuable insights into its performance throughout different training epochs. As we refine our methods and explore further optimization possibilities, these results lay a solid groundwork for future agricultural disease detection research.

We assessed our model's performance by randomly selecting nine images for evaluation and checking its classification accuracy. Our dataset was divided into three sections: training, testing, and validation. Importantly, the testing and validation sets

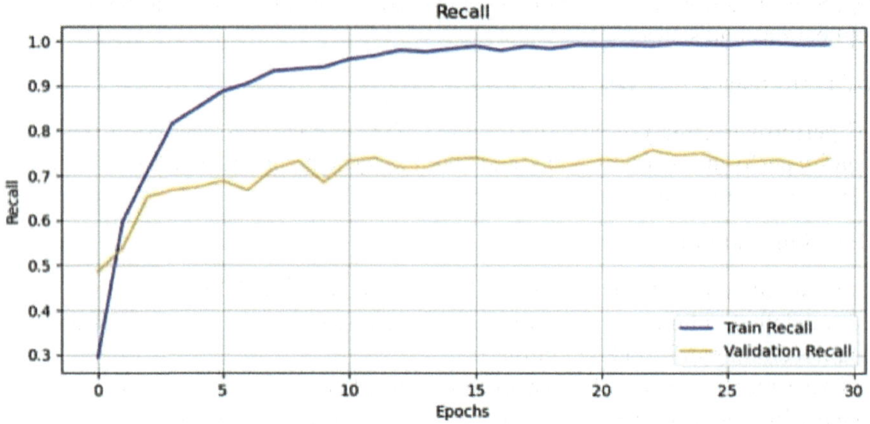

Fig. 9 Recall versus epochs graph for 30 epochs

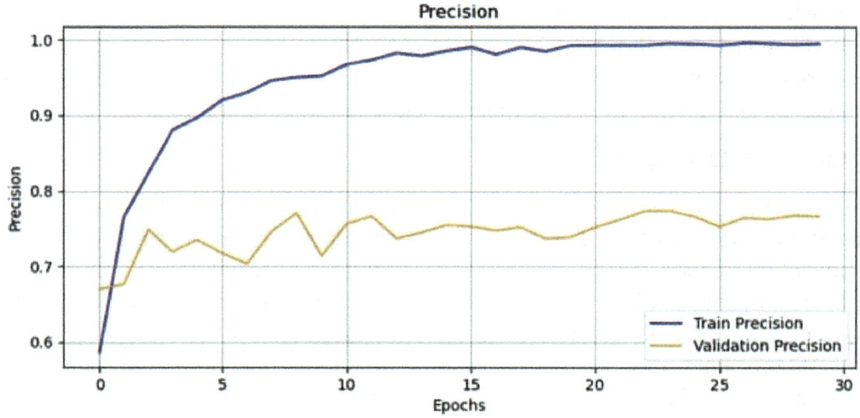

Fig. 10 Precision versus epochs graph for 30 epochs

were entirely separate from the training set to ensure fair evaluation. On average, it took about 70 to 80 min to train the model and less than an hour to test it (Fig. 16).

After analyzing our model's performance metrics, it's valuable to contextualize our findings within the broader landscape of potato leaf disease prediction research. Previous studies have explored various machine learning approaches, including traditional classifiers and deep learning models, to address this agricultural challenge.

Comparatively, our CNN model demonstrates superior performance, achieving an accuracy of 99.51% after 30 epochs of training. This significant improvement underscores the potency of deep learning techniques, mostly CNNs, in accurately classifying potato leaf diseases.

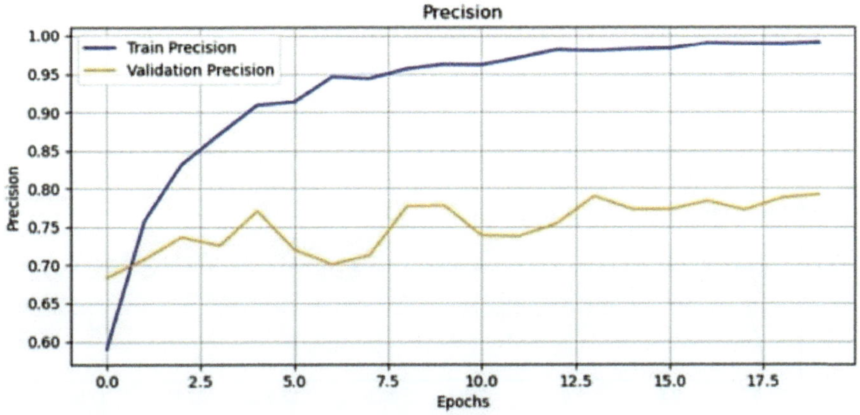

Fig. 11 Precision versus epochs graph for 20 epochs

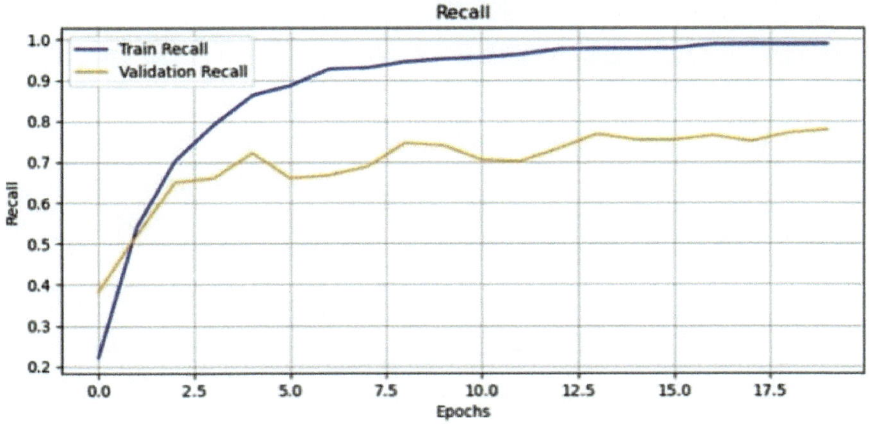

Fig. 12 Recall versus epochs graph for 20 epochs

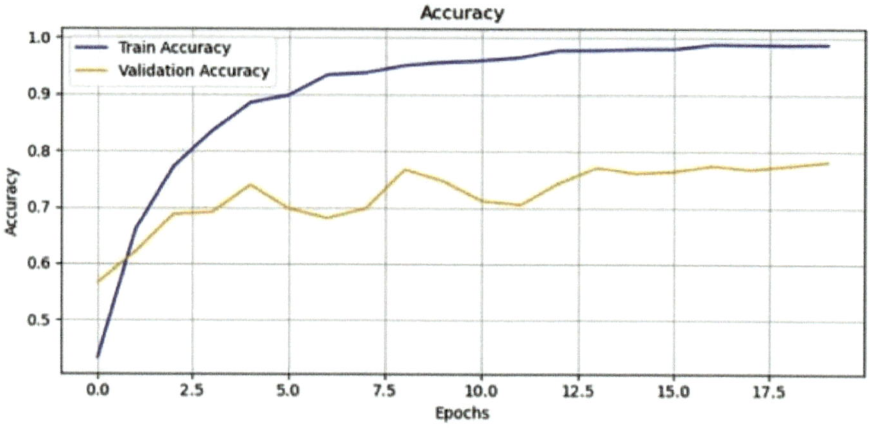

Fig. 13 Accuracy versus epochs graph for 20 epochs

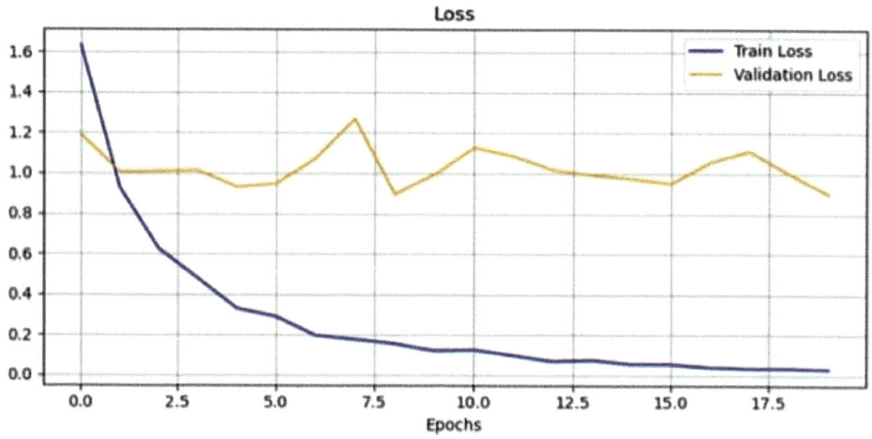

Fig. 14 Loss versus epochs graph for 20 epochs

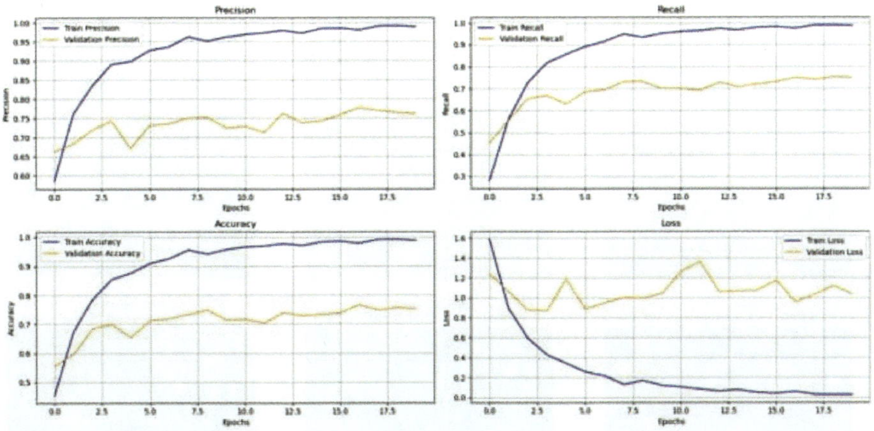

Fig. 15 Validation, precision and loss graph for 20 epochs

Fig. 16 Evaluation images

6 Conclusion

Our study shows a significant enhancement in the field of image classification, especially in the environment of utilizing deep learning techniques for achieving exceptional accuracy, precision, and recall rates. By leveraging the sophisticated architecture of Inception v3, which excels in capturing intricate features from images, and optimizing the training process with the Adam optimization algorithm, we have developed a model that surpasses previous benchmarks in performance. The achievement of 99% accuracy, precision, and recall underscores the reliability and robustness of our model in accurately categorizing images across diverse datasets. This level of accuracy is especially noteworthy in tasks such as disease diagnosis, where the timely and precise identification of conditions can significantly impact patient outcomes and treatment plans. Moreover, our model's success highlights the potential of deep learning approaches, such as CNNs and transfer learning, to address complex image classification challenges across various domains. Whether in healthcare, agriculture, or other sectors, our model's capability to achieve exceptional performance opens up new opportunities for automation, decision support systems, and enhanced data-driven insights. As we look ahead, our findings accelerate further advancements and applications in image classification tasks.

Future research directions may include exploring additional optimization techniques, refining model architectures, and expanding the scope of applications to address emerging challenges and opportunities in image analysis and interpretation. Ultimately, our model appears to be a significant step forward in exploiting the power of deep learning for solving real-world problems and driving innovation in the realm of image classification.

References

1. Purohit, S., Suman, S., Kumar, A., Sarkar, S., Pradhan, C., Chatterjee, J.M.: Comparative Analysis for Detecting Skin Cancer Using SGD-Based Optimizer on a CNN Versus DCNN Architecture and ResNet-50 Versus AlexNet on Adam Optimizer. Walter de Gruyter GmbH, 2021.
2. Liakos, K.G., Busato, P., Moshou, D., Pearson, S., Bochtis, D.: Machine learning in agriculture: a review. Sensors **18**(8) 2018
3. Singh, A., Kaur, H.: Potato plant leaves disease detection and classification using machine learning methodologies. In: IOP Conference Series: Materials Science and Engineering (2021)
4. Oppenheim, D., Shani, G.: Potato disease classification using convolution neural networks. Adv. Anim. Biosci. Elsevier **8**(2), 244–249 (2017)
5. Arshaghi, A., Ashourian, M., Ghabeli, L.: Potato diseases detection and classification using deep learning methods. Multimedia Tools Appl Springer **82**, 5725–5742 (2023)
6. Convolutional Neural Network: An Overview. https://www.analyticsvidhya.com/blog/2022/01/convolutional-neural-network-an-overview/
7. Patil, M.A., Manohar, M.: Potato Leaf Disease Identification using Hybrid Deep Learning Model. In: International Conference on Network, Multimedia and Information Technology (NMITCON) (2023)
8. Transfer Learning. https://www.techtarget.com/searchcio/definition/transfer-learning
9. Chunmao, L.I.U.: Advancing strawberry disease detection in agriculture: a transfer learning approach with YOLOv5 Algorithm. Int J Adv Comp Sci Appl (2024)
10. Shabrina, N.H., Indarti, S., Maharani, R., Kristiyanti, D.A., Prastomo, N.: A novel dataset of potato leaf disease in uncontrolled environment. Data Brief Elsevier **52** (2024)
11. Agarwal, R., Mittal, S., Sharma, A., Hariharan, U.: Enhancing potato disease classification with inception V3-based deep learning model. In: International Conference on Electronics, Materials Engineering & Nano-Technology (IEMENTech) (2023)
12. Advanced Guide to Inception v3. https://cloud.google.com/tpu/docs/inception-v3-advanced
13. Jha, P., Dembla, D., Dubey, W.: Deep learning models for enhancing potato leaf disease prediction: implementation of transfer learning based stacking ensemble model. Multimedia Tools Appl. (2023)
14. Wasalwar, Y.P., Bagga, K.S., Joshi, V.K., Joshi, A.: Potato leaf disease classification using convolutional neural networks. In: International Conference on Emerging Trends in Engineering & Technology—Signal and Information Processing (ICETET—SIP) (2023)
15. Chen, Y., Li, J., Zhang, X., Li, S., Cheng, Y., Fu, X., Li, J., Zhu, L.: Mesenteric adipose-derived exosomal TINAGL1 enhances intestinal fibrosis in Crohn's Disease via SMAD4. J. Adv. Res. (2024)

Leveraging AI and IoT for Enhanced Resource Efficiency in Sustainable Smart Cities

Vishal Jain and Archan Mitra

Abstract The acceleration of urbanization presents unprecedented challenges in managing city resources sustainably. This paper proposes a novel framework utilizing Artificial Intelligence (AI) and the Internet of Things (IoT) to enhance resource efficiency in smart cities, thereby contributing to sustainability. The crux of the framework lies in an integrated AI-IoT system capable of real-time data analysis and decision-making for optimizing resource allocation and consumption. In particular, the study will focus on critical urban systems—energy, water, and waste management—where AI can predict demand patterns, and IoT can provide granular consumption data. Through machine learning algorithms, the framework will analyze diverse datasets, ranging from energy usage to traffic flows, to identify inefficiencies and recommend corrective actions. For instance, in energy management, AI could optimize the distribution of renewable energy resources, while IoT devices monitor energy consumption patterns across the cityscape. The implementation challenges, including data privacy, cybersecurity, and the digital divide, will be critically examined. The potential socio-economic and environmental impacts of the framework's deployment will also be discussed, showcasing how technology can align with the Sustainable Development Goals (SDGs). This paper aims to contribute to the discourse on sustainable urban development, providing actionable insights for policymakers, urban planners, and technology developers. It underscores the synergy between AI and IoT as a driver for innovation in sustainability practices within urban centers.

Keywords Sustainable smart cities · AI-IoT integration · Resource management · Urban sustainability · Machine learning algorithms

V. Jain
Sharda University, Greater Noida, India

A. Mitra (✉)
Amity University Mumbai, Mumbai, India
e-mail: archan6644@gmail.com

B. K. Mishra and A. Rocha (eds.), *Future of AI, IoT, and Sustainability*, Information Systems Engineering and Management 23,
https://doi.org/10.1007/978-3-031-76286-4_12

1 Introduction

1.1 Background

Urbanization is accelerating at a rate that has never been seen before on a global scale, which is defining the demographic and economic landscapes of the future. According to the United Nations, by the year 2050, around 68% of the world's population will be living in urban regions, which will represent a dramatic movement away from rural living and toward urban living [1]. This increasing urban movement presents significant problems for sustainability, notably in terms of the management of critical resources like electricity, water, and garbage. In spite of the fact that urban areas are centers of innovation and economic activity, they are confronted with the formidable challenge of supporting rising populations with limited resources. The management of these resources in a sustainable manner is essential, not only for the purpose of preventing severe environmental degradation, but also for the purpose of ensuring economic stability and quality of life in urban contexts. Energy systems need to advance in order to accommodate greater demand while simultaneously reducing their carbon footprints. Water resources need to be managed in order to prevent scarcity, and waste management systems need to develop in order to handle increased outputs without causing damage to the environment. Because of the complexity of these difficulties, urban planning and resource management need to be approached in a way that is both comprehensive and inventive.

1.2 Rationale

In the realm of urban planning and management, the combination of Artificial Intelligence (AI) and the Internet of Things (IoT) presents a potentially game-changing possibility. In the process of developing urban settings that are more intelligent and responsive, these technologies are at the forefront. According to Hashem et al. [2], artificial intelligence allows for tremendous skills in the areas of predictive analytics and decision-making. These capabilities enable the forecasting of resource needs and the optimization of resource allocation. In the meantime, Internet of Things technologies offer the infrastructure required for real-time data gathering and transfer, which makes it possible to continuously monitor and manage urban systems. The combination of artificial intelligence and the internet of things has the potential to bring about major improvements in resource efficiency. This is demonstrated by the fact that AI is able to assess and make predictions based on enormous datasets, while IoT is able to offer data on resource utilization across the cityscape that is up to the minute. Smart cities are urban environments that make use of digital technology to increase the efficiency of city operations and services, promote connectivity, and reduce costs and resource consumption. This integration serves as the backbone of smart cities.

According to Albino, Berardi, and Dangelico [3], smart cities are an important development that plays a significant role in tackling the problems to sustainability that are faced by urbanization. They provide a road towards urban futures that are more sustainable, efficient, and livable.

1.3 Objectives of the Research

The purpose of this paper is to:

Provide a Novel AI-IoT Framework: Construct and explain a thorough framework that combines AI and IoT to improve resource management in urban environments. This framework intends to connect complex artificial intelligence algorithms with broad Internet of Things networks in order to enable intelligent, real-time analysis and decision-making processes that maximize the utilization of available resources.

Evaluate the Impact of the Framework: The usefulness of the suggested AI-IoT framework in the management of urban resources should be investigated with this project. In order to accomplish this, it is necessary to investigate how artificial intelligence technologies can predict demand patterns across essential resources and how the internet of things can permit detailed monitoring and data procurement on usage. This research will concentrate on important urban systems such as waste management, water management, and energy management.

Conduct an Analysis of the Implementation Implications: It is important to do an analysis of the practical implications of deploying an AI-IoT framework inside urban areas, with a specific emphasis on how well it aligns with sustainability goals. The discussion of the framework's ability to contribute to the Sustainable Development Goals (SDGs), the examination of its socio-economic implications, and the identification of potential impediments to its implementation are all included in this.

Through the accomplishment of these goals, the paper seeks to make a substantial contribution to the ongoing conversation about environmentally responsible urban development. In order to aid policymakers, urban planners, and technology developers in navigating the intricacies of urban sustainability, it intends to give a clear and actionable roadmap. This roadmap will offer insights that have the potential to affect future urban landscapes all over the world.

2 Literature Review

2.1 Current Practices in Urban Resource Management

The term "urban resource management" refers to a wide variety of approaches and technologies that are designed to maximize the utilization, distribution, and long-term viability of essential resources like electricity, water, and garbage. Traditional methods frequently include the utilization of centralized systems for the purpose of monitoring and management. These systems are not only inefficient but also unable to adapt to changes in demand and supply that occur in real time. As an illustration, the management of energy in urban areas has traditionally relied on non-renewable resources, and grid distribution does not dynamically react to fluctuations in demand or supply. This has resulted in inefficiencies and increased carbon emissions [4]. In a similar vein, water management frequently experiences considerable losses as a result of leaks and inefficiencies, which are caused by old infrastructure and a lack of real-time data on usage patterns [5]. According to Kumar, Dahiya, and Bhatia [6], waste management systems frequently confront obstacles in the areas of logistics, inefficiencies in recycling, and inadequate strategies for trash reduction.

2.2 Smart City Technologies

By incorporating cutting-edge information and communication technologies into infrastructure and services, smart city technologies have started to transform urban resource management. This is being accomplished through the integration of these technologies. Particularly noteworthy is the fact that the implementation of Artificial Intelligence (AI) and the Internet of Things (IoT) in urban settings has demonstrated a substantial amount of potential in expanding both efficiency and sustainability. IoT provides the network of sensors and devices that are necessary for collecting real-time data across the urban landscape [7]. Artificial intelligence enables sophisticated data analytics for forecasting resource demand and improving delivery systems. On the other hand, IoT provides the network of sensors and devices. These approaches, on the other hand, do not come without any restrictions. Studies have identified a number of obstacles, including the high initial cost of establishing Internet of Things infrastructures, the difficulty of maintaining and analyzing vast datasets created by these systems, and issues over the security and privacy of the data acquired [2]. Furthermore, there is still a gap in completely integrating these technologies into the urban infrastructures that are already in place, which is frequently the result of impediments that are technical, regulatory, or financial in nature.

2.3 Relevance to Sustainability

Both artificial intelligence and the internet of things are strongly aligned with the sustainability goals that are described in the Sustainable Development Goals (SDGs) of the United Nations. These technologies have the potential to have a direct impact on a number of Sustainable Development Goals (SDGs), including Goal 7 (which focuses on affordable and clean energy), Goal 11 (which focuses on sustainable cities and communities), and Goal 12 (which focuses on responsible consumption and production). According to Vasseur and Dunkels [8], artificial intelligence has the potential to improve the utilization of renewable energy sources by forecasting energy needs and dynamically altering supply. This will result in a reduction in waste and reliance on nonrenewable sources. According to Perera, Qin, Estrella, Reiff-Marganiec, and Vasilakos [9], the Internet of Things has the potential to improve the management of water and waste by giving real-time data on resource utilization. This, in turn, can lead to more sustainable consumption patterns and a reduction in the impact on the environment. There is a need for more comprehensive studies to investigate how these technologies can be effectively scaled and integrated into urban systems to meet sustainability objectives. This is especially important in diverse urban settings that vary significantly in terms of economic, environmental, and social characteristics. Despite these advantages, there is a need for more research to be conducted.

3 Research Methodology

3.1 The AI-IoT Integrated Framework

The suggested framework combines the Internet of Things (IoT) with artificial intelligence (AI) to produce a dynamic system that can improve resource efficiency in urban settings. IoT devices will gather real-time data from diverse urban infrastructures, and AI will interpret this data to make informed decisions about resource management. This layered architecture will govern the interaction between AI and IoT [10]. The following data types are to be used:

Data on Energy Consumption: Sensor and meter readings from the commercial, industrial, and residential sectors.

Water usage statistics: data on consumption from residential and commercial water meters, as well as flow measurements from distribution stations.

Data about trash management, including the amount and kind of waste collected, recycling rates, and operating information from facilities that treat waste.

These data points will be safely sent to a central data management platform, where AI systems will examine and interpret the data to determine the best way to distribute and use resources.

3.2 Data Collection

A network of sensors and smart meters placed across the urban infrastructure will gather data for the AI-IoT framework. These Internet of Things devices will supply constant, real-time data streams that are necessary for the framework to operate efficiently [7]. Particular data sources consist of:

Smart Energy Meters: Used in numerous industries to track electricity use patterns.

Water sensors: Placed strategically throughout the water supply system to monitor flow rates and identify leaks.

Smart Waste Sensors: These sensors are used in waste containers to monitor waste levels and optimize collection routes.

The gathered data will be encrypted before being sent to cloud-based storage platforms, where artificial intelligence (AI) algorithms will process and evaluate it.

3.3 Analytical Techniques

Several machine learning methods will be used by the AI component of the system to forecast demand trends and optimize resource allocation. Among these algorithms are:

Time Series Forecasting: Based on past consumption data, this technique forecasts future resource demand [11].

Regression Analysis: A statistical technique utilized to ascertain the interrelationships among diverse variables that impact resource utilization, including fluctuations in seasons and periods of high demand.

Clustering algorithms: They are used to separate and study how resources are used in various cities, which lets managers make more effective plans [12].

With the use of these algorithms, the AI system would be able to predict outcomes and manage resources effectively, catering to the unique requirements and consumption habits of the metropolitan populace.

3.4 Plan for the Implementation

The following procedures will be followed in a fictional smart city to build the AI-IoT framework:

Infrastructure Assessment: Examine current urban infrastructure to determine what needs to be upgraded in order to integrate IoT devices.

IoT Deployment: Integrate IoT devices and sensors into the trash, water, and energy management systems of the city.

Data Integration: To gather and store data from Internet of Things devices, provide a single platform for data management.

AI Deployment: Use AI algorithms to handle and examine the gathered data, then incorporate the results into systems for managing cities.

Pilot Testing: To improve the system and address any problems, run a pilot test in a certain part of the city.

Full-Scale Implementation: Distribute the system over the entire city while keeping an eye on it and making constant improvements to boost its efficiency.

4 Case Studies

4.1 Energy Management

The distribution and consumption of renewable energy sources can be significantly optimized in the field of energy management with the combination of AI and IoT technology. AI systems effectively integrate renewable energy solutions and improve energy efficiency by using real-time data gathered from IoT devices. For example, sensors and smart meters can offer real-time data on patterns of energy usage; AI algorithms can then use this data to forecast peak demand periods and modify the energy supply appropriately. Furthermore, AI is capable of overseeing the integration of renewable energy sources, such as wind and solar power, making sure that these erratic energy sources are successfully added to the grid without causing instability [13]. In addition to increasing energy efficiency, this dynamic adjustment lessens dependency on nonrenewable energy sources, promoting sustainability objectives.

4.2 Water Management

In order to drastically cut down on waste, the AI-IoT framework is applied to water management with a focus on monitoring and predicting water consumption. The water distribution network's IoT sensors are able to monitor water flow, pressure variations, and possible leaks in real time. Water utilities can dynamically alter water delivery and prevent losses by using AI algorithms that process this data to estimate water demand across various sectors and time periods [14]. Water that is produced and lost before it reaches the consumer is known as non-revenue water, and this precise management helps to remedy it. Preventive maintenance can be triggered by the system's ability to predict peak demand times and possible leak locations, resulting in efficient water distribution and conservation.

4.3 *Waste Management*

AI has the potential to completely transform the effectiveness of garbage collection and recycling procedures in waste management. Important information on the condition and makeup of waste is provided by Internet of Things devices like RFID tags for waste tracking and fill-level sensors in trash cans. AI uses this information to schedule bin collections, manage recycling operations, and optimize garbage collection routes [15]. AI systems, for instance, can identify trends in the production of waste and suggest the best routes for collection, saving time and gasoline. AI can also help with sorting at recycling facilities by anticipating the kinds of materials that will be sent for recycling and modifying sorting systems appropriately, increasing recycling rates and decreasing contamination.

5 Challenges and Considerations

5.1 *Data Privacy*

These frameworks for artificial intelligence and the internet of things create substantial concerns regarding privacy and personal security due to the intensive collection and analysis of data that is key to their performance. There is a growing volume of sensitive personal data being collected as urban systems integrate more Internet of Things devices. This raises concerns over the possibility of unwanted access and misuse of the data. It is necessary to implement stringent privacy policies and data protection safeguards in order to address these issues. On the other hand, according to Roman, Zhou, and Lopez [16], the implementation of sophisticated encryption methods and anonymization procedures can protect data while simultaneously preserving the privacy of users. In addition, the establishment of transparent regulations and compliance standards is extremely important in order to protect individuals' personal information and to foster public confidence in smart city technologies.

5.2 *Online Safety*

Due to the vital nature of the data that they handle and the fact that they are connected to the internet, Internet of Things devices, despite the fact that they are altering urban resource management, present vulnerabilities into urban infrastructure systems. Several different types of cyber attacks have the ability to exploit these vulnerabilities, which could result in the disruption of key services. According to Granjal, Monteiro, and Silva [17], ensuring the cybersecurity of Internet of Things devices requires the implementation of comprehensive security protocols. These protocols include secure booting, access control, and regular software upgrades.

These protocols, among others, are designed to prevent unwanted access and data breaches. Additionally, constant monitoring and the utilization of intrusion detection systems can be of assistance in recognizing and mitigating potential threats in real time, with the purpose of safeguarding the availability and integrity of essential urban infrastructure systems.

5.3 Digital Divide

Concerns over the digital divide are also raised as a result of the urbanization of artificial intelligence and internet of things technologies. In the context of urban populations, this gap refers to the discrepancies that exist between different socioeconomic categories in terms of access to technology and understanding of digital literacy. It is possible that the implementation of smart city technology could disproportionately favor those who have better access to digital resources, which could potentially exacerbate existing inequalities [18]. This could occur if due thought is not given to the matter. For the purpose of addressing this issue, governments need to ensure that everyone has access to technology by establishing public Wi-Fi networks, implementing affordable connectivity options, and implementing community training programs that aim to improve digital literacy. In addition, the participation of community stakeholders in the planning and implementation phases of smart city projects can assist in the customization of solutions to meet a variety of requirements, hence fostering equitable benefits distribution among all urban residents.

6 Impact Analysis

6.1 Socio-Economic Impacts

Urban lifestyles, the labor market, and economic activity are all expected to be significantly impacted socioeconomically by the adoption of an AI and IoT framework. From the standpoint of the labor market, the incorporation of cutting-edge technologies in urban administration may result in the displacement of old roles while also opening up new work opportunities in tech-driven industries. For instance, there will be a rise in the need for network administrators, sustainability coordinators, and experts in AI and data. However, routine and manual jobs might be eliminated, requiring retraining and education measures to handle workforce shifts.

Smarter and more responsive city services will also have a big impact on urban lifestyles. Urban living standards should rise as a result of increased energy efficiency, better waste management, and more effective public transportation. Furthermore, more participatory urban governance models—where residents have more access to

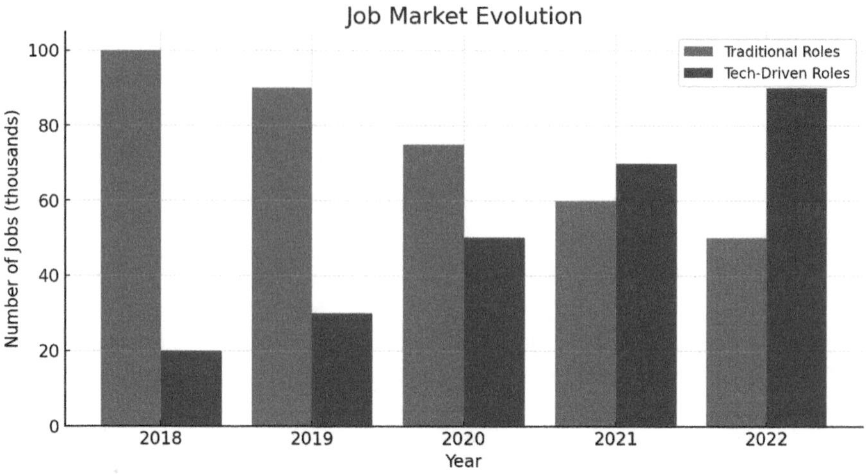

Fig. 1 Job market evolution over the years

and input into city management—may result from the real-time data that IoT devices give.

The AI-IoT framework has the potential to reduce costs for both city administrations and end users. By using resources more wisely, operating expenses can be decreased. These savings can then be transferred to residents, who may see a decrease in utility and public service charges (Fig. 1).

Over the course of five years, this stacked bar chart illustrates the shift in work positions from traditional to tech-driven sectors. It reflects a rise in jobs related to artificial intelligence and the internet of things, while traditional roles have decreased (Fig. 2).

The line graph illustrates gains in a variety of quality-of-life metrics, including air quality, transportation efficiency, and public service satisfaction. These improvements indicate that urban living circumstances have improved as a result of the implementation of innovations in smart city technology.

6.2 Environmental Impacts

It is anticipated that implementing an AI-IoT framework in urban management will have a very good environmental impact. Reduced carbon emissions would be one of the main advantages, which would be attained by increasing the integration of renewable energy sources and optimizing energy use. Because AI algorithms are more precise at managing and forecasting energy consumption, there can be a bigger share of renewable energy sources in the energy mix, which lessens the need for fossil fuels [7].

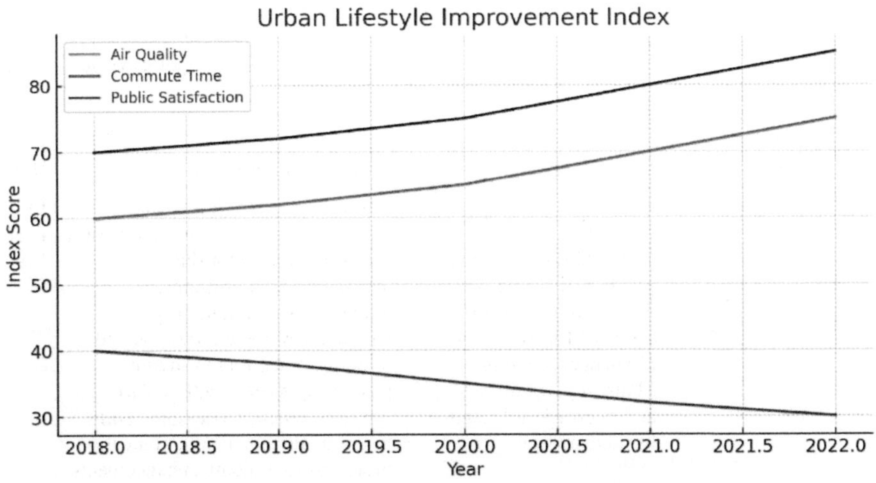

Fig. 2 Urban lifestyle improvement index

Furthermore, the use of smart technology is expected to result in improved resource conservation. Significant water savings can be achieved by using IoT devices to identify leaks and inefficiencies in water usage. Similarly, better waste management techniques made possible by AI can lessen the environmental impact of metropolitan areas by improving recycling procedures and reducing the need for landfills (Fig. 3).

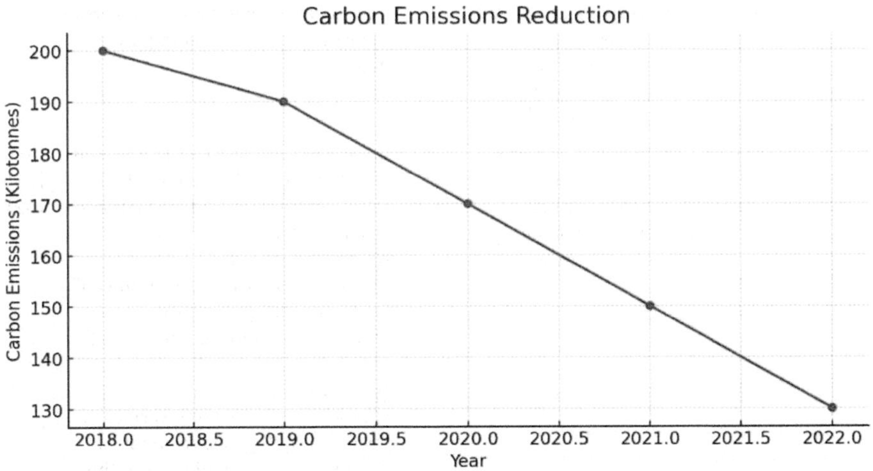

Fig. 3 Carbon emissions reduction

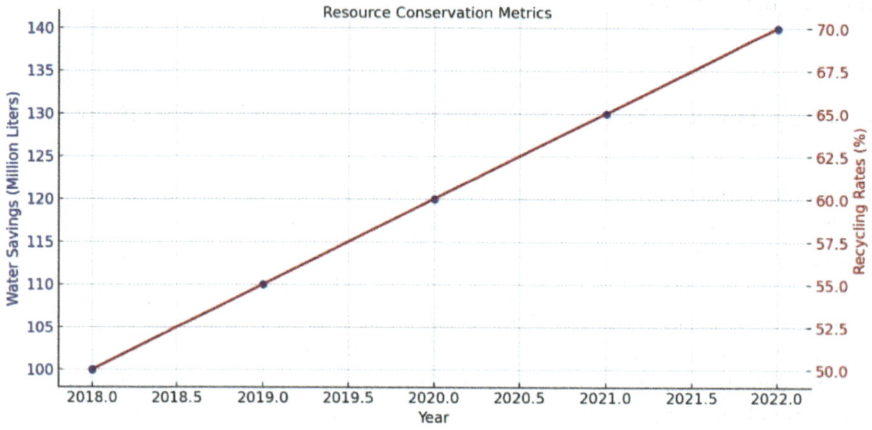

Fig. 4 Resource conservation metrics

This line graph depicts a continuous drop in carbon emissions over the course of five years, emphasizing the environmental benefits of more effective energy management through the framework of the Internet of Things (IoT) and artificial intelligence (Fig. 4).

Increases in water conservation and recycling rates are highlighted by the dual-axis bar graph, which highlights how artificial intelligence and the internet of things can lead to more sustainable resource management methods.

6.3 Alignment with SDGs

Several important Sustainable Development Goals (SDGs) listed by the UN are supported by the AI-IoT architecture. In particular, it helps with:

SDG 7 (Affordable and Clean Energy): The framework contributes to the reduction of energy costs and the promotion of ecologically sustainable energy through energy optimization and increased use of renewable sources.

SDG 11 (Sustainable Cities and Communities): Using cutting-edge technology, the framework makes cities safer, more resilient, and sustainable while also enhancing their infrastructure.

SDG 12 (Responsible Consumption and Production): The framework encourages sustainable production methods and effective resource consumption by streamlining waste management and improving resource efficiency [1] (Fig. 5).

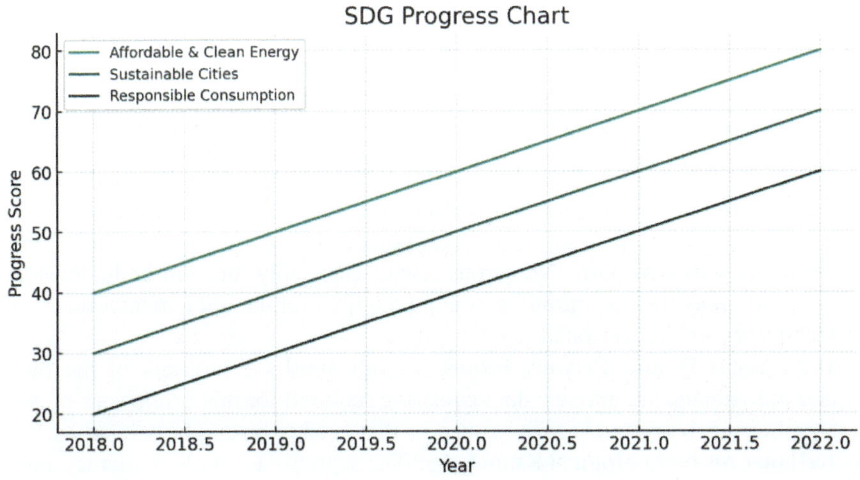

Fig. 5 SDG progress chart

The radar graphic illustrates the progress that has been made in fulfilling certain Sustainable Development Goals. It highlights improvement in areas such as the availability of clean and affordable energy, the development of sustainable cities and communities, and responsible consumption and production.

7 Conclusion

Urban infrastructures have a great chance to improve the sustainability and effectiveness of resource management by incorporating artificial intelligence (AI) and the internet of things (IoT). A novel AI-IoT framework has been developed in this research, with the goal of optimizing the management of vital urban resources like garbage, water, and electricity. With the use of real-time data collection from IoT devices and AI's analytical capabilities, this framework aims to make urban environments smarter, more sustainable, and able to adapt to the changing requirements of its inhabitants.

The case studies on trash, water, and energy management have shown how this paradigm may be used to both increase operational effectiveness and improve urban living standards. The anticipated decreases in carbon emissions and improvements in resource preservation are in line with the objectives of global sustainability, demonstrating how smart city technologies may be used to solve urgent environmental problems.

7.1 Limitations of the Study

Even if the suggested AI-IoT architecture has a lot of potential, there are a few things to be aware of:

- **Technical Difficulties**: It might be difficult and complex to integrate AI and IoT technologies into already-existing urban infrastructures. This requires a great deal of customization and strong cybersecurity measures.
- **Budgetary Restrictions**: Numerous towns, especially those in underdeveloped countries, may find the initial investment required for the implementation of such technology to be incurious.
- **Data Security and Privacy Issues**: To safeguard the interests of the public and individuals, the massive data gathering required for this framework presents significant privacy and security concerns that need to be carefully handled.
- **Reliance on Technological Knowledge**: The degree of technology literacy among consumers of smart city solutions is a critical factor that varies greatly among various demographic groups.

7.2 Future Scope of Research

In the future, AI and IoT have a wide range of applications in urban management. Several topics could be the focus of future research and development:

- **Improving Data Security**: To protect privacy and boost confidence in smart city technology, more sophisticated, secure data transmission and storage solutions should be developed.
- **Cost-Reduction Strategies**: Looking into more economical ways to use these technologies so that a wider range of cities and people can use them.
- **System Interoperability**: Making sure that new technologies can function flawlessly across various platforms and standards and are compatible with current systems.
- **Education Programs**: Encouraging individuals to be technologically literate by incorporating programs that can equip them to make the most of and profit from smart city technologies.

The integration of AI and IoT into urban surroundings can continue to expand, providing even more creative and effective solutions to the problems of contemporary urban living, by overcoming these limits and investigating these future pathways. In addition to adding to the conversation about sustainable urban development, this ongoing work will give politicians, urban planners, and tech developers useful information for building more resilient, livable, and sustainable urban futures.

References

1. United Nations: Revision of world urbanization prospects. Retrieved from https://www.un.org/development/desa/publications/2018-revision-of-world-urbanizationprospects.html (2018)
2. Hashem, I.A.T., Chang, V., Anuar, N.B., Adewole, K., Yaqoob, I., Gani, A., Chiroma, H.: The role of big data in smart city. Int. J. Inf. Manage. **36**(5), 748–758 (2016)
3. Albino, V., Berardi, U., Dangelico, R.M.: Smart cities: definitions, dimensions, performance, and initiatives. J. Urban Technol. **22**(1), 3–21 (2015)
4. Kumar, P., Yamacli, S.: Renewable energy for sustainable urban development: redefining the concept of energization. Energy Procedia **115**, 148–156 (2017)
5. Li, W., Cheng, S., Urban, F.: Smart cities in China. Cem. Concr. Res. **49**, 98–104 (2016)
6. Kumar, A., Dahiya, S., Bhatia, A.: Waste management techniques for improved environmental and public health. J. Environ. Manage. **253**, 105–110 (2018)
7. Bibri, S.E., Krogstie, J.: Smart sustainable cities of the future: an extensive interdisciplinary literature review. Sustain. Cities Soc. **31**, 183–212 (2017)
8. Vasseur, J.P., Dunkels, A.: Interconnecting smart objects with IP: the next internet. Morgan Kaufmann, Burlington (2010)
9. Perera, C., Qin, Y., Estrella, J.C., Reiff-Marganiec, S., Vasilakos, A.V.: Fog computing for sustainable smart cities: a survey. ACM Comput. Surv. **50**(3), 1–43 (2017)
10. Zanella, A., Bui, N., Castellani, A., Vangelista, L., Zorzi, M.: Internet of things for smart cities. IEEE Internet Things J. **1**(1), 22–32 (2014)
11. Taylor, S.J., Letham, B.: Forecasting at scale. Am. Stat. **72**(1), 37–45 (2018)
12. Hastie, T., Tibshirani, R., Friedman, J.: The elements of statistical learning: data mining, inference, and prediction. Springer, New York, NY (2009)
13. Siano, P.: Demand response and smart grids—a survey. Renew. Sustain. Energy Rev. **30**, 461–478 (2014)
14. Garcia, L.M.: Using artificial intelligence to enhance environmental sustainability in water utilities. Water Pract. Technol. **13**(3), 784–792 (2018)
15. Anagnostopoulos, T., Zaslavsky, A., Medvedev, A.: Robust waste collection exploiting cost efficiency of IoT potentiality in smart cities. Int. J. Sens. Wirel. Commun. Control. **6**(1), 1–15 (2016)
16. Roman, R., Zhou, J., Lopez, J.: On the features and challenges of security and privacy in distributed internet of things. Comput. Netw. **57**(10), 2266–2279 (2013)
17. Granjal, J., Monteiro, E., Silva, J.S.: Security for the internet of things: a survey of existing protocols and open research issues. IEEE Commun. Surv. Tutor. **17**(3), 1294–1312 (2015)
18. Van Dijk, J.: The digital divide. Polity Press, Cambridge, UK (2020)

Synergizing MEMS Technology and Artificial Intelligence for Enhanced Sensor Applications

M. Suresh, Sadanand Singh, Rishita Singh, Anirudh Santhosh, and Nibedita Rakshit

Abstract The chapter explores the evolution and integration of Microelectromechanical Systems (MEMS) sensors with artificial intelligence (AI), spanning diverse applications. MEMS technology has revolutionized sensor development, enabling high-performance systems in transportation, healthcare, defense, and consumer products. Sensors detect environmental stimuli and convert them into measurable digital signals, facilitating diverse applications from force and pressure sensing to image and acoustic recognition. Recent trends highlight AI integration with MEMS sensors, enhancing capabilities in edge computing, IoT, and smart devices. Advancements include object recognition in image sensors, speech recognition with MEMS microphones, and gesture-controlled devices. Challenges persist in seamlessly integrating AI sensors for comprehensive environmental monitoring and healthcare applications. The chapter aims to elucidate the fundamentals of MEMS sensors, explore AI integration, and discuss potential benefits and challenges. Through case studies and examples, it provides insights into successful implementations and outlines future directions for advancing sensor technologies through the collaborative synergy of MEMS and AI.

Keywords Micro-electro mechanical systems · Sensors · Artificial Intelligence (AI) · Neural networks · Machine learning algorithms · Deep learning

M. Suresh (✉) · S. Singh · R. Singh · A. Santhosh · N. Rakshit
School of Electrical and Electronics Engineering, VIT Bhopal University, Sehore, Madhya Pradesh, India
e-mail: suresh@vitbhopal.ac.in

B. K. Mishra and A. Rocha (eds.), *Future of AI, IoT, and Sustainability*, Information Systems Engineering and Management 23,
https://doi.org/10.1007/978-3-031-76286-4_13

1 Introduction

Microelectromechanical Systems (MEMS) has undergone Decades-long comprehensive research and development. Beyond the conventional actuators and microfabricated sensors, the field encompasses the combination of Miniature mechanical components and systems integrated with electronics on a shared base or enclosure. This integration achieves high-performance functional systems, playing pivotal roles in transportation, communication, manufacturing, environmental monitoring, healthcare, defense, and a range of consumer goods. Micro-ElectroMechanical Systems, characterized by their small size, offer advantages like reduced weight, size, and energy usage, along with improved speed and precision compared to larger-scale equivalents. The advancement of MEMS depends on fabrication technologies enabling small geometries, precise dimensions, design flexibility, microelectronics interfacing, repeatability, reliability, high yield, and cost-effectiveness. The primary enabling method for MEMS has been integrated circuits (IC) fabrication technology, complemented by specific etching, bonding, and assembly techniques. Microfabrication proves instrumental in batch processing and miniaturizing electromechanical devices and systems to dimensions inaccessible by conventional machining [1, 2].

A sensor identifies input stimuli, such as pressure, force, light, or motion in the surroundings, and generates a detectable digital signal. This signal, often in electric form like voltage or current, is converted for display or transmitted electronically over a network for reading or other applications, reflecting the detected environmental conditions [3–5].

Sensors are categorized based on what they detect, such as force, pressure, flow, temperature, proximity, smoke, gas, alcohol, and more. Each type of sensor produces varying electrical outputs, which can include voltage, capacitance, resistance, current, frequency, and so on.

Sensors are categorized into different groups based on factors such as the measurand, fields of application, conversion principle, energy domain of the measurand, and thermodynamic considerations, among others [3, 6–9]. Some recent trends can be seen in integrating AI with MEMS-based sensors, particularly in image sensors, acoustic sensors, and wearable sensors. Notable advancements include edge computing for rapid response, low power consumption, and high transmission efficiency in IoT. AI algorithms in image sensors focus on object recognition, detection, video tracking, and 3D scene modelling. Integration of MEMS microphones with speech recognition systems is highlighted. The combination of sensors and AI leads to advanced devices like gesture recognition gloves, lip language recognition masks, and robotic manipulators. Multimodal fusion addresses sensor functionality limitations, and neuromorphic computing enhances smart home systems with rapid responses. The emergence of Digital Twins (DT) in smart homes and farms is facilitated by VR, AR, and various sensors, enabling automatic detection and control of activities and environments. The challenge lies in seamlessly combining diverse AI sensors into an integrated system for improved environmental monitoring, immersive interaction, and comprehensive healthcare in the future [10].

The main goals of this chapter are to investigate the basics of MEMS-based sensors and artificial intelligence. It aims to investigate the integration of AI with MEMS sensors, examining its potential benefits, challenges, and applications across various fields. By providing case studies and examples, the review seeks to offer insights into successful implementations, address current challenges, and outline future directions for advancing sensor technologies through the collaborative synergy of MEMS and AI.

2 Fundamentals of the MEMS-Based Sensors

MEMS refers to devices with characteristic lengths ranging from 1 micron to less than 1 mm, integrating electrical and mechanical parts, manufactured using batch-processing technologies for integrated circuits.. MEMS devices have been pivotal in various developmental domains. The crucial components for realizing a specific function within a comprehensive system is miniaturized sensors, actuators, and electronic components. Here are some examples of MEMS-based sensors are discussed as follows:

2.1 Pressure Sensors

Pressure sensors, developed through silicon micromachining technologies, represent early achievements and have emerged as prosperous commercial products. These devices are widely utilized in diverse industrial and biomedical fields. The sensors can adopt various sensing mechanisms, including piezoresistive, capacitive, piezoelectric and resonant technologies [1].

2.1.1 Pressure Sensors Based on Piezoresistive Technology

The piezoresistive effect in silicon is commonly used to manufacture pressure sensors. When pressure causes strain, it changes the silicon band structure, thereby altering the material's electrical resistance. A practical piezoresistive pressure sensor is achieved by producing four resistive sensors positioned at the edges of a thin silicon layer diaphragm, serving as a mechanical device that increases stress and strain at the sensor location. These Four sensing elements are interconnected in a bridge configuration with complementary signals, boosting sensitivity. The sensor's measurable pressure range ranges from 10^{-3} to 10^{-6} Torr, contingent on the design.

2.2 Motion Sensors

Miniaturized inertial sensors include accelerometers and gyroscopes, representing crucial categories of siliconbased MEMS sensors that have achieved successful commercialization. Accelerometers find application in various fields, including automotive safety systems, active suspension, and stability control. They are also utilized in biomedical contexts for activity monitoring, as well as in consumer products like head-mount displays, camcorders, and three-dimensional mice. Gyroscopes, on the other hand, play a role in applications such as automotive ride stability, rollover detection, camcorder stability, and virtual reality.

2.2.1 Accelerometer

An accelerometer consists of a proof mass that is suspended using flexible mechanical supports connected to a stable frame. When an external acceleration occurs, it causes displacement between the support frame and the proof mass. The resulting displacement can lead to a change in stress within the suspension system, which is then identified by piezoresistive sensors to quantify the external acceleration. Additionally, capacitive accelerometers can detect this displacement as a change in capacitance [1, 11].

2.2.2 Gyroscope

The apparatus is comprised of a vibrating mass that is electrostatically driven to resonate along the axis of the drive mode, employing comb fingers. When there is a rotational movement along the vertical axis (z-axis), it induces Coriolis acceleration causes a displacement in the structure along the axis of the sense mode. This deflection causes a modification in the differential sense capacitance value, providing a means to identify and quantify the input angular rotation [1, 12].

The sensing process in MEMS devices involves a mechanism using a polymer on the sensing layer, absorbing specific chemicals to induce changes in stress, mass, or electrical and mechanical characteristics. Detecting mass variations includes measuring resonant frequency or using piezoresistors on the sensing element. Another method uses thermocouples to gauge temperature changes caused by absorbed analytes. Various sensing mechanisms like piezoresistive, capacitive, or optical methods are employed, each with distinct pros and cons. While optical sensing is feasible, it's costlier than capacitive and piezoresistive techniques [13–15].

MEMS sensors find widespread application across diverse industries. In the automotive sector, they contribute to safety, engine performance, comfort, and diagnostics, with examples like airbag and pressure sensors. In aerospace, MEMS devices play pivotal roles in cockpit instrumentation, safety systems, microsatellites, guidance systems, propulsion, and communication technologies. The biomedical field

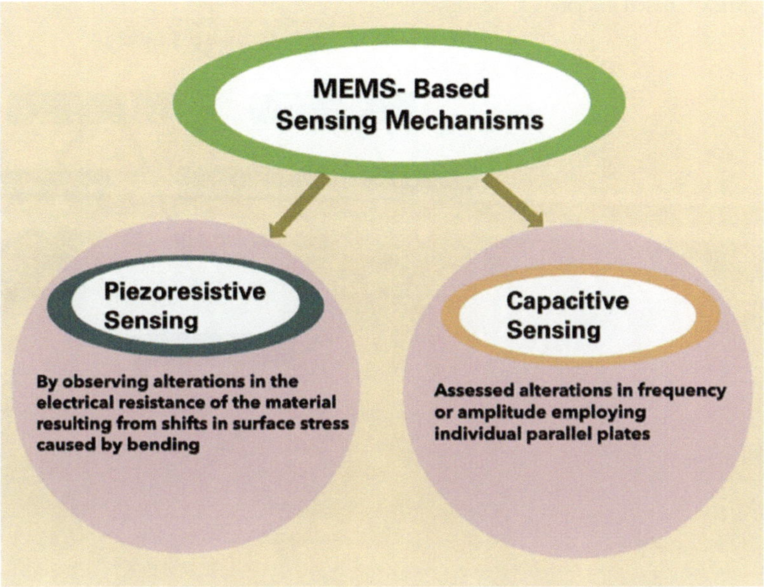

Fig. 1 MEMS based sensing mechanisms: piezoresistive and capacitive sensing

benefits from disposable blood pressure transducers, catheter tip sensors, and various medical monitoring devices. The versatility of MEMS technology extends to revolutionizing safety, efficiency, and functionality in automotive, aerospace, and biomedical applications, showcasing its indispensable role in advancing technology and improving human lives (Figs. 1 and 2).

3 Overview of Artificial Intelligence (AI)

Artificial Intelligence (AI) stands as a transformative force, exerting a profound impact on both historical narratives and future trajectories. Within this context, the essence and significance of AI are expounded upon, highlighting the evolution of its definitions over time. While its origins were rooted in the emulation of human-like behaviours, contemporary interpretations, exemplified by the 2018 European Commission Communication, have broadened its scope considerably. As AI encompasses an array of diverse technologies, the task of assessing its impact becomes inherently complex. To facilitate a meaningful discourse, precision becomes paramount, necessitating a clear distinction between advisory expert systems and intricate algorithms. Moreover, it is imperative to differentiate between speculative conjectures about the future and the tangible societal implications stemming from current AI applications.

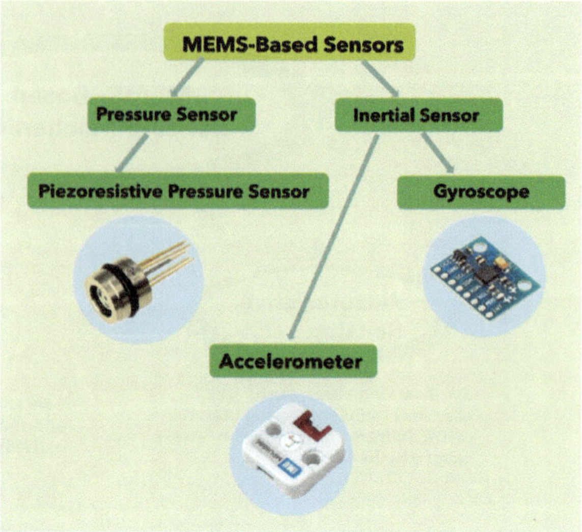

Fig. 2 MEMS based sensors

In the realm of biochemical sensors, various AI algorithms find application, each tailored to enhance sensor functionalities. Among these, supervised learning algorithms such as Linear Regression, Prominent tools include Logistic Regression, K-Nearest Neighbors (KNN), Random Forest (RF), Naive Bayes, Support Vector Machine (SVM), Linear Discriminant Analysis (LDA), Convolutional Neural Network (CNN), Decision Tree (DT) and Neural Network (NN). These algorithms play a pivotal role in optimizing data processing within biosensors, facilitating the extraction of pertinent information and translating it into comprehensible feedback for users, particularly in biomedical applications. Furthermore, the integration of unsupervised learning algorithms, including t-Distributed Stochastic Neighbor Embedding (tSNE) and Principal Component Analysis (PCA) contribute further to the capabilities of biochemical sensors. In essence, the establishment of a robust relationship between raw data and the concentration of target analytes stands as a crucial endeavor, one that underpins the continuous improvement of biosensor performance across diverse domains [16–18] (Figs. 3, 4, 5 and 6).

Fig. 3 Classification of AI algorithms

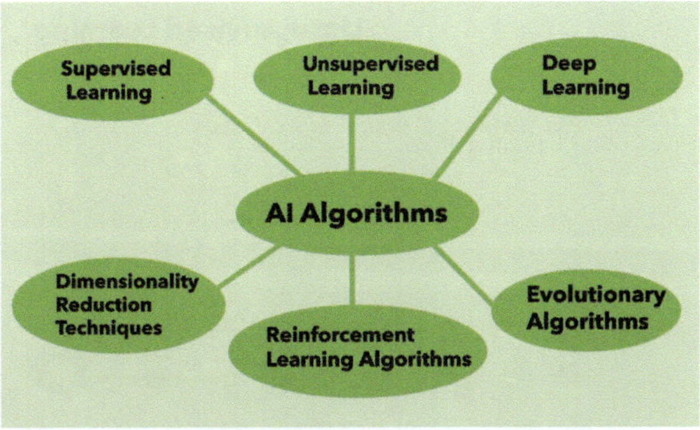

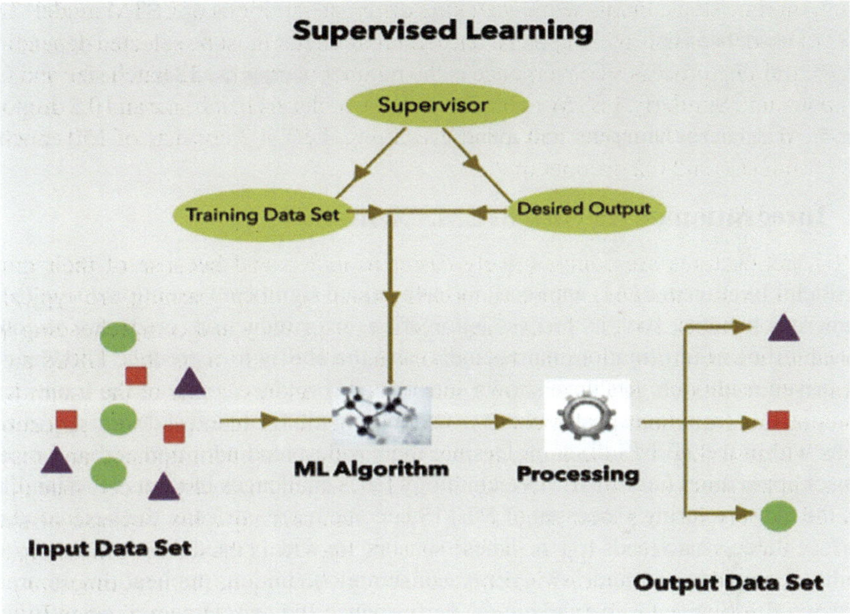

Fig. 4 AI algorithms: supervised learning

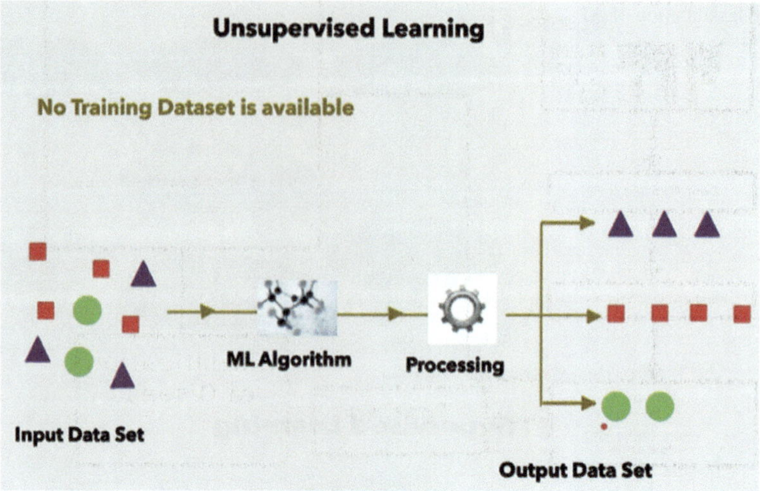

Fig. 5 AI algorithms: unsupervised learning

4 Integration of AI with MEMS Sensor

Artificial Intelligence (AI) applications have gained significant attention due to their numerous benefits, such as process automation, error reduction, predictive maintenance, efficient information management, and the ability to reproduce knowledge. AI-driven methodologies have shown superior performance compared to traditional optimization techniques, particularly in conducting multi-objective design optimizations within the MEMS domain. Despite their widespread adoption across various sensor application fields, MEMS technology faces challenges in terms of reliability. As the manufacturing processes of MEMS become more intricate, the likelihood of surface defects also tends to rise. These sensors are widely used due to their affordability, compact size, and low energy consumption, finding applications in areas like non-destructive testing and human performance analysis. However, despite their impressive qualities, integrating these sensors into condition monitoring systems for rotary machinery has yet to be realized [19, 20].

Monitoring the pulse wave is essential for evaluating health of humans. In their 2018 paper [21], Yu Sun and colleagues presented a wearable multi-sensor system designed for this purpose. The system comprises a measuring unit along with a processing unit. The measuring unit features a thin (0.15 mm) flexible PCB that houses MEMS pressure sensors and a three-axis accelerometer, all strategically placed. It also includes a blood oxygen detector. The processing unit comprises of an amplifier, a filter and a module for power supply. The system successfully detected the waves of pulse from two participants and, during testing, accurately determined pulse wave velocities and calculated the K value related to region of the waves of pulse, adhering to health standards.

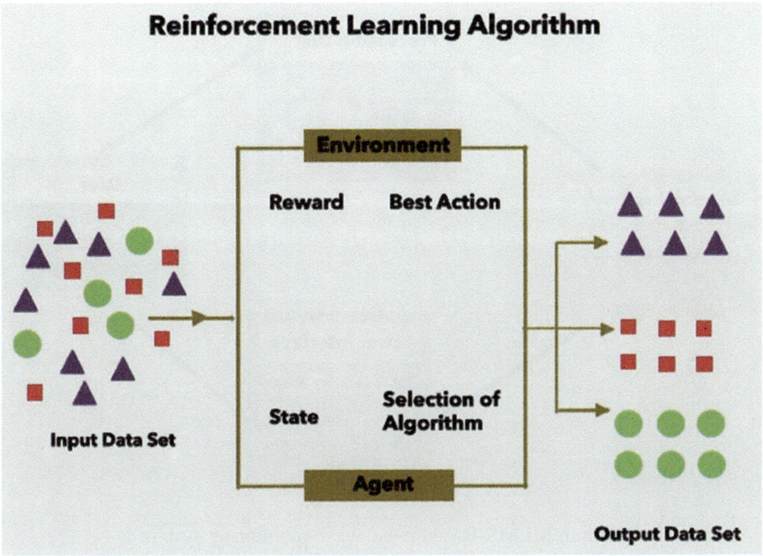

Fig. 6 AI algorithms: reinforcement learning

Several AI algorithms and techniques can be integrated into the proposed design to improve accuracy and achieve superior outcomes. Here are a few illustrative examples:

1. Data Analysis and Interpretation: By employing AI algorithms like machine learning classifiers or deep learning models, the system can effectively analyze collected data from sensors. This analysis can identify patterns, anomalies, and correlations in pulse wave signals, offering valuable insights into the wearer's health status.
2. Signal Processing Optimization: AI techniques can optimize signal processing algorithms within the analog circuit processing unit. For example, neural networks can enhance filtering and amplification processes, resulting in more accurate and reliable pulse wave measurements.
3. Predictive Maintenance: AI algorithms can monitor the health of the monitoring system itself by analyzing sensor data over time. This enables the detection of potential failures or malfunctions, facilitating proactive maintenance and minimizing system downtime.
4. Real-Time Feedback and Alerts: By analyzing pulse wave data, AI algorithms can generate real-time feedback and warnings. For instance, anomaly detection algorithms can identify abnormal pulse wave patterns indicative of potential health risks, triggering timely alerts for intervention or medical attention.

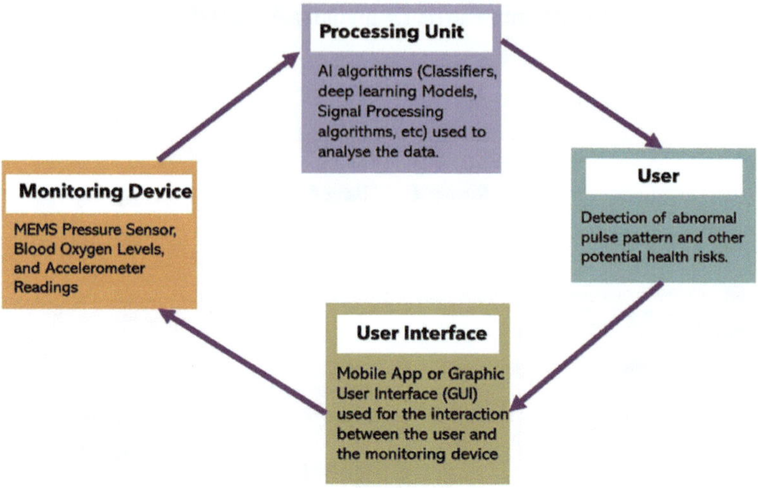

Fig. 7 AI implementation in MEMS-Based pulse wave monitoring system

5. Personalized Health Monitoring: AI algorithms can customize the monitoring system's functionality to individual users based on their unique physiological characteristics and health history. This personalized approach enhances the system's effectiveness in detecting and monitoring specific health conditions.
6. Ongoing Learning and Flexibility: AI algorithms are flexible enough to learn from new data inputs, allowing the monitoring system to continuously improve its performance. This iterative learning process ensures that the system remains up-to-date with evolving health trends and user needs.

These components collaborate to deploy AI algorithms within the pulse wave monitoring device, facilitating sophisticated data analysis, immediate feedback, and personalized health monitoring functionalities (Fig. 7).

5 Synergy of AI and MEMS-Based Sensors

5.1 Streamlining MEMS Sensor Production Process

The research carried out by Itilekha Podder, Tamas Fischl, and Udo Bub [22] delves into the challenges associated with Micro-electromechanical systems (MEMS) manufacturing processes. With hundreds of intricate steps involved, real-time data generated during production poses a significant challenge due to its volume and complexity. Traditional methods struggle to analyse this data effectively, prompting the authors to explore the applicability of Machine Learning (ML) techniques for fault detection and prediction. The study focuses specifically on predicting faults in MEMS motion sensors using various machine learning methods. The employment of statistical analysis helps to understand the root causes, and efforts are made to determine the optimal sub-parameters necessary for achieving maximum performance without over-fitting or redundancy. Overall, the research aims to enhance fault detection and prediction in MEMS manufacturing through advanced data analysis techniques.

5.2 Enhancing the Functionalities of Protective Gears

Helmets are the most widely used industrial protective gear, balancing thermal discomfort with the risk of head injuries. Yan Hao Tan, Hitesh Agrawal and King Ho Holden Li in their paper [23] proposed the concept of utilizing helmet microclimate psychrometry involves employing a machine learning algorithm based on logistic regression and MEMS sensors to assess helmet usage. Experiments involve human subjects and mannequins under various conditions. Ambient-microclimate humidity difference (AMHD) proves effective in detecting helmet usage, strongly associated with the start and cessation. However, using continuously beyond 1-min poses challenges because of the leveling off of the AMHD response. Temperature measurements yield invariant results. The algorithm reliably determines helmet-wearing, offering an improvement over existing methods, enhancing safety through data-driven worker ownership. This research demonstrates the potential of combining analysis and authentic, non-identifiable data to improve safety protocols in industrial settings.

5.3 Enhancing the Precision of MEMS Motion Sensors

The paper [24] discusses the application of Artificial Neural Networks (ANNs) within inertial measurement units for thermal calibration of bias in micro electro-mechanical gyros, crucial for integrated inertial navigation systems. With rising interest in autonomous and personal navigation in GPS-challenging environments, precise calibration is essential. Traditional polynomial fitting methods encounter limitations in modelling abrupt bias changes and sensor hysteresis. To address this, the paper proposes using Back Propagation ANNs, offering adaptive configurations and suitable for mapping thermal bias variations. A comparison between polynomial fitting and ANN compensation demonstrates the better performance of the ANN approach, improving flight attitude accuracy meeting regulatory requirements by 20%. This research underscores the efficacy of ANNs in enhancing calibration accuracy for Micro Electro-Mechanical gyros, crucial for improved navigation system performance. The increasing importance of positioning, navigation, and timing information necessitates. The fusion of inertial navigation systems (INS) and global navigation satellite systems (GNSS) for accurate vehicle positioning. However, GNSS accuracy degrades in challenging signal environments, leading to reliance on standalone INS. To improve INS accuracy, the paper [25] employs advanced deep RNN to model and suppress raw signal noise in micromechanical gyroscope data. By treating gyroscope data as a time series, the proposed method significantly reduces three-axis attitude errors compared to conventional methods, demonstrating the efficacy of mixed deep RNN in enhancing navigation accuracy.

6 Conclusion

In conclusion, this chapter explores the fundamental principles of MEMS-based sensors and their various types, delving into the integration of AI with MEMS technology. The overview of AI highlights its potential applications in enhancing MEMS sensor functionalities across diverse fields such as transportation, communication, healthcare, and defense. Specifically, the inclusion of AI with MEMS sensors opens up new avenues for improving sensor accuracy, streamlining production processes, and enhancing the functionalities of protective gears. Case studies and examples illustrate successful implementations, while addressing challenges and outlining future directions for advancing sensor technologies through collaborative synergy. As MEMS technology continues to evolve alongside AI, the seamless integration of these technologies holds promise for revolutionizing environmental monitoring, immersive interaction, and comprehensive healthcare in the coming years.

References

1. Springer Handbook of Nanotechnology
2. Mehregany, M., Bart, S.F., Tavrow, L.S., Lang, J.H., Senturia, S.D.: Principles in design and microfabrication of variable-capacitance side-drive motors. J. Vac. Sci. Technol. A Vac. Surf. Films **8**(4), 3614–3624 (1990). https://doi.org/10.1116/1.576515
3. Javaid, M., Haleem, A., Singh, R.P., Rab, S., Suman, R.: Significance of sensors for industry 4.0: roles, capabilities, and applications. Sens. Int. **2**. KeAi Communications Co. (2021). https://doi.org/10.1016/j.sintl.2021.100110
4. Huang, L., et al.: One-step rapid quantification of SARS-CoV-2 virus particles via low-cost nanoplasmonic sensors in generic microplate reader and point-of-care device. Biosens. Bioelectron. **171**, 112685 (2021). https://doi.org/10.1016/j.bios.2020.112685
5. Qian, L., Durairaj, S., Prins, S., Chen, A.: Nanomaterial-based electrochemical sensors and biosensors for the detection of pharmaceutical compounds. Biosens. Bioelectron. **175**, 112836 (2021). https://doi.org/10.1016/j.bios.2020.112836
6. Bibby, L., Dehe, B.: Defining and assessing industry 4.0 maturity levels—case of the defence sector. Prod. Plan. Control. **29**(12), 1030–1043 (2018). https://doi.org/10.1080/09537287.2018.1503355
7. Scholl, G., et al.: A8.1—IO-link wireless enhanced sensors and actuators for industry 4.0 networks. In: Proceedings Sensor 2017, AMA Service GmbH, Von-Münchhausen-Str. 49, 31515 Wunstorf, Germany, pp. 134–138 (2017). https://doi.org/10.5162/sensor2017/A8.1
8. Bahl, S., et al.: Biosensors applications in fighting COVID-19 pandemic. Apollo Med. (2020). https://doi.org/10.4103/am.am_56_20

9. Wu, Z., Yang, X., Wu, J.: Conductive hydrogel- and organohydrogel-based stretchable sensors. ACS Appl. Mater. Interfaces **13**(2), 2128–2144 (2021). https://doi.org/10.1021/acsami.0c21841
10. Zhang, Z., Wang, L., Lee, C.: Recent advances in artificial intelligence sensors. Adv. Sens. Res. **2**(8) (2023). https://doi.org/10.1002/adsr.202200072
11. Lu, C., Lemkin, M., Boser, B.E.: A monolithic surface micromachined accelerometer with digital output. In: Proceedings ISSCC' 95—International Solid-State Circuits Conference. IEEE, pp. 160–161. https://doi.org/10.1109/ISSCC.1995.535504
12. Clark, W.A., Horowitz, R., Howe, R.T.: Surface micromachined Z-axis vibratory rate gyroscope. In: 1996 Solid-State, Actuators, and Microsystems Workshop Technical Digest. San Diego, CA USA: Transducer Research Foundation, Inc., pp. 283–287 (1996). https://doi.org/10.31438/trf.hh1996.64
13. Algamili, A.S., et al.: A review of actuation and sensing mechanisms in MEMS-based sensor devices. Nanoscale Res. Lett. **16**(1). Springer (2021). https://doi.org/10.1186/s11671-02103481-7
14. Yazdi, N., Ayazi, F., Najafi, K.: Micromachined inertial sensors. Proc. IEEE **86**(8), 1640–1659 (1998). https://doi.org/10.1109/5.704269
15. Arecco, D.: Analysis and preliminary characterization of a MEMS cantilever-type chemical sensor (2003)
16. Boucher, P.: STUDY panel for the future of science and technology EPRS | European Parliamentary Research Service (2020)
17. Zhou, Z., Xu, T., Zhang, X.: Empowerment of AI algorithms in biochemical sensors. TrAC Trends Anal. Chem. **173**. Elsevier B.V. (2024). https://doi.org/10.1016/j.trac.2024.117613
18. Zhang, K., Wang, J., Liu. T., Luo, Y., Loh, X.J., Chen, X.: Machine learning-reinforced noninvasive biosensors for healthcare. Adv. Healthc. Mater. **10**(17). John Wiley and Sons Inc (2021). https://doi.org/10.1002/adhm.202100734
19. Podder, I., Fischl, T., Bub, U.: Artificial intelligence applications for MEMS-based sensors and manufacturing process optimization. Telecom **4**(1), 165–197. MDPI (2023). https://doi.org/10.3390/telecom4010011
20. Staszewski, W., Jablonski, A., Barszcz, T.: Challenges in application of MEMS sensors for condition monitoring of machinery (2019)
21. Sun, Y., et al.: Wearable pulsewave monitoring system based on MEMS sensors. Micromachines (Basel) **9**(2) (2018). https://doi.org/10.3390/mi9020090
22. Podder, I., Fischl, T., Bub, U.: Smart feature selection for fault detection in the MEMS sensor production process using machine learning methods (2022)
23. Tan, Y.H., Hitesh, A., Li, K.H.H.: Application of machine learning algorithm on memsbased sensors for determination of helmet wearing for workplace safety. Micromachines (Basel) **12**(4) (2021). https://doi.org/10.3390/mi12040449
24. Maudie, T., Hardt, A., Nielsen, R., Stanerson, D., Bieschke, R., Miller, M.: MEMS manufacturing testing: an accelerometer case study (2003)
25. Jiang, C., et al.: A mixed deep recurrent neural network for MEMS gyroscope noise suppressin. Electronics (Switzerland) **8**(2) (2019). https://doi.org/10.3390/electronics8020181

Connected Realities: Exploring IoT's Technological Marvels and Varied Applications

Sri Ramya Siraparapu and S. M. A. K. Azad

Abstract The swift advancement of the Internet of Things from a theoretical concept to practical reality has deeply affected numerous sectors, including process control systems. This paper extensively explores the dynamic landscape of IoT technologies, particularly their integration into streaming data analytics for process control applications. By seamlessly integrating crucial components such as sensors, connectivity protocols, cloud computing, and artificial intelligence, IoT facilitates efficient device connectivity and data exchange, fundamentally revolutionizing traditional process control methodologies. Leveraging machine learning techniques within IoT-driven streaming data analytics empowers industries with real-time decision-making capabilities, predictive maintenance strategies, and optimized industrial processes, thereby extracting actionable insights for enhanced operational efficiency and cost savings. Despite encountering challenges like data security concerns and interoperability issues, continuous advancements in connectivity, edge computing, and sensor technologies promise to address these obstacles and fully unleash the potential of IoT-driven streaming data analytics in process control systems. Looking forward, emerging trends in this domain include the adoption of smart infrastructure, autonomous vehicles, wearable healthcare devices, and immersive augmented reality experiences, indicating a future characterized by unparalleled connectivity and innovation in process control. In conclusion, the convergence of IoT and ML in streaming data analytics signifies a transformative shift, enabling industries to achieve heightened efficiency, elevate product quality, and tackle global challenges, thereby laying the foundation for a future that is more sustainable and resilient.

Keywords Internet of Things (IoT) · Streaming data analytics · Process control systems · Machine learning · Industrial IoT

S. R. Siraparapu
School of Electronics Engineering, VIT AP University, Amaravati, India

S. M. A. K. Azad (✉)
Department of Electronics and Communication Engineering, Sri Venkateswara College of Engineering, Tirupati, Andhra Pradesh, India
e-mail: freedom_smak@rediffmail.com

B. K. Mishra and A. Rocha (eds.), *Future of AI, IoT, and Sustainability*, Information Systems Engineering and Management 23,
https://doi.org/10.1007/978-3-031-76286-4_14

1 Introduction

1.1 Evolution of IoT Technologies

The Internet of Things (IoT) encompasses a network of interconnected physical devices outfitted with sensors, software, and various technologies, allowing them to communicate and share data via the internet. [1]. Figure 1 illustrate the interconnected nature of IoT systems, displaying the various components integral to their operation. These components work unruffled seamlessly to aid IoT devices to sense, communicate, and interact with their environments. The environment represents the physical surroundings where IoT devices operate, while actuators convert digital signals into physical actions within that environment [2]. Microcontrollers, embedded within IoT devices, control their functions, enabling them to process data and execute tasks. These IoT devices, often referred to as "things," are tangible objects fitted with sensors and linked to the internet to facilitate data exchange. Users interact with IoT devices through interfaces, which could include physical interfaces like buttons or touchscreens, as well as digital interfaces such as mobile apps or web-based dashboards. The Internet serves as the infrastructure facilitating communication and data exchange between IoT devices and other connected systems, enabling seamless connectivity. Users, whether individuals or entities, play a crucial role in interacting with and managing IoT devices and systems. IoT devices generate data, and organizations manage and store this data in data management and repositories systems, where they can process, analyze, and utilize it for insights and decision-making. Communication interfaces, comprising protocols and technologies, enable communication between IoT devices and external systems, ensuring interoperability and compatibility [3].

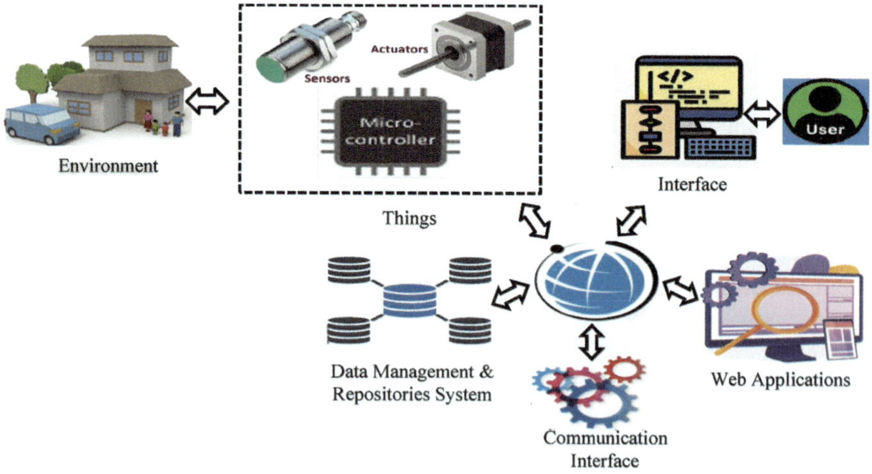

Fig. 1 Anatomy of IoT systems

Web applications provide software interfaces accessible via the internet for monitoring and controlling IoT devices, offering real-time insights and management capabilities. Through bridging the physical and digital realms, IoT offers organizations automation, task simplification, and data-driven insights, fundamentally transforming operations and enhancing connectivity, efficiency, and safety. While the concept of IoT has been around since the 1990s, recent advancements in technology, including affordable sensors, cloud computing platforms, and artificial intelligence, have made it more feasible and widespread. IoT devices encompass a wide range of objects, from everyday household appliances to complex industrial tools, each equipped with a Unique Identifier (UID) [4] and capable of transmitting data autonomously. The types of IoT applications vary based on their specific usage, including consumer IoT for everyday appliances, commercial IoT in healthcare and transportation, military IoT (IoMT) for military applications, industrial IoT (IIoT) in manufacturing and energy sectors, and infrastructure IoT for smart city connectivity. Figure 1 illustrates the interplay between industrial IoT (IIoT) and consumer IoT, facilitated by an intelligent network system, to maintain digital connectivity across large and small-scale networks. In the setting, IoT is prevalent in the retail sector, while IIoT [5] is pivotal in the industrial sector, playing a crucial role in enhancing manufacturers' operational efficiency.

1.2 Importance of IoT in Modern Applications

IIoT represents a cutting-edge subset of IoT, integrating networked-devices and technologies with industrial automation applications as shown in Fig. 2. Traditionally, industrial-wireless-technologies relied on ad-hoc-solutions, but with IIoT, cost-effective technologies like Wireless HART, ISA100.11a are developed with increased adaptableness [6]. IIoT presents compelling opportunities by connecting various sensors to the internet, offering significant potential for lifestyle enhancement [7].

Recognized as an optimal approach for improving production performance, IIoT merges IT&OT domains, leading to cost reductions and enhanced machine-to-machine communication. Notably, IIoT applications facilitate the improvement of Smart Factories, where sensor-embedded devices are connected over the internet, enabling real-time data access and advanced visibility of operations. Environmental monitoring in industrial settings, facilitated by IoT technologies, ensures a safe work environment by detecting smoke contaminants, and monitoring air and water quality. The data is transmitted via Bluetooth to a centralized server aimed at analysis [7]. Additionally, IIoT applications such as smart-grid systems, industrial-automation, and smart-communication leverage vast amounts of data to increase productivity efficiently and cost-effectively. The importance of IoT deceptions in its capability to bridge physical and digital worlds, offering numerous benefits to organizations, such as automation, simplification of tasks, and data-driven insights IoT devices function by sensing activities in the physical world, with integrated components like CPUs and network adapters, often configured and managed through software

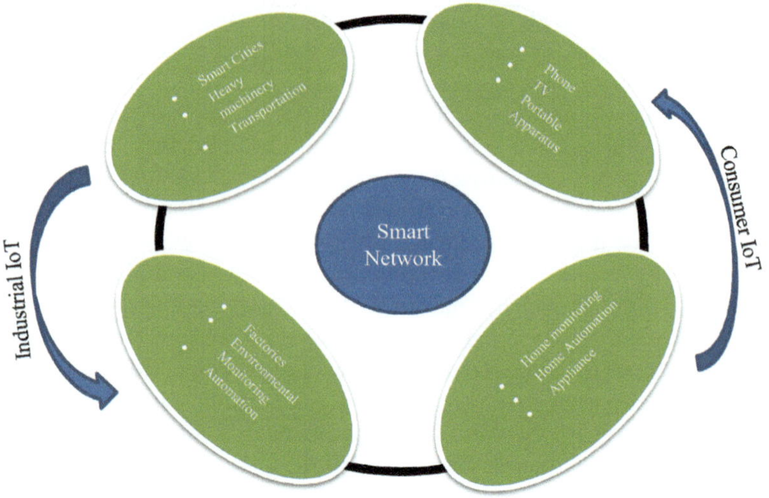

Fig. 2 Internet of Things and industrial Internet of Things

applications or integrated web servers[8]. Examples of IoT devices encompass home security systems, fitness trackers, industrial safety and security systems, augmented reality glasses, and motion detection sensors. These devices collectively enhance connectivity, efficiency, and safety across various fields.

2 Foundational Technologies

Foundational technologies [9] form the backbone of IoT, enabling seamless connectivity and data processing. These technologies include sensors and actuators for data-collection and control, connectivity protocols for communication, embedded systems for specialized tasks, cloud computing for data storage and analysis, edge computing for real-time processing at the network edge, and ML and AI for advanced analytics and decision-making[10]. Together, these technologies drive the utility and effectiveness of IoT-systems across various domains.

2.1 Sensors and Actuators

Sensors and actuators are fundamental components of the IIoT, revolutionizing industries worldwide. Sensors, acting as the "senses" of the IIoT, collect vital data from the physical environment, including temperature, pressure, and position. This data facilitates informed decision-making and proactive maintenance [11]. Actuators,

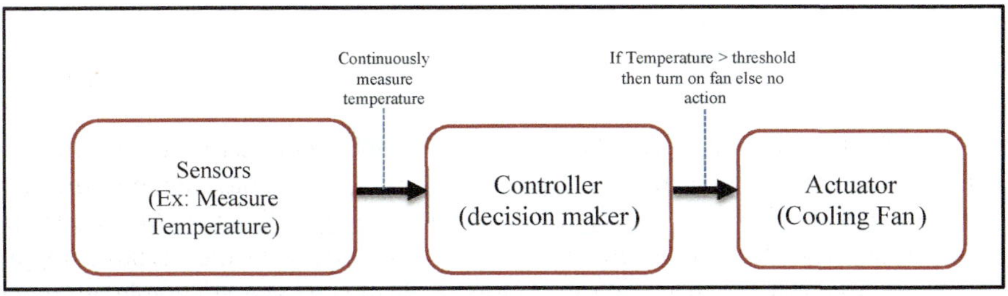

Fig. 3 IoT environmental control system

the "muscles" of the IIoT, manipulate machines and processes based on received signals, exerting control over the physical world as shown in Fig. 3. The integration of sensors and actuators is pivotal for harnessing the full potential of IIoT systems. Establishing seamless communication between these components is essential. Protocols like Modbus, OPC-UA, and MQTT facilitate data transmission, while industrial communication standards such as CAN and Ethernet enhance system versatility. Specialized interfaces like ADCs and DACs mediate between sensors and actuators, ensuring seamless data transfer and compatibility [12]. Effective interfacing ensures reliable operation and optimization of IIoT systems, enabling industries to deliver accurate, real-time data for intelligent decision-making. By embracing standardized protocols and specialized interfaces, industries move towards a more connected and automated future.

2.2 Connectivity Protocols

Communication protocols form the backbone of IoT systems, as depicted in Fig. 4, enabling network connectivity and seamless integration with applications. These protocols facilitate the exchange of data among devices over the network. Diverse protocols address various communication aspects, and when combined, they form a protocol suite. Implemented in software, they constitute a protocol stack and are essential in data encoding and addressing schemes.

Data encoding protocols define rules for encoding data, ensuring compatibility and efficient transmission across devices and networks. Addressing schemes protocols establish addressing schemes that permit devices to locate and communicate with

Fig. 4 IoT protocol stack

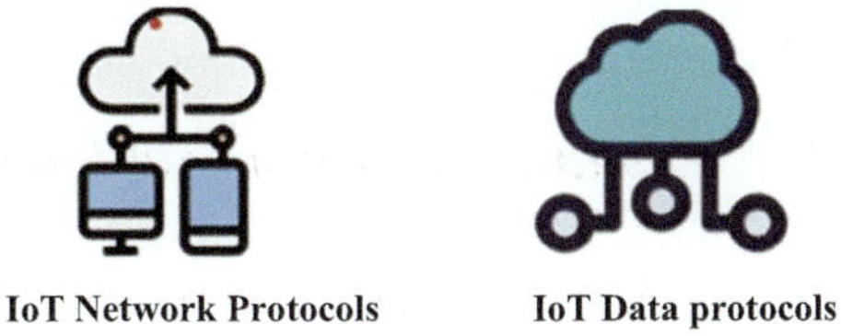

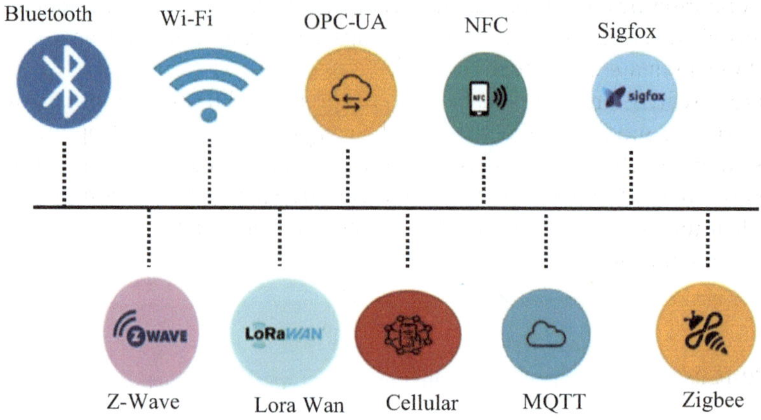

Fig. 5 Connectivity protocols

each other within the network [13]. The landscape of IoT communication protocols shown in Fig. 5 encompasses a diverse array of options tailored to specific needs and environments. These protocols facilitate seamless connectivity and data exchange, crucial for the functioning of IoT ecosystems. From ubiquitous standards like Wi-Fi and Bluetooth to specialized solutions like Zigbee and LoRaWAN [14], each protocol offers unique advantages suited to different applications. For instance, Wi-Fi, with its high data rates and widespread adoption, excels in local area environments like homes and businesses. Bluetooth, particularly its lowenergy variant, serves as a linchpin for personal devices and wearables due to its low power consumption and scalability. Zigbee stands out for its robustness and scalability, making it a preferred choice for industrial settings requiring low-power operations. MQTT, OPC-UA, and LoRaWAN cater to specific niches, offering lightweight messaging, industrial-grade communication, and long-range connectivity, respectively. Similarly, NFC and Sigfox address use cases requiring short-range, contactless communication and wide-area coverage with minimal power consumption, respectively [15]. Each protocol contributes to the diverse IoT ecosystem, enabling innovative solutions across industries such as health-care, smart-cities, and agriculture, and paving the way for a connected future.

2.3 *Cloud Computing*

Cloud Cloud computing allows users to access applications and resources through the internet, providing flexibility and convenience [16]. It encompasses several characteristics and service models.

Characteristics:

- Broad network access: Users can access resources from any location via the internet.
- On-demand self-service: Users can provision resources without requiring intervention from the service provider.
- Rapid scalability: Resources can be quickly adjusted to meet changing demands.
- Measured service: Usage of resources can be monitored, controlled, and reported for transparency and accountability.
- Pay-per-use: Users pay only for the resources they consume, typically on a subscription or usage basis.

Service Models:

- Infrastructure as a Service (IaaS): Offers online services like physical/virtual machines, servers, networking, storage, and data center space on a pay-per-use basis. Major providers include Google Compute Engine, Amazon Web Services, and Microsoft Azure. Examples include web hosting and virtual machines.
- Platform as a Service (PaaS): Provides a cloud-based environment with all the tools needed to support the complete lifecycle of web-based applications. It eliminates the need to manage underlying hardware and software provisioning. Examples include App Cloud and Google App Engine.
- Software as a Service (SaaS): Delivers applications over the internet as a service, removing the need for users to install and maintain software locally. Users access applications via the internet, relieving them from managing complex software and hardware. SaaS providers handle security, availability, and performance. Examples include Google Docs, Gmail, and Office.

2.4 *Edge Computing*

Edge-computing facilitates the intelligent processing of data at the edge of IoT networks as shown in Fig. 6, enabling faster and more effective operations by analyzing and filtering data closer to where it is generated. This approach addresses challenges like Latency, bandwidth constraints, and securityissues linked to conventional cloud-computing [17]. By pushing computational tasks to the network edge, edge computing enhances scalability, reduces latency, and ensures real-time processing. Unlike cloud computing, which centralizes services in remote data-centers, edge-computing distributes services closer to end-users, improving accessibility and reducing transmission times. It leverages high bandwidth and low latency to offer diverse applications, making it advantageous for IoT scenarios where rapid responses are crucial. While both edge and cloud computing have their strengths, edge computing excels in IoT applications due to its ability to provide adequate computing power, storage, and quick responses [18]. Additionally, edge computing can adapt to the dynamic nature of IoT devices, leveraging them to extend its framework and cater to distributed computing needs effectively.

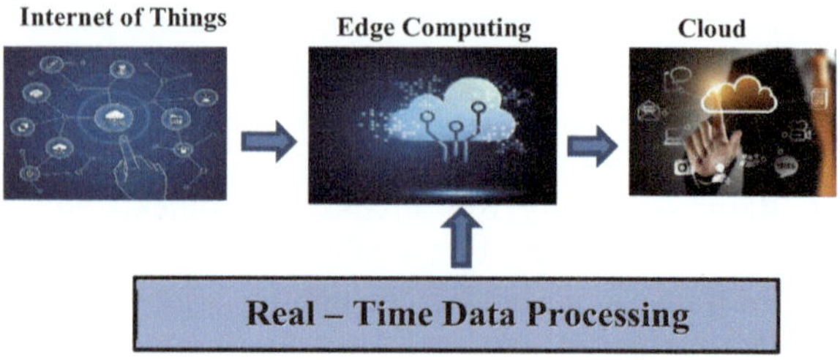

Fig. 6 Edge and cloud computing integration in IoT networks

Edge-computing and cloud-computing are vital in IoT, yet they differ in data processing locations and application suitability. Edge computing processes data near its source, enhancing speed and security, ideal for real-time applications like predictive maintenance [19]. Conversely, cloud computing processes data remotely in data centers, facilitating large-scale processing and dynamic resource allocation, making it ideal for tasks like big data analytics and machine learning.

2.5 Machine Learning and AI

ML and AI are fundamental technologies underpinning the IoT, enabling the extraction of actionable insights from the massive volumes of data generated by IoT devices [20]. ML algorithms analyze data to identify patterns, trends, and correlations, facilitating predictive maintenance, real-time decisionmaking, anomaly detection, energy efficiency optimization, and personalized user experiences within IoT systems. On the other hand, AI enhances IoT capabilities by offering predictive analytics, natural language processing, and cognitive computing, enabling automation, predictive analytics, and improved operational efficiency across various industries. These technologies find applications in smart buildings, industrial processes, healthcare, transportation, agriculture, smart cities, and retail, transforming processes and enhancing user experiences [21] as shown in Fig. 7. As ML and AI continue to evolve, their integration with IoT promises to create intelligent systems capable of autonomously responding to evolving conditions and fostering additional advancements and improvements in efficiency.

3 Applications of IoT

The Internet has undoubtedly been a transformative invention, enabling global connectivity and revolutionizing how we interact with each other through computers and smartphones. However, the emergence of the IoT represents a new phase in this technological evolution [22]. IoT extends connectivity beyond just people, allowing everyday objects to sense, communicate, and even act upon data, ushering in a new

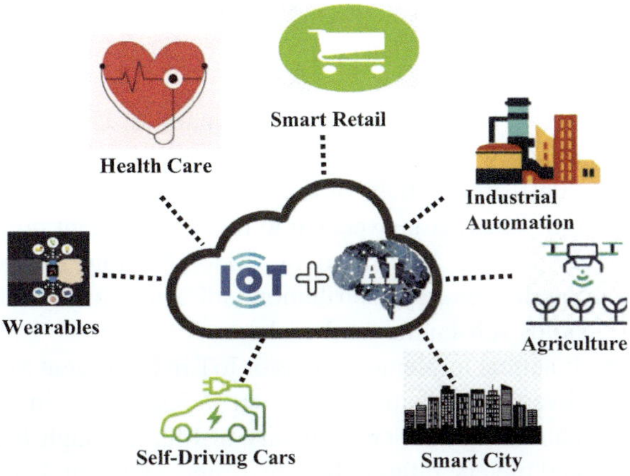

Fig. 7 IoT applications across industries

era of innovation and efficiency. When combined with other disruptive technologies like 5G, automation, and machine learning, IoT has the potential to fundamentally reshape our lives and the way we conduct business. Let us delve into some practical real-world applications of IoT like smart-cities, IIoT, healthcare, agriculture, smart-homes, retail and supplychain that are already making a positive impact.

3.1 Smart Cities

Smart cities rely on Information and Communication Technologies (ICT) to implement sustainable development initiatives amidst rapid urbanization. Through IoT devices and data-driven insights, cities aim to enhance citizen experiences, improve resource management, and streamline service delivery. The necessity for smart cities stems from various factors like resource optimization, enhanced quality of life, sustainability, economic development, IoT in smart cities market dynamics, real-world applications of IoT in smart cities, smart traffic-management, smart-parking, public-safety, and wastemanagement [23].

- Resource Optimization: IoT, AI, and big data analytics enable cities to optimize-energy, water, transportation resources, leading to cost savings and increased efficiency.
- Enhanced Quality of Life: Data and technology empower cities to deliver better healthcare, education, and public safety services, while also facilitating citizen engagement.
- Sustainability: Smart cities focus on sustainability by decreasing energy usage, advocating for renewable energy options, and minimizing waste to address climate change and enhance resilience.

- Economic Development: By attracting businesses, creating jobs, and improving operational efficiency, smart cities drive economic growth and foster local development.
- IoT in Smart Cities Market Dynamics: Urbanization remains an ongoing global trend, with millions relocating to cities weekly. Projections suggest that by 2040, around 65% of the world's population will be urban residents. The smart cities market is expected to reach significant proportions, with a projected value surpassing $6 trillion by 2030. Notably, the smart utility segment, encompassing energy distribution, infrastructure management, and water treatment, holds substantial market share.
- Practical Implementations of IoT in Intelligent Smart-cities: IoT technology finds diverse applications in smart cities, improving efficiency, reducing costs, and enhancing residents' quality of life. Through IoT-powered devices like sensors and meters, cities collect real-time data to update decision-making and optimize-infrastructure and services.
- Smart Traffic Management: Beyond optimizing-traffic, flow and reducing-congestion, advanced smart traffic management systems can incorporate predictive analytics to anticipate traffic patterns and proactively adjust signals and routes. Additionally, integrating with autonomous vehicle technology can further enhance efficiency and safety on roadways.
- Smart Parking: Expanding smart parking solutions can involve integrating with mobile payment platforms to enable cashless transactions and reservation systems for guaranteed parking spots. Furthermore, implementing dynamic pricing based on demand can incentivize turnover and optimize parking utilization in high-demand areas.
- Public Safety: In addition to monitoring for security threats, IoT-enabled public safety systems can incorporate facial recognition technology for identifying wanted individuals or missing persons. Integrating with emergency response systems can enable automated alerts and dispatch of first responders in critical situations.
- Waste Management: Extending IoT solutions in waste management may involve deploying smart bins equipped with compaction technology to maximize capacity and reduce collection frequency. Implementing blockchain technology can enhance transparency and traceability in waste management processes, facilitating better accountability and resource allocation.
- Utility Management: Beyond monitoring energy, water, and lighting usage, utility management systems can incorporate demand-response mechanisms to optimize resource allocation in real-time. Integrating with smart grid technology enables bi-directional communication between utilities and consumers, facilitating more efficient energy distribution and consumption.
- Environmental Monitoring: Expanding environmental monitoring efforts can involve deploying mobile sensor networks to monitor pollution levels in real-time across a wider geographic area. Integrating with weather forecasting systems enables predictive modeling of air and water quality, facilitating early intervention and mitigation of environmental risks.

- Public Transport: In addition to optimizing operations, smart public transport systems can offer personalized journey planning and real-time updates to passengers via mobile apps. Implementing contactless fare payment systems and integrating with multi-modal transportation networks can enhance accessibility and convenience for commuters.

3.2 Industrial IoT

IIoT encompasses enhanced efficiency, workplace observations, and modifications within industrial settings. By 2030, it's projected that global GDP could rise to $15 trillion when IIoT is installed worldwide [24]. Here are some key applications of IIoT:

- Manufacturing: IIoT facilitates the emergence of fresh business opportunities and market dynamics within the manufacturing sector, driving what is known as Industry 4.0. By prioritizing customer data and gathering insights from diverse sources, IIoT empowers manufacturers to devise actionable solutions. This leads to enhanced responsiveness to customer needs, with reduced response times and decreased product waste [24].
- Agriculture: IIoT plays a pivotal role in agriculture by enabling the gathering of data on factors like rainfall, humidity, soil composition, and pest presence [25]. This data improves farming techniques, reduces waste, and minimizes efforts required for crop management. For instance, farmers can monitor soil temperature and moisture content and utilize IoT-derived data for precision fertilization programs. Collaborations such as Toyota Tsusho's partnership with Microsoft to innovate fish farming tools exemplify how IoT technologies can optimize farming processes [26].
- Military Applications: In military contexts, IoT technologies are harnessed for investigation, inspection, and other battle-related purposes. Sensors, robots, biometrics, vehicles, and smart technologies deployed on the battlefield enhance military capabilities and operations.
- Internet of Battlefield Things (IoBT): Initiated by the U.S. Army Research Laboratory (ARL), IoBT aims to empower army soldiers through IoT technologies. ARL collaborates with universities, industries, and Army researchers to enhance the conceptual integration of IoT technologies into Army operations, thereby boosting military effectiveness.
- Ocean of Things (OoT): DARPA's Ocean of Things (OoT) project aims to extend the IoT into vast oceanic regions for gathering, monitoring, and analyzing environmental and vessel activity data. By deploying low-cost, floating-enabled sensors across the ocean, DARPA intends to create a real-time monitoring system for maritime activities, leveraging the extensive coverage of the Earth's surface by oceans.

These applications of IIoT highlight its potential to revolutionize various sectors, from manufacturing and agriculture to military operations and maritime surveillance.

3.3 Healthcare

IoT-based smart healthcare applications encompass a range of functionalities tailored to diverse needs. These applications leverage IoT characteristics to innovate healthcare delivery significantly. Here's an overview of key examples:

- Remote Monitoring and Disease Detection: Continuous remote monitoring via wearable devices is crucial for early disease detection. For example, Wu et al. [27] introduced an ECG data monitoring system utilizing a bipotential device attached to a user's t-shirt, transmitting data to a smartphone via Bluetooth. Similarly, Agustine et al. [28] developed an alert system notifying medical professionals of oxygen level anomalies, enabling timely interventions.
- Smart Treatment and Smart Surgical Environment: IoT plays a crucial role in cancer treatment, integrating stages like chemotherapy and radiotherapy. Heshmat et al. demonstrated a system storing laboratory test data on a cloud-based server, facilitating remote consultations and prescription monitoring [29]. In surgical training, IoT enables solutions like virtual realitybased frameworks for realistic simulations, fostering global collaboration among surgeons [30]. Robotic systems such as the Da Vinci system and the Flex® robotic system further advance surgical capabilities.
- Virtual Health Platform and Assistant: Intelligent healthcare introduces mHealth platforms for remote monitoring and telemedicine services. Examples include stress level tracking systems using wearable sensors [31, 32]. Virtual assistants aid patients in translating medical terms and providing relevant information to physicians, optimizing patient management and care delivery efficiency.
- Smart Hospitals and Pharmaceutical Industry: IoT-driven smart hospitals enhance patient care and streamline operations. These hospitals employ technologies for patient monitoring, staff management, and inventory tracking. In the pharmaceutical sector, IoT optimizes inventory management, anti-counterfeiting measures, and drug production, benefiting patients with improved access to medical services and enhanced doctor-patient interactions. Wearable devices facilitate real-time monitoring in clinical trials, particularly for lung diseases [33].

3.4 Smart Homes

A smart home employs internet-connected devices to enable remote monitoring and control of appliances and systems like lighting and heating, providing homeowners with security, comfort, convenience, and energy efficiency [34]. Also known as home automation or domotics, smart home technology allows users to manage smart devices via a dedicated app on their smartphone or other networked devices. As part of the Internet of Things (IoT), smart home systems and devices collaborate, exchanging usage data and automating actions according to the homeowners' preferences. A smart home comprises interconnected devices and appliances that work together to form a remotely controllable network. Various aspects of household living [35], from lighting to kitchen appliances, have been transformed by smart alternatives.

- Smart TVs: Apart from accessing content through applications, smart TVs can integrate with other smart devices in the home ecosystem. For example, they can synchronize with smart lighting systems to adjust ambient lighting according to the content being viewed, providing a more immersive viewing experience.
- Smart Lighting Systems: In addition to adjusting lighting based on occupancy and daylight, smart lighting systems can also integrate with home security systems. They can simulate occupancy when residents are away by turning lights on and off at varying intervals, enhancing home security.
- Smart Thermostats: Beyond temperature optimization, smart thermostats can analyze external factors such as weather forecasts to adjust heating and cooling settings preemptively. This proactive stance not only improves comfort but also maximizes energy efficiency.
- Smart Door Locks and Garage Door Openers: Smart locks can integrate with home automation platforms to trigger specific actions based on user-defined scenarios. For instance, unlocking the front door can automatically disarm the security system and adjust lighting to welcome the homeowner.
- Smart Security Cameras and Systems: In addition to monitoring for security threats, smart security cameras can integrate with voice assistants to provide hands-free control. Users can command the camera to display live feeds on compatible smart displays or TVs.
- Smart Pet and Lawn Care: Beyond automated feeding and watering, smart pet care devices can track pet activity and health metrics. Similarly, smart lawn care The systems can adapt watering schedules according to current weather conditions, maximizing water efficiency.
- Smart Kitchen Appliances: Smart kitchen appliances can integrate with meal planning and grocery delivery services. They can suggest Recipes tailored to the ingredients you have on hand, create shopping lists, and even place orders for necessary items.
- Smart Household Monitors: Household monitors can provide insights into energy consumption patterns, enabling users to identify opportunities for energy savings. They can also integrate with utility providers to offer real-time pricing information, allowing users to adjust usage during peak demand periods.
- Smart Plugs: Beyond remote control, smart plugs can integrate with home automation platforms to create custom automation routines. For example, a smart plug connected to a coffee maker can be programmed to start brewing coffee automatically when the alarm goes off in the morning.

3.5 Supply Chain Management

IoT presents a diverse array of applications in supply chain management (SCM); fundamentally transforming the way businesses handle logistics and operations [36]. Here are the key areas where IoT is revolutionizing SCM:

- Asset-Tracking and Monitoring: Leveraging IoT-enabled sensors and GPS-trackers provides businesses with real-time visibility into the location and status of assets such as vehicles, containers, and packages. This heightened visibility enhances accountability and allows for more efficient management of the entire supply chain.
- Inventory Optimization: IoT-devices deployed in warehouses and storage facilities offer continuous updates on inventory levels, facilitating proactive stock management to minimize stockouts and optimize order fulfillment processes.
- Proactive Maintenance: By deploying IoT sensors on machinery and equipment, businesses can gather data on their performance and health, enabling predictive maintenance. This proactive approach reduces downtime and mitigates repair costs by addressing issues before they escalate.
- End-to-End Visibility: IoT technologies provide comprehensive visibility across the supply chain, offering real-time insights into product movement, potential delays, and disruptions. This transparency empowers businesses to make informed decisions and effectively manage risks.
- Environmental-Monitoring: IoT-sensors equipped to monitor-temperature, humidity, and other environmental issues ensure the integrity and quality of sensitive goods during transportation and storage, such as pharmaceuticals or food items.
- Condition Monitoring: IoT devices track and monitor the condition of goods throughout transit, detecting factors like shock, vibration, temperature, and humidity. This proactive monitoring helps prevent damage and ensures product integrity.
- Optimized Routing: By harnessing IoT-data coupled with advanced analytics and AI algorithms, businesses can optimize transportation routes and schedules. This optimization leads to reduced fuel consumption, lower transportation costs, and expedited delivery times.
- Warehouse Automation: IoT-driven automation enhances warehouse operations by orchestrating tasks and maximizing space utilization. Automated processes such as picking, packing, and sorting minimize errors and increase operational efficiency.
- Supplier Management: IoT data enables businesses to evaluate supplier performance in real time, monitor adherence to delivery schedules, and assess vendors based on data-driven insights. This enhanced visibility facilitates more effective supplier management practices.
- Supply Chain Analytics: The abundant data produced by IoT devices provides valuable understanding into the performance of supply chains. By analyzing this

data, companies can pinpoint areas needing enhancement and enact tactics to improve the overall efficiency and effectiveness of operations.

4 IoT in Process Control Systems

The concept of the IoT encompasses the expansive networks of edge devices, commonly referred to as sensors, which are interconnected to the internet. These devices are tasked with gathering, analyzing, and sharing data, thereby enhancing communication and collaboration among physical objects across a multitude of industries [37]. From manufacturing and smart homes to transportation, health-care, and agriculture, IoT revolutionizes the way devices interact and operate within their respective environments. Its scalability and versatility make it a pivotal technology in shaping the future of connected eco-systems, facilitating seamless data exchange and enabling innovative applications across diverse sectors [38].

4.1 Streaming Data Analytics

Real-time data streaming entails the continuous collection and ingestion of data sequences from multiple sources, followed by real-time processing to derive significance and glean insights from the data [39] as shown in Fig. 8. The various components involved is discussed below.

- Source: Hundreds or thousands of devices or applications produce significant amounts of continuous data swiftly. Instances include mobile devices, web applications (clickstream), application logs, IoT sensors, smart devices, and gaming applications.
- Data Streaming: Effortlessly connect with over 15 AWS services (such as Amazon API Gateway, AWS IoT Core, Amazon CloudWatch, etc.) to securely and consistently capture ongoing data streams from thousands of devices.
- Stream Storage: Opt for a storage solution tailored to your scaling, latency, and processing requirements, such as Amazon Kinesis Data Streams, Amazon Data Firehose, or Amazon Managed Streaming for Apache Kafka (Amazon MSK).

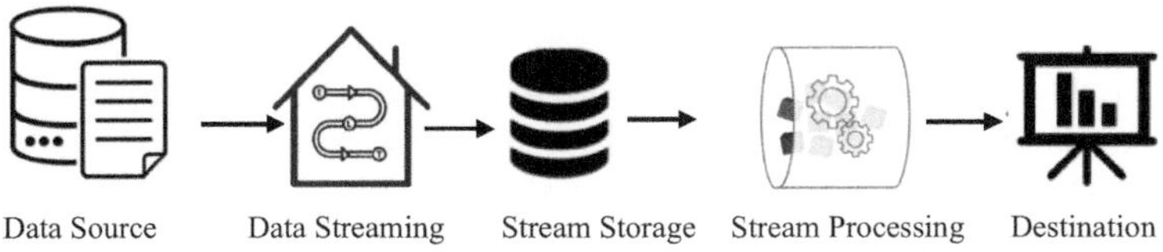

Fig. 8 Components of real time data streaming

- Stream Processing: Select from a variety of services, ranging from simple solutions requiring minimal setup to transform and continuously deliver data (such as Amazon Data Firehose) to advanced, customized real-time applications and machine learning integration using services like Amazon Managed Service for Apache Flink and AWS Lambda.
- Destination: Channel streaming data to fully integrated data lakes, data warehouses, and analytics services for further examination or long-term storage, including options like Amazon S3, Amazon Redshift, Amazon OpenSearch Service, and Amazon EMR.

4.2 Machine Learning Integration

Leveraging machine learning in IoT enables the unveiling of concealed insights within data streams, facilitating swift and automated responses while enhancing decision-making processes [40]. By delving into vast volumes of data through sophisticated algorithms, machine learning discerns hidden patterns and anomalies, augmenting intelligence by processing image, video, and audio inputs. This integration empowers businesses to predict future trends, detect anomalies, and streamline processes through automated actions derived from statistical analyses, thereby supplementing or replacing manual procedures with more efficient systems. Companies harnessing ML for IoT unlock predictive-capabilities across a spectrum of use-cases, enabling them to glean new insights and elevate automation levels. Through this approach, organizations can seamlessly ingest and standardize data, construct machine-learning models, and deploy them across cloud, edge, and device environments [41].

Cumulocity IoT Machine Learning expedites the real-time scoring of IoT data within Cumulocity IoT Streaming Analytics, as depicted in Fig. 9. Embedded in the platform's visual analytics builder, Cumulocity IoT Streaming Analytics presents a "Machine Learning" building block [42], enabling users to implement specified machine learning models for scoring live data. This functionality provides a user-friendly environment for effortlessly incorporating machine learning models into streaming analytics workflows.

4.3 Real-Time Decision-Making

The thought of data as a strategic asset has gained substantial traction in recent years, owing to its invaluable insights that drive informed decision-making and progress for organizations and individuals alike as shown in Fig. 10. Data plays a crucial role in various aspects such as decisionmaking, problem-solving, innovation, personalization, performance optimization, risk management, scientific advancements, and governance. Leveraging data empowers entities to navigate the complexities of the

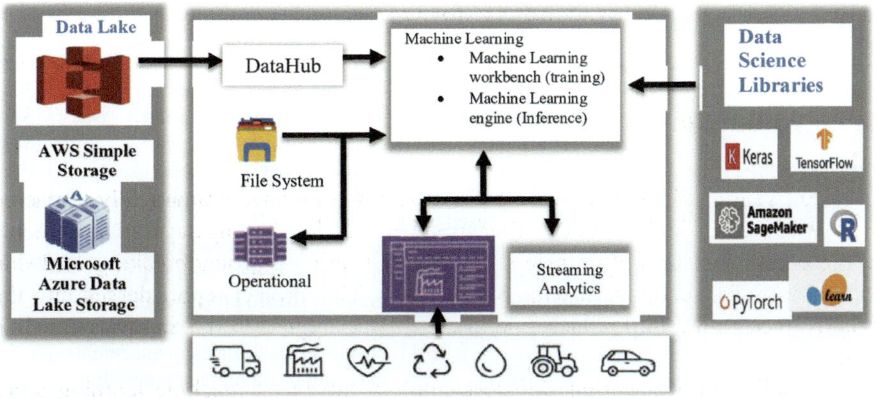

Fig. 9 Cumulocity IoT machine learning architecture overview

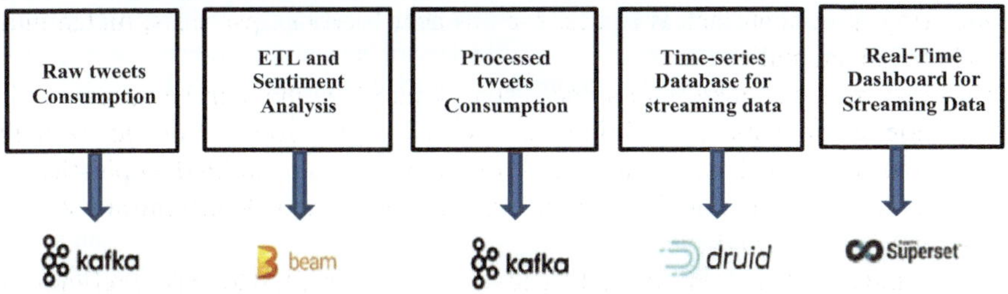

Fig. 10 Real-time streaming analytics for faster decision-making

modern world and achieve meaningful progress. Organizations are increasingly prioritizing rapid, data-driven decision-making across critical business units, employing emerging technologies like AI, ML and IoT [43], and cloud computing to revolutionize operations and stay competitive. With the surge in data volume from sources like IoT sensors, social media, web, and mobile apps, there's a growing need for real-time, data-driven decision-making using streaming data analytics. This article highlights the design of a platform-agnostic, real-time streaming analytics engine capable of processing live data feeds from Twitter to conduct real-time sentiment analysis of users. The application design focuses on key challenging characteristics such as low latency, high availability, and horizontal scalability to ensure efficient processing and timely responses [44] are mentioned in Table 1.

5 Challenges in IoT Implementation

- Security and Privacy

Table 1 Advancements, solutions, and challenges in industrial IoT (IIoT) technologies

Challenges	Areas of Application	Potential solutions/remarks	Ref
Real-time performance	• Stream processing-systems designed for managing substantial amounts of data from IoT sensors are essential for IoT based big data processing • Managing smart grids and microgrids through IoT technology	• Introducing a real-time data collection system featuring a variety of sensors for measuring accelerometer, gyroscope, temperature, and humidity • Create a develop a prototype and integrate a distributed stream processing system with IoT to improve stream processing results • Examine three IoT communication protocols to evaluate enhancements in realtime communication within an IoT-enabled microgrid	[30] [37] [38]
Energy efficiency	• Regarding the positioning and allocation of resources within UAVs • Developing energy-efficient methods for wide-scale access in 5G networks and environmentally conscious cellular standards • Utilizing LoRaWAN technology in IoT devices with a focus on optimizing energy consumption	• Create and adapt an algorithm to allocate subchannels to devices systematically, aiming to efficiently reduce interference • Describing how the power usage at the MAC layer influences the battery life of devices in an IoT-enabled setup • There are significant considerations regarding energy efficiency across different layers, such as the medium access control and network layers of the LoRa standard that require addressing	[39] [40] [32]
Security and safety of IIoT devices	• IIoT manufacturing procedures • Industrial IoT • IIoT necessitating security measures	• Recognize challenges and research prospects related to scalability, interoperability, and dependability, crucial aspects of numerous distributed systems • Determine the vulnerable points of an embedded system using commonly employed devices in the field • Introduce a classification of solutions encompassing various artificial intelligence methods to address security challenges within IIoT environments	[41] [33] [42]

(continued)

Table 1 (continued)

Challenges	Areas of Application	Potential solutions/remarks	Ref
Coexistence and interoperability	• Industrial IoT applications requiring compatibility and mutual operation • 5G networks integrated with IoT technology • Creating networks based on IoT technology	• Concentrate on the challenges related to the necessity for mutual operation and compatibility • Offer perspectives on the future potential and challenges concerning the integration of AI into 5G networking within the IoT ecosystem, particularly focusing on coexistence and interoperability • Emphasizing the persistent problems and hurdles in this field, while suggesting a framework that utilizes various approaches to improve interoperability across all levels concurrently	[43] [34] [35]
Limited research on OT and ICT convergence	• Mining industry applications utilizing IoT technology • Systems that combine Operational Technology (OT) and Information Technology (IT)	• Recognize a significant obstacle related to vertical fragmentation arising from technical disparities among various systems and devices. • Implementing the OPC unified using architecture as the interoperable interface between devices and systems is the most effective approach to meeting the demands for highquality services at the Operational Technology (OT) layer	[44] [45]

A major obstacle in IoT implementation is guaranteeing the security and privacy of interconnected devices and their data exchange. Challenges like inadequate encryption, insufficient testing and updates, and susceptibility to network attacks present substantial risks. Furthermore, worries about unsecured data transmission and privacy breaches underscore the importance of implementing strong security measures and privacy protections to safeguard sensitive information. [45].

- Interoperability

Interoperability poses a significant challenge in IoT deployment, as various devices and systems must communicate and share data seamlessly. The lack of standardized protocols can result in compatibility issues, impeding the integration and efficiency of IoT solutions. Implementing common standards and protocols is crucial to addressing interoperability obstacles and facilitating smooth connectivity throughout IoT ecosystems. [46].

- Scalability

Scalability is another critical challenge in IoT implementation, given the exponential growth in connected devices and data volume. Efficient data management, network capacity, and device management become increasingly complex as IoT ecosystems expand. Adopting scalable architectures and efficient data handling techniques is crucial to accommodate the growing demands of IoT deployments and ensure optimal performance and efficiency [47].

- Data Management

Efficiently managing data is a primary hurdle in IoT adoption, given the substantial volumes of data produced by interconnected devices. Challenges like storing, processing, and analyzing data necessitate strong strategies and technologies to address the scale, speed, and diversity of IoT data. Employing effective data management approaches is critical for extracting meaningful insights, maintaining data precision, and facilitating informed decision-making in IoT settings. [48].

6 Future Trends and Directions

Continued advancements in connectivity, such as the widespread adoption of 5G, coupled with the rise of edge computing, are set to revolutionize the IoT landscape. These developments promise faster, data that are transmission that is more reliable and enable real-time communication, fostering the seamless operation of IoT applications across diverse environments. Concurrently, advancements in sensor technologies will drive innovation, allowing for precise data collection across various domains. The future of IoT will witness the emergence of novel applications spanning smart agriculture, autonomous vehicles, healthcare, and more, addressing societal challenges and enhancing efficiency. Interconnected ecosystems will be the norm, facilitating collaboration among devices, platforms, and services across different domains, ultimately reshaping industries and societies globally. Embracing these trends will reveal unique levels of connectivity, efficiency, and innovation in the digital age.

7 Conclusion

In conclusion, the discourse on IoT technologies and applications underscores its transformative potential across various sectors, from smart cities to healthcare and industrial IoT. As IoT rapidly evolves from a conceptual framework to a tangible reality, essential components like sensors, connectivity protocols, cloud computing, and artificial intelligence play pivotal roles in enabling seamless device connectivity and data exchange. However, challenges such as security, interoperability, scalability, and data management persist, necessitating robust solutions to ensure the reliability and integrity of IoT systems and data. Looking ahead, advancements in connectivity, edge computing, and sensor technologies, coupled with emerging applications, will continue to drive innovation in IoT, ushering in an era characterized by unprecedented connectivity and efficiency. By embracing these trends and leveraging the transformative power of IoT, organizations and individuals can unlock new opportunities, enhance decision-making processes, and address global challenges, marking the onset of a more connected and intelligent future.

References

1. Abdul-Qawy, A.S., et al.: The Internet of Things (IoT): An overview. Int. J. Eng. Res. Appl. **5**(12), 71–82
2. Al Mamun, M.A., Yuce, M.R.: Sensors and systems for wearable environmental monitoring toward IoT-enabled applications: A review. IEEE Sens. J. **19**(18), 77717788
3. Al-Ali, A.R., et al.: Role of IoT technologies in big data management systems: A review and smart grid case study. Pervasive Mob. Comput., 101905 (2024)
4. Madakam, S., Takahiro, U.: Industrial Internet of Things (IIoT): principles, processes and protocols. In: The Internet of Things in the Industrial Sector: Security and Device Connectivity, Smart Environments, and Industry 4.0, pp. 35–53 (2019)
5. Gupta, V.P.: Smart sensors and industrial IoT (IIoT): A driver of the growth of industry 4.0. In: Smart Sensors for Industrial Internet of Things: Challenges, Solutions and Applications, pp. 37–49 (2021)
6. Kiesel, R., van Roessel, J., Schmitt, R.H.: Quantification of economic potential of 5G for latency critical applications in production. Procedia Manuf. **52**, 113–120
7. Xu, Y., et al.: A blockchain-based nonrepudiation network computing service scheme for industrial IoT. IEEE Trans. Ind. Inform. **15**(6), 3632–3641
8. Javaid, M., et al.: An integrated outlook of Cyber–Physical Systems for Industry 4.0: Topical practices, architecture, and applications. Green Technol. Sustain. **1**(1), 100001 (2023)
9. Hanes, D., et al.: IoT fundamentals: Networking technologies, protocols, and use cases for the Internet of Things. Cisco Press (2017)
10. Salhaoui, M.: Smart IoT monitoring and real-time control based on autonomous robots, visualrecognition and cloud/edge computing services (2021)
11. Kanagachidambaresan, G.R., Bharathi, N.: Sensors and protocols for industry 4.0
12. Ladegourdie, M., Kua, J.: Performance analysis of OPC UA for industrial interoperability towards industry 4.0. IoT **3**(4), 507–525 (2022)
13. Zikria, Y.B., et al.: Internet of things (IoT): Operating system, applications and protocols design, and validation techniques. Futur. Gener. Comput. Syst. **88**, 699–706
14. Alliance, LR.: Wi-Fi & LoRaWAN® Deployment Synergies (2019)
15. Zanaj, E., et al.: Energy efficiency in short and wide-area IoT technologies—a survey. Technologies **9**(1), 22 (2021)
16. Srivastava, A.: A detailed literature review on cloud computing. Asian J. Technol. Manag. Res. **4**(2)
17. Zhou, Z., et al.: Edge intelligence: Paving the last mile of artificial intelligence with edgecomputing. In: Proceedings of the IEEE **107**(8), 1738–1762
18. Du, J., et al.: Machine learning for 6G wireless networks: Carrying forward enhanced bandwidth, massive access, and ultrareliable/low-latency service. IEEE Veh. Technol. Mag. **15**(4), 122–134
19. Salman, O., et al.: Edge computing enabling the Internet of Things. In: 2015 IEEE 2nd world forum on Internet of Things (WF-IoT). IEEE (2015)
20. Bzai, J., et al.: Machine learning-enabled Internet of Things (IoT): Data, applications, and industry perspective. Electronics **11**(17), 2676

21. Ramana, T.V., et al. (eds).: Artificial intelligence and machine learning for smart community: Conceptsand Applications (2024)
22. Greengard, S.: The Internet of Things. MIT press (2021)
23. Ullah, A., et al.: Smart cities: The role of Internet of Things and machine learning in realizing a data centric smart environment. Complex Intell. Syst. **10**(1), 1607–1637
24. Tiwari, A., et al.: Privacy issues & security techniques in big data. In: 2019 International Conference on Computing, Communication, and Intelligent Systems (ICCCIS). IEEE (2019)
25. Müller, J.M., Kiel, D., Voigt, K.-I.: What drives the implementation of industry 4.0? The role of opportunities and challenges in the context of sustainability. Sustainability **10**(1), 247
26. Kiel, D., Arnold, C., Voigt, K.-I.: The influence of the Industrial Internet of Things onbusiness models of established manufacturing companies–a business level perspective. Technovation **68**, 4–19 (2017)
27. Wu, T., Redouté, J.-M., Yuce, M.: A wearable, low-power, real-time ECG monitorfor smart t-shirt and IoT healthcare applications. In: Advances in body area networks I: Post-conference proceedings of bodynets 2017. Springer International Publishing (2019)
28. Agustine, L., et al.: Heart rate monitoring device for arrhythmia using pulse oximeter sensor basedon android. In: 2018 International Conference on Computer Engineering, Network and Intelligent Multimedia (CENIM). IEEE (2018)
29. Heshmat, M., Shehata, A.-R.S. (2018) A framework about using internet of things for smartcancer treatment process. In: Proceedings of the International Conference on Industrial Engineering and Operations Management (2018)
30. Cecil, J., et al.: An IoMT based cyber training framework for orthopedic surgery using next generation internet technologies. Inform. Med. Unlocked **12**, 128–137 (2018)
31. Estrin, D., Sim, I.: Open mHealth architecture: an engine for health care innovation. Science **330**(6005), 759–760 (2010)
32. Akmandor, A.O., Jha, N.K.: Keep the stress away with SoDA: Stress detection and alleviation system. IEEE Trans. Multiscale Comput. Syst. **3**(4), 269–282
33. Geller, N.L., Kim, D.-Y., Tian, X.: Smart technology in lung disease clinical trials. Chest **149**(1), 22–26 (2016)
34. King, J.: Energy impacts of smart home technologies. Report A1801 (2018)
35. Lobaccaro, G., Carlucci, S., Löfström, E.: A review of systems and technologies for smart homes and smart grids. Energies **9**(5), 348 (2016)
36. Rejeb, A., Keogh, J.G., Treiblmaier, H.: Leveraging the internet of things and blockchain technology in supply chain management. Futur. Internet **11**(7), 161 (2019)
37. DaCosta, F., Henderson, B.: Rethinking the internet of things: a scalable approach to connecting everything. Springer Nature (2013)
38. Vyas, B.: Java-powered AI: Implementing intelligent systems with code. J. Sci. Technol. **4**(6), 1–12 (2023)
39. Biem, A., et al.: Real-time analysis and management of big time-series data. IBM J. Res. Dev. **57**(3/4), 8–1
40. Nazir, A., et al.: Advancing IoT security: A systematic review of machine learning approaches for the detection of IoT botnets. J. King Saud Univ. Comput. Inf. Sci., 101820 (2023)
41. Adel, A.: Unlocking the future: fostering human–machine collaboration and driving intelligent automation through industry 5.0 in smart cities. Smart Cities **6**(5), 2742–2782 (2023)
42. Oktug, F., Yaslan, Y., Gulacar, H.: A prediction module for smart city IoT platforms. Transp. Power Grid Smart Cities (Commun. Netw. Serv.), 269–290 (2018)
43. Husain, G.: Harnessing big data analytics for enhanced machine learning algorithms. J. Environ. Sci. Technol. **2**(2), 21–35 (2023)
44. Mutlag, A.A., et al.: Enabling technologies for fog computing in healthcare IoT systems. Futur. Gener. Comput. Syst. **90**, 62–78 (2019)
45. Sisinni, E., et al.: Industrial internet of things: Challenges, opportunities, and directions. IEEE Trans. Ind. Inform. **14**(11), 4724–4734 (2018)
46. Dave, D.M.K., Mittapally, B.K.: Data integration and interoperability in IoT: Challenges, strategies and future direction

47. El Kafhali, S., Salah, K: Efficient and dynamic scaling of fog nodes for IoT devices. J. Supercomput. **73**, 5261–5284 (2017)
48. Diène, B., et al.: Data management techniques for internet of things. Mech. Syst. Signal Process. **138**, 106564 (2020)

Sustainable Electric Vehicle Market Segmentation Using Deep Learning Techniques

Nrusingha Tripathy, Debahuti Mishra, Sarbeswara Hota, Subrat Kumar Nayak, and Gobinda Chandra Das

Abstract In order to implement market segmentation as a key strategy for promoting the widespread use of emergent agility technologies such as electric vehicles (EVs). The adoption of electric vehicles (EVs) is predicted to rise substantially in the near future due to their low operating costs and minimal emissions. It will therefore greatly increase interest in further academic research in the future. This study aims to combine the approaches of "perceived benefits-attitude-intention" in order to examine and identify many categories of potential EV purchasers based on psychographic, behavioral, and socioeconomic aspects. In this work, different LSTM model variants, GRU, and Bi-LSTM models were utilized. We are taking the electric vehicle population dataset from Kaggle competition. According to the report, three different young groupings for segmenting consumers and markets behavioral, psychographic, and demographic are thought to be forming relationships with purchasers of electric vehicles (EVs). The implications are outlined, which can provide researchers and decision-makers some helpful guidance on how to encourage the use of EVs in light of the expanding sustainable transportation industry. We compared the models with their Mean Absolute Error (MAE), Root Mean Squared Error (RMSE), and Mean Squared Error (MSE). The Bi-LSTM model gives a MAE score of 0.417, RMSE score of 1.254, and MSE score of 1.573, which is minimal as compared to other models.

Keywords Greener ways of transportation · Market segmentation · Electric vehicles · Intention to embrace electric vehicles · Cluster analysis

N. Tripathy (✉) · D. Mishra · S. Hota · S. K. Nayak
ITER College Rd, Bhubaneswar, Odisha, India
e-mail: nrusinghatripathy654@gmail.com

G. C. Das
Department of CSE, KL (Deemed to be) University, Vijayawada, India

B. K. Mishra and A. Rocha (eds.), *Future of AI, IoT, and Sustainability*, Information Systems Engineering and Management 23,
https://doi.org/10.1007/978-3-031-76286-4_15

1 Introduction

1.1 Scope and Purpose

The rapid expansion of the transportation sector and its corresponding emissions are a focus for authorities seeking to ensure environmental sustainability. Consequently, it is crucial to understand the determining elements of transport emissions. Given the growing emissions from transportation, electric vehicles are essential [1]. A significant portion of Europe's CO_2 emissions 22% in the UK come from road-based transportation. Concerns about climate change have led to agreements among EU member states to reduce emissions by 80% by 2050, with the goal of alleviating atmospheric CO_2 levels at 450 parts per million and preventing global warming from exceeding 2 degrees Celsius. Several industries participate in the endeavor, with the transport industry aiming to cut emissions by 95%. It also depends heavily on oil, which creates issues with resource depletion and supply security.

1.2 Process Overview

Finally, vehicle-related urban pollution results in health issues. Because of this, creating oil-free and low-carbon transportation options is regarded as essential. Three approaches are being investigated to address this issue: increasing the economy of existing cars, using biofuels, and implementing electric powertrains. However, absolute oil independence and zero tailpipe emissions technology would likely be required in the long run due to the anticipated growth in passenger numbers. Battery electric vehicles (BEVs), meet these two requirements. Their idea is straightforward: the Internal Combustion Engine Vehicle (ICEV) and the tank are replaced with an electric motor that runs on a battery, and the car is hooked into a charging station when it's not in use [2, 3]. Ease of use. They have a number of benefits, including excellent acceleration, minimal tailpipe emissions that improve local air quality, and the ability to be charged overnight with inexpensive energy generated by any kind of power plant, including renewable sources. But even with these benefits, BEVs still have a lot of obstacles to overcome. The range of these cars is constrained since energy storage is currently costly and requires time for battery recharge.

Prior to any market penetration, a charging spot infrastructure must be established, and the associated expenditure is crucial. Concerns about main resource depletion have also been highlighted about a few battery components. BEVs may have a negative effect on the grid. But the hardest problem is that societal acceptability of these cars is the last major obstacle that needs to be overcome before BEVs can become widely available [4]. Customers are hampered by their high construction costs, and their low operating costs are seldom advertised. The biggest obstacle is most likely "range anxiety," which arises from customers' concern that they won't be able to finish their trip since the range is less than in traditional cars and charging takes time.

In essence, we use three different kinds of deep learning models: GRU, Bi-LSTM, LSTM, and variants of the LSTM model. In evaluation to other models, the Bi-LSTM model's accuracy score is greater.

2 Literature Review

In urban areas, public transit is offered via several means of transportation. Large vehicles like buses and trams, as well as specific components like engines, wheels, and tyres, are all featured, along having significant networks such as the rail and road systems [5]. Despite having different functions, each of these parts contributes to the overall operation of "public transport". In this part, the development and spread of different modes of transport throughout different cities in Europe are further explored. All that will be assessed is transport equipment. Dates will be mentioned with the events if practical.

In an effort to attain sustainability, the rapidly growing digital economy and society are looking for suitable solutions that will guarantee sustainable growth. This holds true for logistics as well as the fields of economics and economics [6]. Since over 90% of cars worldwide are powered by oil, there is a discernible trend towards the desire to power cars using alternate energy sources. Consequently, there has been an increase in interest in electric cars, or EVs. Contemporary methods of vehicle propulsion, such electrically powered engines, may offer a solution that is in line with the sustainable development agenda. Considering how important the issue of growing global air pollution is, using electric vehicles might help accomplish sustainable development goals [7]. Since electric automobiles don't release any hazardous chemicals during operation, including carbon dioxide or nitrogen oxides, they help to slow down the growth of urban pollution and the greenhouse effect. In addition, there may be benefits associated with owning, using, and purchasing an EV, such tax rebates, bus lanes access, or free parking in urban areas [8]. The function of electrification in public transportation, personal transportation devices (such electric scooters), electric bicycles, and electric vehicles are the most researched topics in the field of electric vehicle markets. Most research on electric vehicles raises questions like these:

- battery life and how to optimize it
- Ecological and environmental effect;
- charging speed;
- EV pricing policy;
- charging stations for electric vehicles;
- legal laws and holders' facilitations;
- commercial strategies in electric mobility.

However, from the standpoint of the economic sciences, the most important aspects are those that influence a consumer's ultimate decision to purchase an electric vehicle [9]. When deciding which item to buy, price is the primary factor that matters the most and has the most influence. This also applies to electric cars. A customer's greater propensity to purchase a certain automobile model may also be influenced by other variables such as vehicle range, flexibility (including acceleration), trunk size, and brand prestige. In this work The Bi-LSTM model yields low scores likened to other models, with an MAE of 0.417, an RMSE of 1.254, and an MSE of 1.573.

3 Methodology

Preparing the data for use in a deep learning model is a crucial first step. The results might not be correct if the data wasn't first tidy or in the precise arrangement for the model [10]. The data must be prepared in the format that is required beforehand. It is used to handle the dataset's noise, doubling, and missing values. Data pre-processing comprises activities such as attribute scaling, dataset segmentation, and import. The pre-processed data is necessary for the model to have greater accuracy [11]. It describes the procedures for transforming, integrating, and cleansing data in order to get it ready for analysis Congestion issues may arise during electric vehicle (EV) integration into the electrical grid if EVs operate on a free charging strategy rather than being controlled and supervised by an analytical structure like the one proposed in this study [12]. Failure to employ this kind of plan may result in significant expenditures being needed for grid reinforcement. A few distribution grids have looked at this problem in preliminary ways. Figure 1 shows different electric vehicle brands in our dataset. The impact of EVs' conventional MV distribution system is discussed in this part, along with an enhanced charging technique designed to facilitate the amalgamation of additional EVs into the same lattice without requiring infrastructural improvements [13]. Figure 2 shows the overall car price in our dataset.

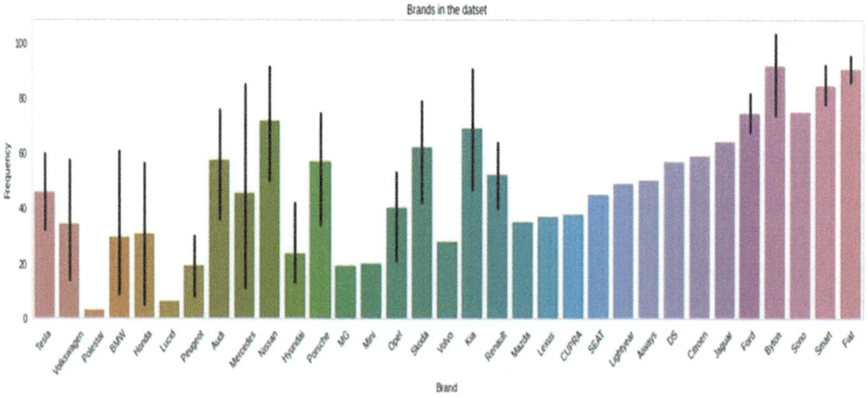

Fig. 1 Different EV brands in the dataset

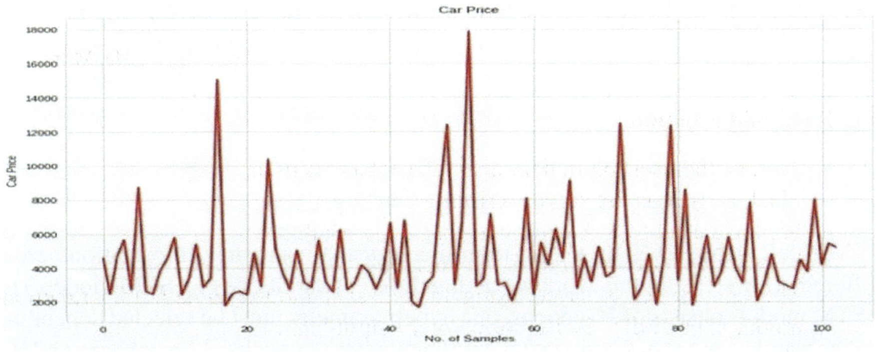

Fig. 2 Overall car price

3.1 *Model Analysis*

EVs provide data from a variety of sources, including exterior meteorological conditions, charging stations, and onboard sensors. Pre-processing makes it easier to combine several data sources into a single format, enabling thorough analysis and judgment to maximize EV efficiency, range, and charging tactics. Different deep learning model predicted result

a) **Long Short-Term Memory (LSTM)**

Time series forecasting problems like EV charging behaviour prediction and energy consumption pattern prediction may be effectively handled by LSTM networks since they are well-suited to handle sequential data. In sequential data, long-term dependencies may be captured by LSTMs, unlike standard feedforward neural networks. This is important for jobs involving EV prediction, since behaviour in the future may be greatly influenced by prior charging characteristics. LSTMs were developed to improve the performance of traditional RNNs in jobs containing consecutive or time-series data by addressing the vanishing gradient issue that often arises with them [14]. Figure 3 shows the LSTM architecture. LSTMs have two main components: the cell state (*Ct*) and the hidden state (*ht*).

Fig. 3 LSTM architecture

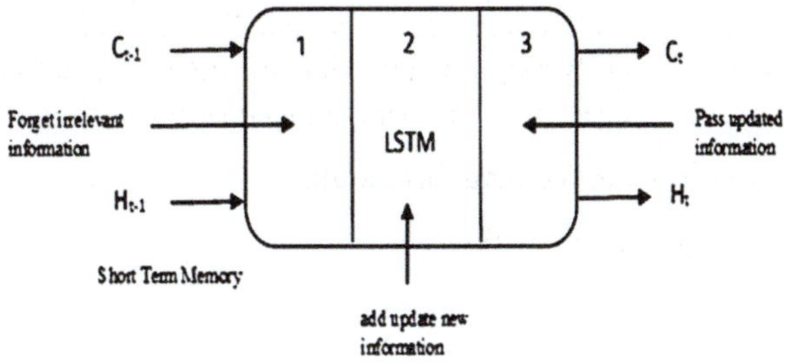

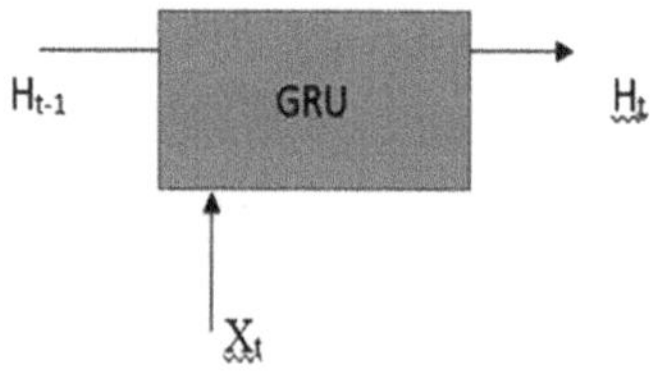

Fig. 4 GRU architecture

The cell state serves as a long-term memory that can carry information across different time steps. In this work we taking different variations of LSTM model. The LSTM model consists of 50 epochs, one hyperparameter must be selected depending on the training process's convergence is the number of epochs, 32 batch size and 0.2 dropout unit. Similarly, LSTM 1 consists of 100 epochs, 64 batch size and 0.2 dropout unit to regularize the model and avoid overfitting. LSTM 2 consists of 150 epochs, 128 batch size and 0.2 dropout unit.

b) **Gated Recurrent Unit (GRU)**

GRU architectures are comparatively easier to understand because of their more straightforward structure, whereas models based on deep learning are typically regarded as "black boxes." Enhancing model transparency and comprehending the variables affecting EV charging behaviour can both benefit from this. GRUs are a variant of traditional RNNs and were introduced to address some of the issues like the vanishing gradient problem [15]. They are ideal for learning from sequential input because they include a gating mechanism that lets them updated and resets their hidden states selectively. Unlike the LSTM, which has a distinct cell state (Ct), the GRU only has a hidden state (Ht). They train more quickly because of their straightforward structure. It gets the hidden state Ht-1 from the former timestamp t-1 and an input Xt at each individual timestamp t. Following it, the next timestamp is shifted to the current hidden state, Ht. The simple GRU architecture given in Fig. 4.

c) **Bidirectional LSTM (Bi-LSTM)**

A Bidirectional LSTM (Bi-LSTM) is an allowance of the traditional Long Short-Term Memory (LSTM) architecture. While a standard LSTM processes input sequences in a forward manner, a Bi-LSTM processes the sequences that go both forward and backward. Because of its ability to process input in both directions, the network is better able to identify long-term reliance in sequential data by collecting data from both past and future contexts [16]. Bidirectional LSTMs are particularly useful for tasks where context from both past and future is important, such as natural language processing (NLP) tasks, speech recognition, and other sequence-to-sequence problems. The bidirectional nature allows the model to capture dependencies that may be missed by a unidirectional LSTM. Bi-LSTM models concurrently record dependencies from previous and subsequent time steps. This implies that, when it comes to EVs, the model can both examine past data and predict future trends (such impending charging demands or traffic patterns), which will result in a deeper comprehension of EV usage trends.

3.2 Data-Preprocessing

The quality of data gathered from numerous sensors and sources in EVs is improved by pre-processing techniques including cleaning, normalization, and transformation. This guarantees precise and trustworthy data for analysis and decision-making, which is essential for maximizing electric vehicle performance and efficiency. Deeper insights into the behaviour and performance of EVs may be gained by pre-processing data in order to create additional characteristics or variables from it. To improve comprehension and control of battery health and usage, information such as battery status of charge (SoC) and charging trends may be obtained from the raw data obtained from EVs.

a) **Data Cleaning**

The gathered data is small and is utilized for both grouping and visualization. The process makes use of Python modules like NumPy, Pandas, Scikit-Learn, and SciPy, and repeatable results are guaranteed. Data cleansing ensures that identities, formats, and units are consistent across sources, which makes it easier to integrate disparate datasets. As a result, businesses may gain benefits from a completer and more cohesive picture of their data. Data cleansing guarantees the accuracy and dependability of the data employed in market analysis, including sales numbers, client demographics, and metrics related to vehicle performance. As a result, stakeholders are able to decide wisely by having a thorough grasp of customer preferences and industry trends. EV producers can effectively examine market demand, vehicle performance statistics, and customer feedback when they have access to clean data. This aids in pinpointing problem areas and maximizing next product development endeavors to satisfy client demands and legal requirements [17].

b) **Exploratory Data Analysis (EDA)**

Exploratory data analysis (EDA) is a crucial step in the data inspection process that involves looking at and analysing data to find trends, gain understanding, and understand the underlying structure. EDA helps analysts and data scientists to formulate hypotheses, discover relationships, and guide the selection of appropriate modelling techniques. We begin the exploratory data analysis by analyzing a portion of the data that was collected without the use of principal components and a portion of the dataset that was created by combining all of the available data. PCA is a statistical technique that converts a set of associated characteristic information into a set of linearly independent features by orthogonal transformation [18]. The newly altered qualities are the Principal Components. The procedure aids in decreasing the dimensionality of the data to lower the cost of any machine learning operation.

c) **Scree Plot**

Scree plot is a popular technique for graphically representing the number of PCs that should be kept. According to Fig. 5, shows the eigenvalues for each and every PC in a simple line segment plot. It displays the factors on the x-axis and the eigenvalues on the y-axis. It always displays a falling gradient. The majority of scree plots have

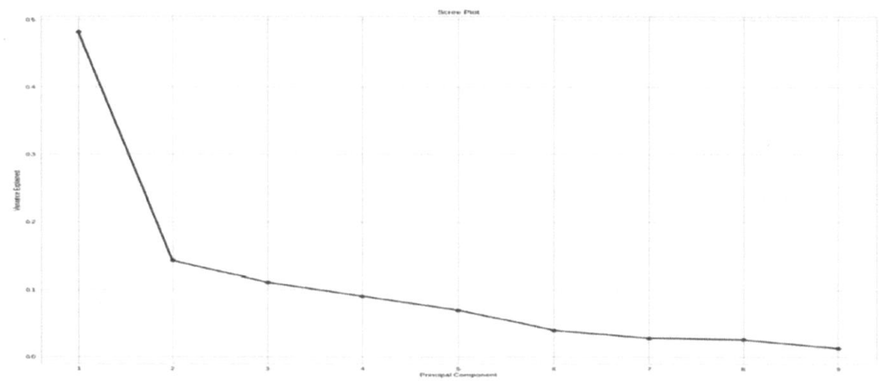

Fig. 5 Scree plot for our dataset

roughly the same shapes: they start elevated on the left, drop quickly, and then level off. This is due to the fact that most of the variability is often explained by the first component, a reasonable amount by the next few components, and a minor portion of the total variation by the subsequent components [19]. The scree plot criterion chooses all components immediately before the line flattens out by searching for the "elbow" in the curve. The variance plot's proportion: At least 80% of the variation clarified by the chosen PCs.

d) **Extracting Segments**

Agglomerative hierarchical clustering is a special kind of clustering that uses this particular tactic. The agglomerative centralized technique of clustering starts with each point as a separate cluster and works its way up by linking points to clusters based on distances. We find the optimal number of clusters for clustering in a hierarchy by using a dendrogram, a tree-like diagram that shows the classifications of cluster splits and unites [20]. If two clusters in a network merge, the dendrogram will unite them; the height of the join will be determined by the gap between the clusters. We may ascertain the finest number of clusters by examining the hierarchical structure of the dendrogram, as illustrated in Fig. 6.

e) **Elbow Method**

A well-liked technique for reckoning out the elbow approach is the idyllic numeral clusters. The process consists of finding the k clusters for which the change in WSS starts to decrease by calculating each cluster's Within-Cluster-Sum of Squared Errors (WSS). The explanation for the elbow approach is because the variance varies fast for a small number of clusters, then slows down and forms an elbow in the curve [21]. The elbow point is the maximum number of clusters we can use in our clustering method. The elbow point is highlighted by the function itself. Which is attained as given in Fig. 7. The function also tells us how long it takes to generate models for various numbers of clusters through the green line.

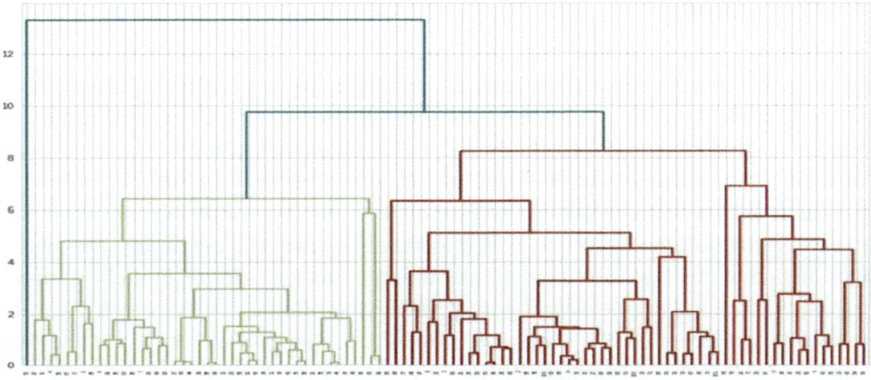

Fig. 6 Dendrogram plot for our dataset

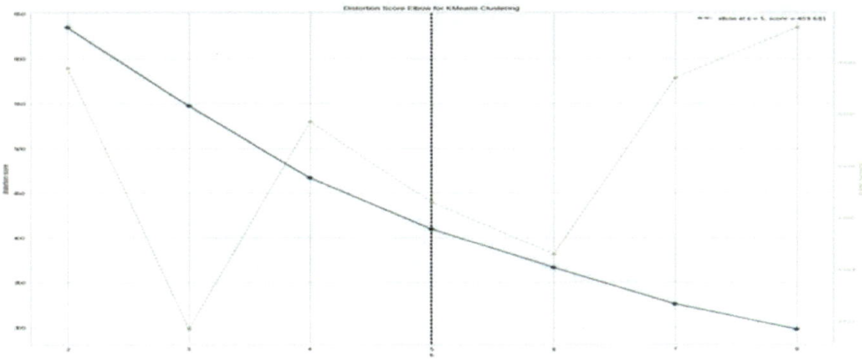

Fig. 7 Evaluating the clusters using distortion and elbow method

f) **Clustering**

One of the most popular exploratory data analysis methods for determining the data's structure is clustering [22]. It can be defined as the method of finding segments in the data so that data elements in one cluster (identical subgroup) are substantially diverse from one another and data points in another cluster are almost identical. Stated otherwise, finding homogenous subgroups in the data is our aim. When evaluating the individual data points within each cluster, the Euclidean or correlation similarity metric is as high as feasible. Which similarity metric to use will depend on the specific use case. Two approaches to cluster analysis are sample-based (finding subgroups of features based on samples) and feature-based (verdict subgroups of samples based on features) [23].

4 Discussion

Operators of electric vehicle (EV) rental sites are paying greater attention to customer service demands as their client base continues to grow at a rapid pace. However, EV operators are unable to concurrently address the demands of every consumer with a particular product or service because to the wide diversity of customer wants [24]. Research on market segmentation of explicit or implicit consumer groups is essential to raising the quality of service as it may reveal market characteristics and internal workings that can be used to create marketing plans and service models tailored to specific target audiences [25]. Figure 8 shows the rapid charge of electric vehicles which fall under inr (10e3). Figure 9 shows the scatter plot of *range_km* and brand and Fig. 10 shows the heatmap to show the correlation of the data.

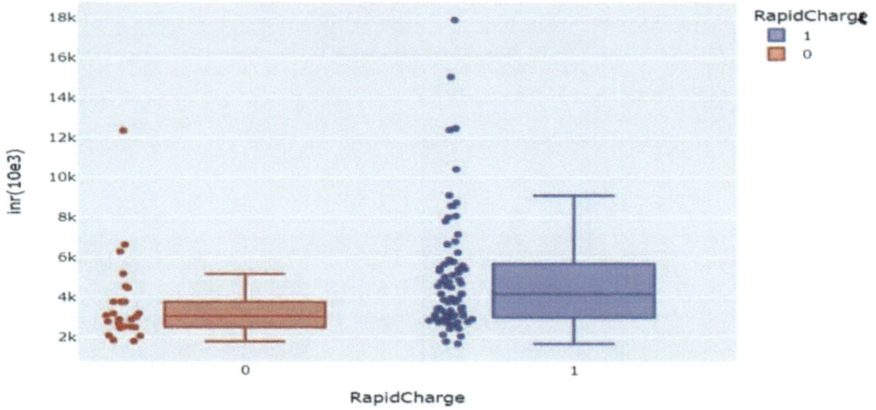

Fig. 8 Rapid charge in which the vehicles fall under *inr* (10e3)

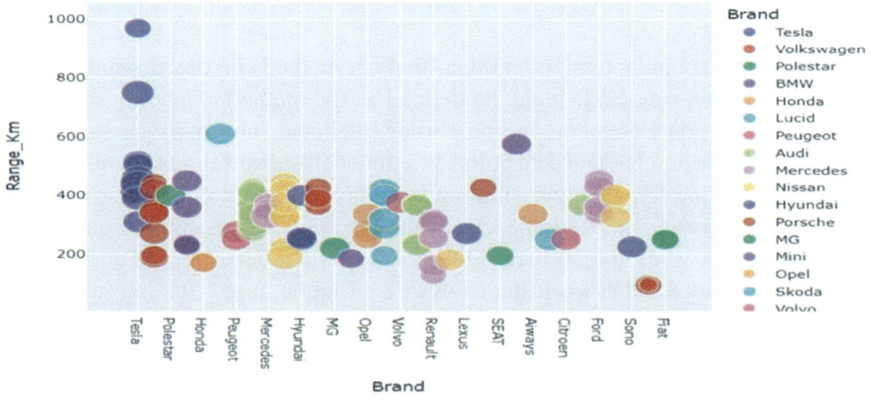

Fig. 9 Scatter plot of *Range_Km* and brand

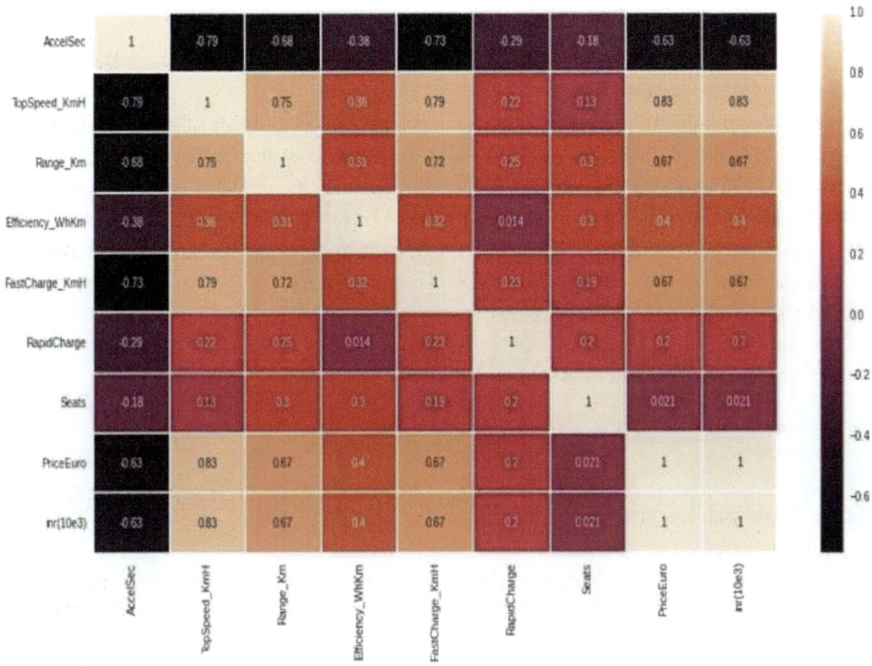

Fig. 10 Heatmap to show the correlation of the data

A table that shows the linking is called a correlation matrix. It works well with variables that show a straight-line link to one another coefficients pertaining to several variables. The matrix uses the heatmap in the graphic below to show the correlation between all possible pairings of values. A strong relationship is typically defined as two variables with a correlation coefficient value greater than 0.7.

5 Result Analysis

Battery-powered vehicles alleviate a lot of the negative externalities caused by traditional internal combustion engines. There are no harmful pollutants released into the atmosphere by electric vehicles. Electric vehicles can minimize pollution in the places they are driven in, even in situations where the energy source that was originally used to create the power was not fully environmentally friendly. This is one of their key advantages. More than 80 percent of the concentration of toxins in certain cities comes from automotive emissions. Figure 11 shows the residual histogram plot.

Table 1 shows the model appraisal on the basis of MSE, MAE, RMSE score. LSTM1 has one dense layer, 50 epochs, and 32 batch sizes; LSTM2 has two dense

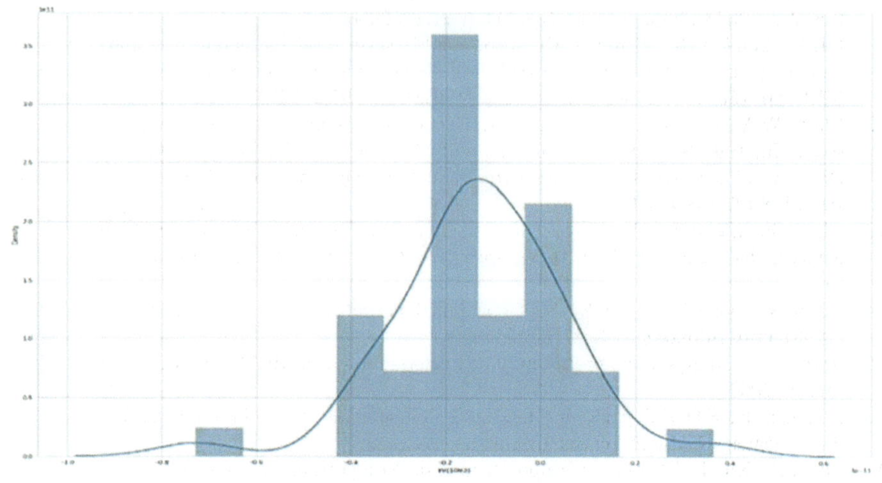

Fig. 11 Residual histogram plot

layers, 100 epochs, and 64 batch sizes; and LSTM3 has three dense layers, 128 batch sizes, and 200 epochs. These are the three distinct architectures of the LSTM model that we used in this study. In this study, we estimated three separate parameters: MAE, MSE, and RMSE. Based on the results shown in Table 1, The Mean Absolute Error (MAE), Root Mean Squared Error (RMSE), and Mean Squared Error (MSE) of each model were compared. The Bi-LSTM model yields low scores compared to other models, with an MAE of 0.417, an RMSE of 1.254, and an MSE of 1.573. Figure 12 shows the Model performance analysis through bar diagram.

Table 1 Model assessment on the basis of MSE, MAE, RMSE

Model name	MSE	MAE	RMSE
LSTM	4.854	1.667	2.203
LSTM 1	4.469	1.417	2.114
LSTM 2	5.153	2.316	2.270
GRU	5.021	2.134	2.241
Bi-LSTM	1.573	0.417	1.254

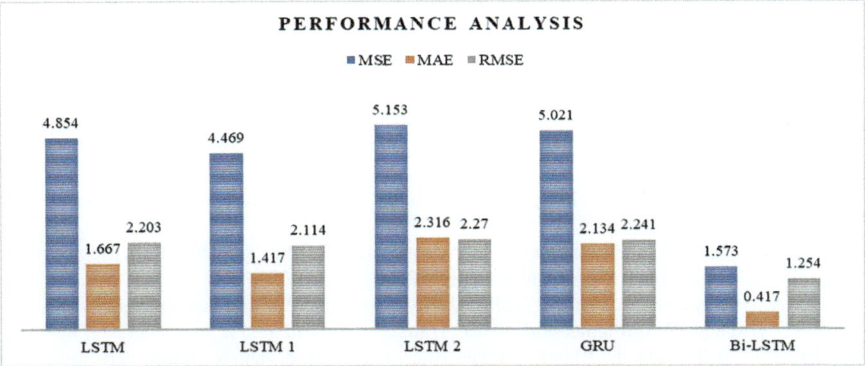

Fig. 12 Model performance analysis

6 Conclusion

In the near future, the administration and operation of the electric power infrastructure will face significant challenges due to the growing usage of EVs. In order to optimize the hidden benefits of electric vehicles and mitigate any possible drawbacks, a new structure for the market and technology would need to be created. By gradually utilizing electric vehicles, the general public should be introduced to the idea of electric mobility. With the support of this initial stage, grid machinists will be able to enhance the planning, management, and supervision of the electrical system while also gaining the opportunity to gain knowledge from their mistakes.

We employ three distinct deep learning model types: LSTM, GRU, Bi-LSTM, and LSTM model adjustments. We used the Kaggle competition's electric car population dataset for this study. The Bi- LSTM has an MAE score of 0.417, an MSE score of 1.573, and an RMSE value of 1.254. In comparison to other models, the accuracy score of the Bi-LSTM model is higher. Not only the electric vehicle market expanding, which is greatly valued, but it is also necessary considering the mounting amounts of carbon emissions in the atmosphere. The primary barrier to the general adoption of electric vehicles is their cost, as gasoline-powered vehicles are more accessible, more functional, and less expensive. Since the global population greatly affects the realization and success of our business, we want to use environmental awareness campaigns and mass marketing to encourage and empower people to drive electric cars. If every motorist converted to an electric vehicle, the surrounding environment might be completely free of dangerous toxins. Considering all the harmful consequences of pollution, driving an electric car every day instead of only occasionally could have a substantial positive impact on your health.

References

1. Tripathy, N., Hota, S., Satapathy, P., Nayak S.K., Prusty, S.: An empirical analysis of electric vehicle in urban transportation market using deep-learning techniques. In: 2023 IEEE 3rd International Conference on Sustainable Energy and Future Electric Transportation (SEFET). IEEE, pp. 1–5 (2023)
2. Shimizu, H.: The new possibility of electric vehicles. IATSS Res. **17**(2), 24–33 (1993)
3. Saputra, Y.M., Nguyen, N.D., Hoang, D.T., Vu, X.T.X., Dutkiewicz, E., Chatzinotas, S.: Federated learning meets contract theory: economic-efficiency framework for electric vehicle networks. IEEE Trans. Mob. Comput. **21**(8), 2803–2817 (2022)
4. Liu, N., Hu, X., Ma, L., Yu, X.: Vulnerability assessment for coupled network consisting of power grid and EV traffic network. IEEE Trans. Smart Grid **13**(1), 589–598 (2022)
5. Shimizu, H., Harada, J., Bland, C.: The role of optimized vehicle design and power semiconductor devices to improve the performance of an electric vehicle. In: Proceedings of International Symposium on Power Semiconductor Devices and IC's, pp. 8–12 (1995)
6. Tripathy, N., Hota, S., Mishra, D.: Performance analysis of bitcoin forecasting using deep learning techniques. Indones. J. Electr. Eng. Comput. Sci. **31**(3), 1515–1522 (2023)
7. Zhang, L., Cheng, L., Alsokhiry, F., Mohamed, M.A.: A novel stochastic blockchain-based energy management in smart cities using V2S and V2G. IEEE Trans. Intell. Transp. Syst. **24**(1), 915–922 (2023)
8. Bollerslev, J., Andersen, P.B., Jensen, T.V., Marinelli, M., Thingvad, A., Calearo, L., Weckesser, T.: Coincidence factors for domestic EV charging from driving and plug-in behavior. IEEE Trans. Transp. Electrification **8**(1), 808–819 (2022)
9. Tajjour, S., Chandel, S.S.: A comprehensive review on sustainable energy management systems for optimal operation of future-generation of solar microgrids. Sustain. Energy Technol. Assess. **1**(58), 103377 (2023)
10. Cao, Y., Wang, H., Li, D., Zhang, G.: Smart online charging algorithm for electric vehicles via customized actor critic learning. IEEE Internet Things J. **9**(1), 684–694 (2022)
11. Tripathy, N., Hota, S., Prusty, S., Nayak, S.K.: Performance analysis of deep learning techniques for time series forecasting. In 2023 International Conference in Advances in Power, Signal, and Information Technology (APSIT). IEEE, pp. 639–644 (2023)
12. ElGhanam, E., Sharf, H., Hassan, M.S., Osman, A.: On the coordination of the charging demand of roadway powered electric vehicles. In: 2022 IEEE 2nd International Conference on Sustainable Energy and Future Electric Transportation (SeFeT), pp. 1–6 (2022)
13. Tveit, M.H., Sevdari, K., Marinelli, M., Calearo, L.: Behind-the-meter residential electric vehicle smart charging strategies: Danish cases. In: 2022 International Conference on Renewable Energies and Smart Technologies (REST), vol. 1, pp. 1–5 (2022)
14. Tripathy, N., Hota, S., Mishra, D., Satapathy, P., Nayak, S.K.: Empirical forecasting analysis of bitcoin prices: a comparison of machine learning, deep learning, and ensemble learning models. Int. J. Electr. Comput. Eng. Syst. **15**(1), 21–9 (2024)
15. Dong, M., Mazin, H.E., Xu, W.: Data segmentation algorithms for a time-domain harmonic source modeling method. In: 2009 IEEE electrical power and energy conference (EPEC), pp. 1–6 (2009)
16. Li, F., Yang, H., Gao, X., Han, H.: Towards IoT-based sustainable digital communities. Intell. Converg. Netw. **3**(2), 190–203 (2022)
17. Tripathy, N., Balabantaray, S.K., Parida, S., Nayak, S.K.: Cryptocurrency fraud detection through classification techniques. Int. J. Electr. Comput. Eng. (IJECE) **14**(3), 2918–26 (2024)
18. Usman, O.: Renewable energy and CO_2 emissions in G7 countries: does the level of expenditure on green energy technologies matter? Environ. Sci. Pollut. Res. **30**(10), 26050–62 (2023)
19. Kim, B., Paik, M., Kim, Y., Ko, H., Pack, S.: Distributed electric vehicle charging mechanism: a game-theoretical approach. IEEE Trans. Veh. Technol. **71**(8), 8309–8317 (2022)
20. Kumar, B.A., Kennedy, R.: High-density electric vehicle charging using wide bandgap semiconductors. SoutheastCon **2022**, 491–495 (2022)

21. Krevor, S., De Coninck, H., Gasda, S.E., Ghaleigh, N.S., de Gooyert, V., Hajibeygi, H., Juanes, R., Neufeld, J., Roberts, J.J., Swennenhuis, F.: Subsurface carbon dioxide and hydrogen storage for a sustainable energy future. Nat. Rev. Earth Environ. **4**(2), 102–18 (2023)
22. Nižetić, S., Arıcı, M., Hoang, A.T.: Smart and sustainable technologies in energy transition. J. Clean. Prod. **20**(389), 135944 (2023)
23. Rane, N.: Contribution of ChatGPT and other generative artificial intelligence (AI) in renewable and sustainable energy. SSRN 4597674 (2023)
24. Tripathy, N., Nayak, S.K., Prusty, S.: A comparative analysis of Silverkite and inter-dependent deep learning models for bitcoin price prediction. Front. Blockchain **28**(7), 1346410 (2024)
25. Kumar, C.M., Singh, S., Gupta, M.K., Nimdeo, Y.M., Raushan, R., Deorankar, A.V., Kumar, T.A., Rout, P.K., Chanotiya, C.S., Pakhale, V.D., Nannaware, A.D.: Solar energy: a promising renewable source for meeting energy demand in Indian agriculture applications. Sustain. Energy Technol. Assess. **1**(55), 102905 (2023)

Energy Management for Household Integrated with Renewable and Electric Vehicle

T. J. Deepika, L. Avinash, P. Sai Teja, and A. Kishor Reddy

Abstract In the realm of sustainable energy solutions, the integration of renewable energy and electric vehicle (EV) storage in household energy management presents a significant advancement. This paper introduces a novel energy management system that synergizes solar power, grid energy, and an EV battery for optimized energy utilization. We analyze three operational scenarios: a base case relying on grid power and EV battery, a case incorporating solar power without energy sell-back, and an advanced scenario with renewable integration and energy sell-back capability. The results demonstrate enhanced energy efficiency and cost savings, showcasing the system's potential in transforming residential energy management towards greater sustainability.

Keywords Renewable energy integration · Electric vehicle storage · Household energy management · Solar power · Energy efficiency · Sustainable solutions

1 Introduction

In recent years, the paradigm of energy consumption and management in households has undergone a significant transformation. This shift is primarily driven by two factors: the escalating global demand for energy and the urgent need to address environmental concerns associated with traditional energy sources. As a result, there has been a growing emphasis on incorporating renewable energy sources into the domestic energy matrix. Concurrently, the advent of electric vehicles (EVs) has introduced a novel dimension to household energy systems, both as consumers and potential storage units for electrical energy.

This paper delves into the realm of household energy management, particularly focusing on the integration of renewable energy sources and the strategic utilization of

T. J. Deepika (✉) · L. Avinash · P. Sai Teja · A. Kishor Reddy
EEE Department, St. Peter's Engineering College, Hyderabad, India
e-mail: deepikatj@gmail.com

B. K. Mishra and A. Rocha (eds.), *Future of AI, IoT, and Sustainability*, Information Systems Engineering and Management 23,
https://doi.org/10.1007/978-3-031-76286-4_16

electric vehicle battery storage. The traditional household energy systems, predominantly reliant on grid power, are increasingly proving to be inadequate in meeting the contemporary challenges of sustainability and energy efficiency. Moreover, the dynamic nature of household energy demands, coupled with the variable supply from renewable sources, necessitates a more sophisticated and adaptable approach to energy management [1].

The integration of renewable energy sources, such as solar and wind [2], into household energy systems offers a promising avenue to enhance sustainability and reduce the carbon footprint. However, this integration is not without challenges. The intermittent nature of these sources and their dependency on environmental conditions necessitate a robust mechanism to ensure a consistent and reliable energy supply. Herein lies the potential of electric vehicles, which can serve a dual role - as a consumer of household energy and as a mobile storage unit that can feed energy back into the household system [3].

The proposed energy management system in this paper aims to create a harmonious balance between these various elements. By leveraging advanced control algorithms and smart energy management strategies, the system is designed to optimize the use of renewable energy, minimize reliance on the grid, and reduce overall energy costs. This approach not only aligns with the global drive towards sustainable energy practices but also enhances the efficiency and flexibility of household energy systems.

In summary, this paper presents an innovative approach to household energy management, addressing the contemporary challenges and leveraging the opportunities presented by renewable energy sources and electric vehicles. The ensuing sections will detail the methodology, implementation, and results of this approach, providing insights into its efficacy and potential implications for future energy management practices.

2 Literature Survey

In The advancement of household energy management systems integrating renewable energy and electric vehicles (EVs) represents a pivotal stride towards sustainable living. This survey encapsulates significant research contributions in this domain, elucidating the progression from foundational concepts to contemporary innovations.

2.1 Evolution of Household Energy Management

Elastic Energy Management Algorithms: Powroźnik [4] introduced the concept of an elastic energy management (EEM) algorithm. This algorithm, comprising EEM1 for periods of higher energy prices and EEM2 for peak demand times, leverages the Greedy Randomized Adaptive Search Procedure (GRASP) for optimizing appliance power consumption. This research laid the groundwork for intelligent energy management in households [5], particularly in the integration of renewable energy sources (RES).

Integration of Renewable Energy Sources: Denholm et al., in their study [6], further delved into the necessity of energy storage to manage the intermittency of renewable sources like solar and wind power. This research highlighted the importance of integrating storage solutions to enhance the efficiency and reliability of household renewable energy systems.

2.2 *Incorporating Electric Vehicles in Energy Management*

Vehicle-to-Grid Systems: The potential of EVs as energy storage was expounded by Kempton et al. [7]. Their research presented EVs not only as consumers of energy but also as a crucial component in household energy systems, capable of storing and supplying energy back to the grid.

Optimizing Hybrid Power Systems: Aditi [8] explored the stability of hybrid power systems integrating wind, solar, and hydro energy sources. The study emphasized the role of advanced controllers like fuzzy logic and artificial neural networks in stabilizing and optimizing these systems.

2.3 *Smart Energy Management and Multi-Criteria Design*

Smart Grid Technologies: Fang et al. [9] illustrated the evolution of smart grid technologies, including smart meters and IoT devices. These advancements are instrumental in the real-time management and optimization of energy flows in households with integrated renewable sources and EVs.

Multi-Criteria Hybrid Systems: Naghizadeh [10] proposed a multi-criteria approach for designing hybrid renewable energy systems (HRES). This approach, considering factors like cost, power supply probability, and energy loss, underscores the complexity and necessity of versatile designs in modern energy systems.

2.4 *Challenges and Future Directions*

Identifying Gaps and Proposing Solutions: Despite these advancements, several gaps remain in household energy management. The lack of adequate storage solutions, environmental concerns, and the complexity of certain control algorithms are notable challenges. Moreover, the integration of EVs in these systems, as proposed in our study, offers a novel solution to these challenges, promising enhanced efficiency and sustainability.

This literature survey reveals a trajectory of increasingly sophisticated approaches to household energy management. From the initial focus on integrating renewable sources to the incorporation of EVs and the development of smart, multi-criteria systems, the research landscape underscores a commitment to sustainable and efficient energy solutions. Our proposed paper builds upon these foundations, introducing innovative methodologies to optimize energy management in households.

3 Proposed Household Energy Management

This section outlines our proposed energy management system, designed to integrate renewable energy sources and electric vehicle (EV) storage capabilities into household energy systems. The proposal aims to fill existing gaps in energy management, optimize resource allocation, and enhance overall system efficiency.

3.1 Hybrid System—Block Diagram

The proposed hybrid system constitutes a multifaceted approach to household energy management, comprising:

- Solar Panels and Wind Turbines: These renewable sources provide a significant portion of the household's energy needs, reducing reliance on the grid and lowering carbon emissions.
- Electric Vehicle (EV) Battery: Serving as a dynamic energy storage unit, the EV battery can store excess energy generated from renewable sources or from the grid during off-peak hours.
- Grid Connection: Acts as a supplementary energy source and a backup in case of insufficient renewable energy generation.
- Energy Management Controller (EMC): A central unit equipped with intelligent algorithms to manage energy flow between the grid, renewable sources, EV battery, and household appliances.

A schematic block diagram (Fig. 1) illustrates the interconnectedness of these components and the energy flow pathways.

4 Overview of Proposed Energy Management

The proposed system addresses several limitations in current household energy management frameworks:

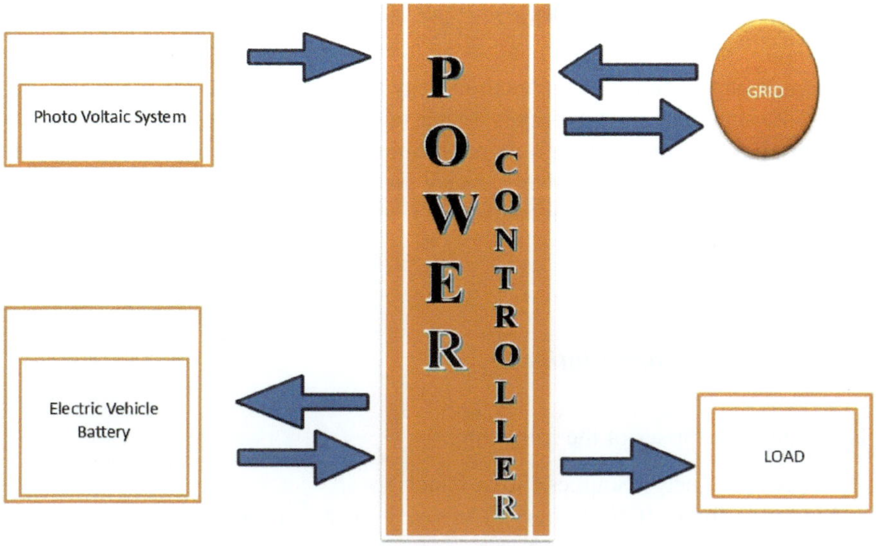

Fig. 1 Block diagram of hybrid system

- ***Versatile Energy Sourcing***: Unlike conventional systems, our approach combines solar and wind energy with grid power and EV battery storage, ensuring a more reliable and sustainable energy supply.
- ***Dynamic Energy Allocation***: The system's EMC utilizes smart algorithms to dynamically allocate energy based on real-time consumption patterns, renewable energy availability, and grid tariffs.
- ***Optimized EV Battery Use***: By integrating the EV battery as both a consumer and a storage unit, the system enhances the overall efficiency and offers a novel solution to energy storage challenges.

4.1 Objectives of the Proposal

The objectives of the proposed system are multi-fold:

- ***Maximizing Renewable Energy Use***: To substantially increase the household's reliance on renewable energy, thereby advancing sustainability goals.
- ***Reducing Grid Dependence***: Minimizing the need for grid power, especially during peak demand periods, to lower energy costs and enhance self-sufficiency.
- ***Cost Efficiency***: Optimizing energy consumption patterns and storage to reduce overall electricity costs.
- ***System Reliability and Adaptability***: Ensuring a stable and flexible energy supply that can adapt to varying household needs and environmental conditions.

5 Proposed Methodology

In this section, we delineate the methodology underpinning our proposed household energy management system. This system is tailored to optimize energy consumption in a household setting that integrates renewable energy sources and an electric vehicle (EV) battery. To comprehensively assess the system's efficacy, we consider three distinct case scenarios, each reflecting different operational conditions and energy source combinations.

5.1 Problem Formulation

The system comprises of the following:

- Renewable Energy Sources (Solar Panels).
- Electric Vehicle (EV) Battery: Utilized both as an energy storage unit and a supplementary energy source.
- Grid Connection: Acts as a primary or backup energy source.
- Energy Management Controller: A smart system to manage and optimize energy distribution and consumption.

The primary aim is to maximize the utilization of renewable energy, minimize grid dependence, and efficiently utilize the EV battery for energy storage and supply. This objective necessitates the dynamic allocation of energy based on the state of charge (SoC) of the EV battery, the household energy demand, and the availability of solar power.

5.2 Cases for Analysis

We evaluate the proposed system under three different scenarios, each with its own set of conditions and challenges:

- ***Case 1: Base Case (No PV and No Selling Back to Grid)***: This case serves as a control scenario, allowing us to benchmark the system's performance without renewable energy integration. It highlights the baseline efficiency and cost implications in a traditional energy setup.
- ***Case 2: With Renewable (No Selling Back to Grid)***: This scenario assesses the impact of adding renewable energy sources while still maintaining a level of reliance on the grid. It demonstrates the benefits and challenges of integrating solar power into the household energy mix.
- ***Case 3: With Renewable and Selling Back to Grid***: This case explores the dynamics of a more complex energy management system that not only incorporates renewable energy but also engages in energy trading with the grid. It is

crucial for understanding the potential for cost savings and energy efficiency in a more interconnected and smart grid environment.

Each of these cases is crucial for a holistic evaluation of the proposed energy management system. They allow us to analyze the system's adaptability, efficiency, and economic viability under varying operational conditions and energy availability scenarios. The outcomes from these case studies will offer insights into optimizing household energy systems for future sustainable and smart homes.

6 Control Algorithm for Energy Management

In this section, we present the step-wise algorithms for the three distinct case scenarios outlined in the methodology. These algorithms are designed to optimize the energy management in a household integrated with renewable energy sources and an electric vehicle (EV) battery.

Case 1: Base Case (no PV and no Selling Back to Grid)

Description: This scenario represents a basic setup where the household relies solely on the grid and the EV battery. There is no contribution from photovoltaic (PV) solar panels, and no energy is sold back to the grid.

Algorithm: Following is the step-wise algorithm for the base case:

Step 1: Initialize Parameters:

- Set ev_soc (state of charge of EV battery).
- Define load (household energy consumption).
- Specify price (cost of energy from the grid).
- Set maxsoc (maximum SoC) and minsoc (minimum SoC) for the EV battery.

Step 2: Calculate EV Load:

- If ev_soc < 40, calculate additional load to charge EV to maxsoc.
- Determine totalload as the sum of household and EV load.

Step 3: Calculate Energy from Grid:

- Compute energy required from the grid as grid = totalload.
- Calculate cost as cost = grid * price.

Step 4: Output Results:

- Display energy purchased from grid and total cost.

Case 2: With Renewable (no Selling Back to Grid)

Description: In this case, the household energy system includes PV solar panels, alongside grid and EV battery sources. However, excess energy generated from the PV system is not sold back to the grid but can be stored in the EV battery.

Algorithm: Following is the step-wise algorithm for case2 with renewables:

Step 1: Initialize Parameters:

- Set ev_soc (state of charge of EV battery).
- Define load (household energy consumption).
- Specify price (cost of energy from the grid).
- Set maxsoc (maximum SoC) and minsoc (minimum SoC) for the EV battery.
- Set pv (solar power generation) for the hour.

Step 2: Calculate Total Load and Generation:

- *If ev_soc < 40, calculate additional EV load.*
- *Determine totalload (household + EV load).*
- *Calculate totalgen as solar power generation (pv)*

Step 3: Calculate Energy from Grid:

- *If totalgen > totalload, no grid interaction is required.*
- *Otherwise, calculate energy required from the grid as grid = totalload − totalgen.*
- Calculate cost as cost = grid * price.

Step 4: Output Results:

- *Display energy purchased from grid, total cost and savings*

Case 3: With Renewable and Selling Back to Grid

Description: This advanced scenario integrates PV solar panels with the option to sell excess energy back to the grid. It represents a fully-fledged renewable energy system with a bidirectional interaction with the grid.

Algorithm: Following is the step-wise algorithm for case2 with renewables:

Step 1: Initialize Parameters:

- Set ev_soc (state of charge of EV battery).
- Define load (household energy consumption).
- Specify price (cost of energy from the grid).
- Set maxsoc (maximum SoC) and minsoc (minimum SoC) for the EV battery.
- Set pv (solar power generation) for the hour
- Specify sprice (price for selling energy back to the grid).

Step 2: Calculate Total Load and Generation:

- *If ev_soc < 40, calculate*
- *evload = maxsoc − ev_soc*
- *totalload = load + evload*
- *totalgen = pv.*
- *Else calculate*
- *evgen = ev_soc − minsoc*
- *totalgen = pv + evgen*
- *totalload = load*

Step 3: Calculate Energy from Grid:

- *If totalgen > totalload, calculate energy sold to grid as grids = totalgen-totalload.*
- *Otherwise, calculate energy required from the grid as grid = totalload − totalgen.*
- Calculate cost of selling = grids * sprice.
- Calculate cost of purchasing = grid * price
- Compute net cost and profit.

Step 4: Output Results:

- *Display energy purchased from grid, total cost and savings*

7 Results and Discussion

This section presents the results obtained from the implementation of the proposed energy management system under three different cases. Each case is analyzed based on the system data, the results obtained, and a comparative analysis is provided.

7.1 System Data

The following parameters are considered as input parameters for the system:

- EV State of Charge (Ev_soc)
- Household Load (Load)
- Maximum and Minimum SoC for EV Battery (Maxsoc and Minsoc)
- Electricity Buying and Selling Prices.

The input data for each of the 3 cases is summarized in Table 1.

Table 1 Input data—case-wise

Case	EV state of charge (Ev_soc)	Household load (Load)	Maximum SoC (Maxsoc)	Minimum SoC (Minsoc)	Buying price (Rs/kWh)	Selling price (Rs/kWh)
1	35 kW	100 kW	95 kW	25 kW	5	–
2	35 kW	100 kW	95 kW	25 kW	5	–
3	85 kW	100 kW	95 kW	25 kW	5	8

7.2 Results Obtained—Case-Wise

- **Case 1:** Base Case (No PV and No Selling Back to Grid) was analysed with an input of Ev_soc = 35 kW, Load = 100 kW, Maxsoc = 95 kW, Minsoc = 25 kW, Buying Price = 5 Rs/kWh.

  The obtained Energy purchased from grid = 160kW, Cost = 800 Rs and Profit = 0 Rs.

- **Case 2:** With Renewable (No Selling Back to Grid) is analysed for the input of Ev_soc = 35 kW, Load = 100 kW, PV = 60 kW, Maxsoc = 95 kW, Minsoc = 25 kW, Price = 5 Rs/kWh.

  An output of Energy purchased from grid = 100kW, Cost = 500 Rs, Profit = 300 Rs is obtained.

- **Case 3:** With Renewable and Selling Back to Grid was analysed for an input of Ev_soc = 85 kW, Load = 100 kW, PV = 60 kW, Maxsoc = 95 kW, Minsoc = 25 kW, Buying Price = 5 Rs/kWh, Selling Price = 8 Rs/kW.

  The output obtained was Energy purchased from grid = 0 kW, Cost = 0 Rs, Energy sold to grid = 20, Income from selling = 160 Rs and Profit = 960 Rs.

7.3 Results Analysis—Case-Wise

The results from the three cases indicate a significant variation in energy consumption patterns, costs, and profit generation.

- ***Case 1*** represents a traditional energy setup with complete reliance on the grid. The absence of renewable energy sources and the inability to sell back to the grid leads to higher energy costs and no profit generation.
- ***Case 2*** shows the impact of integrating renewable energy sources (PV) into the system. The reliance on the grid is reduced, leading to a decrease in energy costs and a moderate profit. This case highlights the benefits of adding renewable energy sources to household energy systems.

Table 2 Comparative analysis of case-wise results

Case	Details	Energy purchased from grid (kW)	Cost of purchasing (Rs)	Energy sold to grid (kW)	Income from selling (Rs)	Profit (Rs)
1	No PV and No selling back to grid	160	800	–	–	0
2	PV + No selling back to grid	100	500	–	–	300
3	PV + Selling back to grid	0	0	20	160	960

- ***Case 3*** demonstrates the most efficient and profitable scenario. The integration of renewable sources, combined with the ability to sell excess energy back to the grid, results in zero energy purchases and a substantial profit. This case underscores the potential of a fully integrated renewable energy system with bidirectional grid interaction.

7.4 Results Analysis—Comparative

A comparison of the results obtained in the 3 cases is given in Table 2.

From the table, the transition from Case 1 to Case 3 illustrates a clear trajectory towards increased sustainability and economic benefits.

The inclusion of renewable energy sources significantly reduces the dependence on grid power, thereby lowering the overall energy costs.

The ability to sell back to the grid in Case 3 not only eliminates the cost of purchasing energy but also generates substantial profit, showcasing the economic advantage of a well-integrated energy management system.

These results affirm the effectiveness of the proposed energy management system in enhancing the efficiency, sustainability, and economic viability of household energy systems. The integration of renewable energy sources and the strategic use of EV batteries for energy storage and supply are pivotal in achieving these outcomes.

8 Conclusion

The research presented in this paper underscores the significant potential of integrating renewable energy sources and electric vehicles (EVs) into household energy management systems. By examining different case scenarios, we have demonstrated how multimodal operation can lead to enhanced energy efficiency, cost savings, and a reduced environmental footprint.

The integration of renewable energy sources, notably solar power, in conjunction with the strategic use of EV batteries for energy storage and supply, presents a sustainable and efficient approach to energy management. The results from the three case studies clearly indicate that the shift from traditional grid dependence to a more diversified energy mix can significantly lower energy costs and even generate profit.

Our proposed system, leveraging smart technologies and advanced control algorithms, effectively balances energy supply and demand. This balance is crucial for minimizing reliance on fossil fuels and reducing greenhouse gas emissions, thereby contributing to environmental sustainability. The ability to sell excess energy back to the grid, as observed in the most advanced case scenario, further enhances the economic viability and appeal of such integrated systems.

In conclusion, the incorporation of renewable energy and EVs into household energy systems represents a forward-thinking approach to energy management. It aligns with global sustainability goals while offering practical benefits to consumers. As technology advances, these systems will become increasingly sophisticated, making them more accessible and beneficial for a broader range of users, and contributing significantly to the transition towards a more sustainable and efficient energy future.

References

1. Tezer, T., Yaman, R., Yaman, G.: Evaluation of approaches used for optimization of stand-alone hybrid renewable energy systems. Renew. Sustain. Energy Rev. **73**, 840–853 (2017)
2. Eke, R., Kara, O., Ulgen, K.: Optimization of a wind/PV hybrid power generation system. Int. J. Green Energy **2**(1), 57–63 (2005)
3. Elliston, B., MacGill, I., Diesendorf, M.: Least cost 100% renewable electricity scenario s in the Australian national electricity market. Energy Policy **5**(9), 270–282 (2013)
4. Powroźnik, P.: Household energy management. Appl. Sci. (2001)
5. Palensky, P., Dietrich, D.: Demand side management: demand response, intelligent energy systems, and smart loads. IEEE Trans. Industr. Inf. **7**(3), 381–388 (2011)
6. Denholm, P., et al.: The role of energy storage with renewable electricity generation. NREL/TP-6A2-47187 (2010)
7. Kempton, W., Tomić, J.: Vehicle-to-grid power fundamentals: calculating capacity and net revenue. J. Power. Sources **144**(1), 268–279 (2005)
8. Aditi: A review paper on hybrid power system with different controllers and tracking methods. Int. J. Eng. Res. Technol. (2016)
9. Fang, X., Misra, S., Xue, G., Yang, D.: Smart grid—the new and improved power grid: a Survey. IEEE Commun. Surv. Tutor. **14**(4), 944–980 (2012)
10. Naghizadeh, R.A.: Multi-criteria optimal sizing of hybrid renewable energy systems including wind, photovoltaic, battery and hydrogen storage with ε-constraint method. IET J. (2017)

Implementation of Skin Disease Identification and Analysis Through Deep Learning

Debasree Mitra, Antara Das, Aradhya Rai, Ritoja Banerjee, and Kanhaiya Kumar

Abstract The "Skin Care Analysis System using Deep Learning" is an innovative project aimed at revolutionizing personalized skincare diagnostics and recommendations. This web-based system employs a robust deep learning model, constructed with Keras and Convolutional Neural Networks (CNN), capable of accurately classifying 9 different skin diseases. Users are engaged through an intuitive web interface featuring user authentication, a survey module, and image upload capabilities. The system analyzes uploaded facial images, providing real-time predictions for skin conditions such as acne, melanoma, Eczema, and more. Additionally, a novel bruise detection feature enhances the system's utility, enabling users to identify various types of skin discolorations. Benefits include early disease detection, personalized skincare recommendations, and user empowerment through insights into skin health. The system aims to reduce healthcare costs, promote skin health awareness, and provide a convenient, accessible, and data-driven platform for skincare analysis. Future enhancements, such as skin evolution analysis and foundation match recommendations, further position the project at the forefront of technology-driven skincare solutions. This project signifies a significant advancement in the intersection of technology and healthcare, offering users a comprehensive and user-friendly approach to personalized skincare.

Keywords Skin disease · Deep learning · CNN · Model evaluation · Accuracy

D. Mitra (✉)
Department of Computer Science and Engineering, Adamas University, Kolakata, India
e-mail: debasree.mitra2005@gmail.com

A. Das
Tata Consultancy Services, Kolkata, India

A. Rai
Apmosys, Navi Mumbai, India

R. Banerjee
Tata Consultancy Services, Bhubaneswar, India

K. Kumar
Indian Space Research Organisation, Bengaluru, India

B. K. Mishra and A. Rocha (eds.), *Future of AI, IoT, and Sustainability*, Information Systems Engineering and Management 23,
https://doi.org/10.1007/978-3-031-76286-4_17

1 Introduction

In recent years, strides in computer vision technology have dramatically impacted numerous sectors, including surveillance and healthcare. Among these advancements is the creation of an online face detection skin test—a groundbreaking tool that integrates facial recognition with dermatology. This project report delves into the origins, mechanics, importance, and prospective influence of this innovative online tool.

The online face detection skin test employs advanced machine learning algorithms, especially convolutional neural networks (CNNs), to detect and analyze facial skin conditions with precision in real-time. By utilizing a vast database of images featuring various skin types and conditions, the system becomes adept at identifying a wide array of dermatological issues, such as acne, eczema, psoriasis, melanoma, and other skin anomalies.

This system is designed with a user-friendly interface that can be accessed via web or mobile applications, enabling users to take high-resolution images of their facial skin using their device's camera. After the image is submitted, sophisticated algorithms analyze it, pinpointing specific areas of interest and identifying any visible irregularities or skin conditions.

The online face detection skin test stands out for its potential to make preliminary skin health assessments widely accessible. By offering an easy-to-use and accessible platform, individuals can perform initial self-assessments of their skin condition regardless of their location, without the immediate need for a dermatological consultation. This tool can serve as an early screening method, prompting users to seek professional medical advice when necessary.

Furthermore, the use of machine learning algorithms ensures that the system continually improves and becomes more accurate over time. With each new image and diagnosis, the system learns and adapts, improving its capability to precisely identify and classify a broader variety of skin conditions.

The implementation of an online skin disease detection application holds great promise for improving dermatological care, empowering individuals to take proactive steps toward their skin health. But, it has some limitations, ethical considerations, and the necessity of professional medical validation before making definitive diagnoses or treatment recommendations.

In summary, this project report explores the development and implications of the online face detection skin test, shedding light on its mechanics, significance, and the potential it holds to transform the field of dermatological care and self-assessment (Fig. 1).

a

	Deaths	Average Age at Death (years)	YPLL 75	
			Average	Total
All causes of death	2,596,993	73.2	7.9	20,391,598
Skin disease	22,953	68.2	10.8	248,656
Percentage	0.90%			1.20%

b

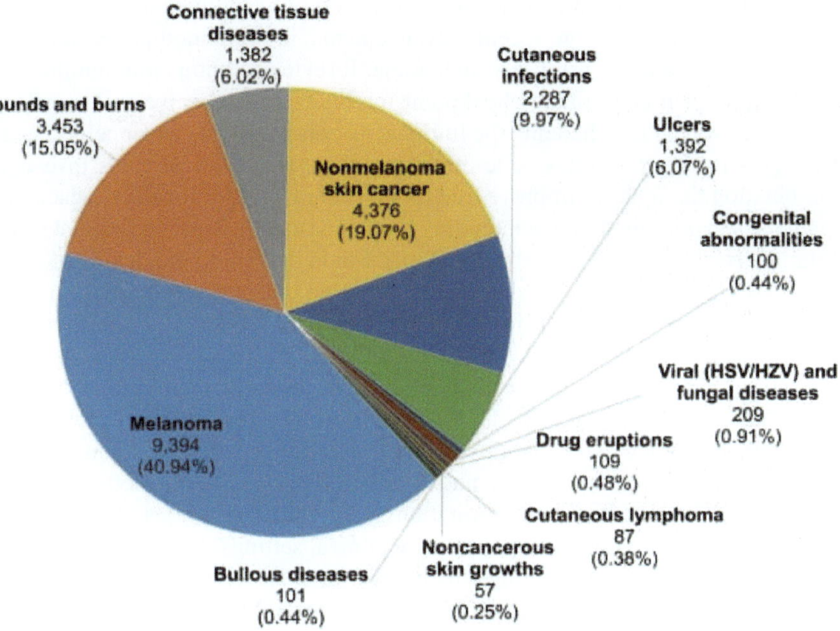

Fig. 1 A brief worldwide statistical data about different skin disease

2 Literature Survey

This paper explores the genetic factors associated with melanoma, focusing on genetic predispositions and susceptibility. It delves into the role of specific gene mutations and variations in increasing the risk of melanoma development. The study uses genomic analysis techniques to identify key genetic markers and pathways linked to melanoma. Furthermore, it highlights the significance of genetic counseling and testing for individuals who have a family history of melanoma to evalute their risk and take preventive measures. "Genetic Markers and Molecular Profiling in Skin Cancer Diagnostics": Focused on the molecular aspects, this paper could highlight specific genetic markers, mutations, or molecular profiling techniques associated with various skin cancers. It might discuss the diagnostic potential of these markers in early detection, prognosis, and personalized treatment strategies. "Comparative

Analysis of Biopsy Techniques for Skin Cancer Confirmation": Such a paper could compare different biopsy methods in confirming skin cancer diagnoses. It might assess the accuracy, reliability, and practicality of techniques such as punch biopsy, shave biopsy, or incisional biopsy, considering factors like lesion size, location, and histopathological examination. "Integration of Multimodal Approaches for Precise Skin Cancer Diagnosis": This research might focus on combining multiple diagnostic approaches, such as dermoscopy, imaging, molecular profiling, and clinical data, to improve overall accuracy in diagnosing skin cancer. It might discuss how a multimodal approach can enhance diagnostic certainty and patient care [1].

This paper focuses on recent advancements in immunotherapy for treating advanced skin cancers, primarily melanoma. It reviews various immunotherapeutic approaches, such as immune checkpoint inhibitors and adoptive cell transfer, and their effectiveness in enhancing the immune response against cancer cells. The study highlights the success rates, side effects, and ongoing clinical trials investigating combination therapies to improve treatment outcomes. Additionally, it discusses the challenges in immunotherapy, including resistance mechanisms and patient selection criteria. Overview of Skin Tests: The paper would delve into the details of different skin tests used in diagnosing skin cancer. It might discuss how dermoscopy allows for magnified views of skin lesions, enabling dermatologists to examine them closely for signs of malignancy. Biopsy procedures, which involve taking tissue samples for laboratory analysis, would also be explained in terms of their role in confirming cancerous growth. Evaluation of Test Accuracy: The research paper would likely discuss the accuracy and reliability of these tests in detecting skin cancer. It might cover the sensitivity, specificity, and overall efficacy of each method, providing insights into their strengths and limitations. Clinical Applications: The paper might discuss how these skin tests are applied in clinical settings. It could detail protocols for using these tests, determining which tests are most suitable for specific types of skin lesions, and how healthcare professionals interpret the results to make accurate diagnoses [2].

This paper examines the role of environmental factors in increasing the risk of skin cancer. It discusses the impact of ultraviolet (UV) radiation from sunlight exposure as the primary environmental risk factor for skin cancer development, emphasizing its role in DNA damage and carcinogenesis. Additionally, the study explores other environmental factors such as chemical exposures, pollution, and lifestyle habits contributing to skin cancer incidence. It provides recommendations for preventive measures, including sun protection strategies and reducing exposure to environmental carcinogens. Skin cancer is a common type of cancer worldwide, which can be detected using various methods, including skin tests, are used for its diagnosis. A research paper focused on skin tests for skin cancer would likely cover different types of skin tests, their effectiveness, and their role in identifying and diagnosing skin cancer. The summary of such a research paper might include: Introduction: The paper would introduce skin cancer as a significant health concern and discuss the importance of early detection in improving outcomes. It would highlight the various types of skin tests available for diagnosing skin cancer, such as dermoscopy, biopsy, and imaging techniques like ultrasound or MRI. Challenges and Future Directions: The summary might touch upon challenges faced in using skin tests for skin cancer diagnosis, such as limitations in differentiating certain types of lesions or variability in interpretation among clinicians. It could also suggest potential advancements or future directions in skin test technology or integration of complementary diagnostic tools to enhance accuracy and efficiency. Conclusion: Summarizing the findings, the conclusion might reiterate the importance of skin tests in diagnosing skin cancer, emphasize their contributions to early detection, and highlight areas for further research or improvement [3].

This research paper focuses on targeted therapy for skin cancers other than melanoma, such as squamous cell carcinoma (SCC) and basal cell carcinoma (BCC). It reviews the molecular pathways involved in these skin cancers and discusses targeted therapies such as hedgehog pathway inhibitors for BCC and epidermal growth factor receptor or EGFR, inhibitors for SCC. The study evaluates the efficacy, safety, and limitations of these targeted therapies, emphasizing their role as alternative treatments in advanced or recurrent non-melanoma skin cancers. Diagnosis: This research paper examines the role of molecular and genetic markers in skin tests for skin cancer. It may discuss specific biomarkers, genetic mutations, or gene expression profiles associated with different types of skin cancer and their potential as diagnostic tools. Comparative Study of Biopsy Techniques in Skin Cancer Diagnosis: This paper compares different biopsy methods, such as punch biopsy, shave biopsy, and excisional biopsy, in terms of diagnostic accuracy, tissue adequacy for analysis, patient comfort, and scarring. It might provide insights into the optimal biopsy technique based on lesion characteristics and location. "Evaluation of Non-Invasive Imaging Techniques in Skin Cancer Diagnosis": This paper could delve into the assessment of non-invasive imaging methods like dermoscopy, reflectance confocal microscopy, or optical coherence tomography. It might discuss their efficacy, accuracy, and limitations in diagnosing different types of skin cancer lesions without the need for invasive procedures. "Role of Artificial Intelligence in Skin Lesion Classification and Diagnosis": This research paper may explore the application of artificial intelligence and machine learning algorithms in analyzing skin lesion images for cancer detection. It might evaluate the performance of Artificial Intelligence models in distinguishing between benign and malignant skin lesions and discuss their potential impact on clinical practice [4].

3 Background Study

The Skin Care Analysis System is a cutting-edge AI-powered web application which utilizes deep learning technologies, specifically Convolutional Neural Networks or CNNs, to analyze facial images for comprehensive skin evaluation. The system integrates a user survey to gather essential details about skin type, concerns (such as acne, pimples, wrinkles), and allergies. After completing the survey, users upload a

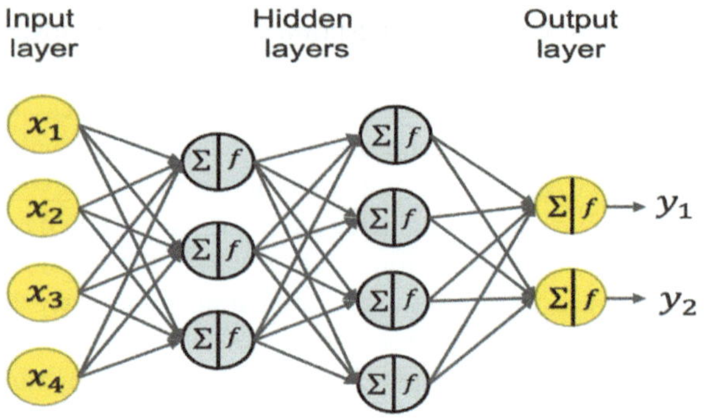

Fig. 2 Neural network

photo, which the model then thoroughly analyzes. Based on the survey responses and image assessment, the system provides personalized skincare recommendations.

3.1 Computer Vision

The Skin Care Analysis System is a cutting-edge AI-powered web application. The system integrates a user survey to gather essential details about skin type, concerns (such as acne, pimples, wrinkles), and allergies [5]. After completing the survey, users upload a photo, which the model then thoroughly analyzes. Based on the survey responses and image assessment, the system provides personalized skincare recommendations.

3.2 Deep Learning

Deep learning is a subset of machine learning that stands out for its exceptional ability to discovering complex patterns and connection between data. Our model is trained using Convolutional Neural Networks (CNNs), which are widely used for object recognition and categorization. CNNs are a subclass of neural networks that focuses in processing grid-like data, such as images [7] (Fig. 2).

3.3 Convolutional Neural Network (CNN)

We have used a CNN model in our project as CNNs are widely used for object recognition and categorization. CNNs are a type of neural networks which specialize in handling grid-structured data, like images. A digital image made up of pixels organized in a grid, where each pixel contains a value that represent brightness and color. Similar to the human brain's visual processing, CNN neurons respond to stimuli within their receptive fields and are interconnected to cover the entire visual field. CNNs first detect basic patterns such as lines and curves, and after that it recognizes more complex patterns such as faces and objects as the layers progress. A typical CNN model includes three layers:

I. convolutional layer,
II. pooling layer,
III. fully connected layer [6, 7].

The first layer, convolutional layer detects features in the input image using a set of learnable filters. The pooling layer is used to reduce the spatial dimensions of the feature maps while retaining important features. The fully connected layer maps high-level abstractions learned by preceding layers into the final output space (Figs. 3 and 4).

- **Convolution Layer**

Detects features in the input image using a set of learnable filters (Fig. 5).

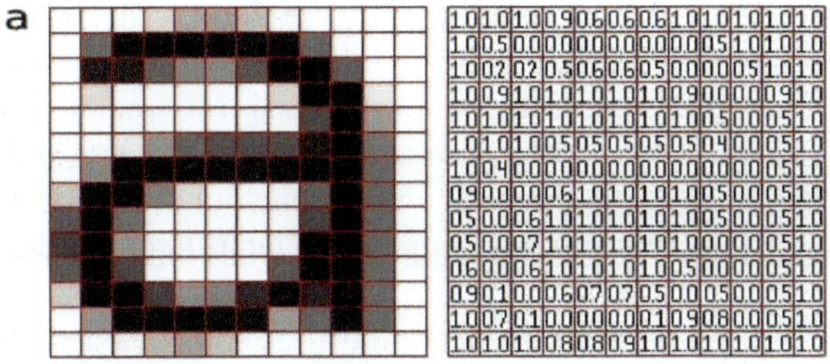

Fig. 3 Representation of image as a grid of pixels

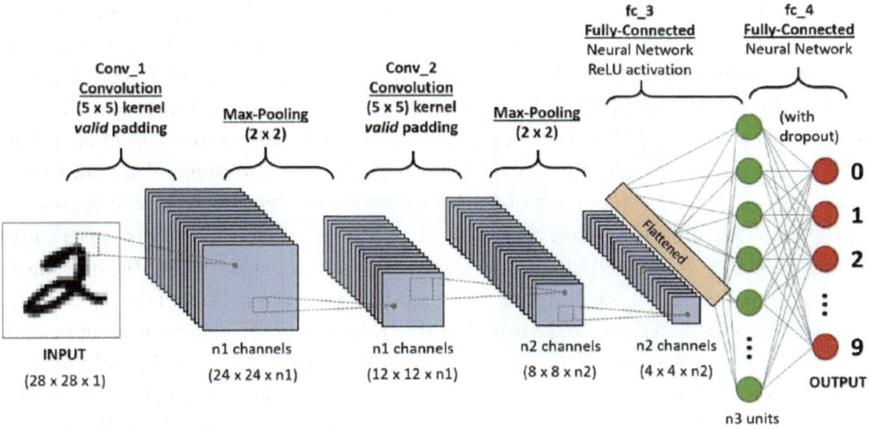

Fig. 4 Layers in CNN with example

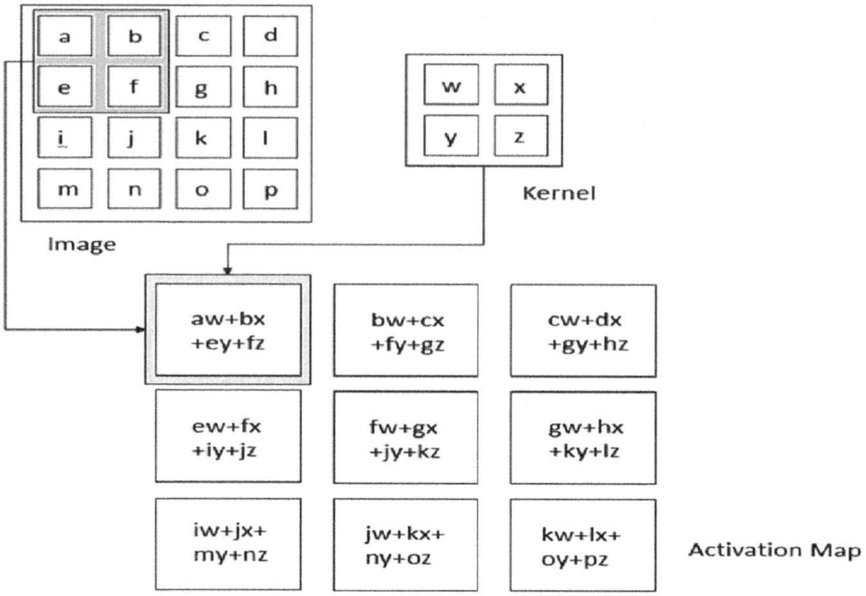

Fig. 5 Convolution operation

- **Pooling Layer**

Reduces the spatial dimensions of the feature maps while retaining important features (Fig. 6).

- **Fully Connected Layer**

A Fully Connected (FC) layer, or a dense layer, play a crucial role in cnn. In an FC layer, each neuron is connected to every neuron in both the preceding and succeeding

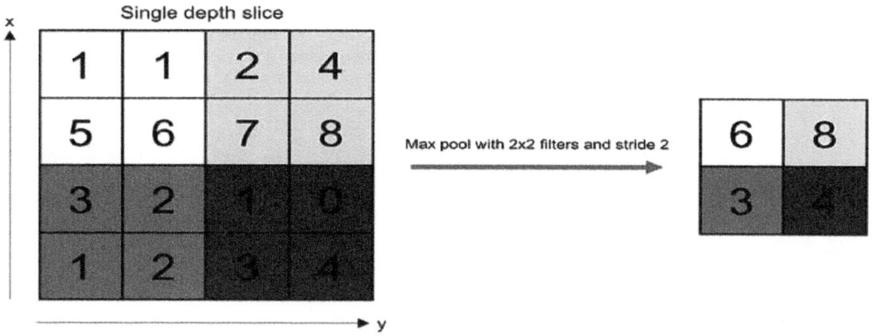

Fig. 6 Pooling operation

layers. This full connectivity allows the network to learn complex global patterns and relationships within the data.

The primary function of an FC layer is to map high-level abstractions learned by preceding layers into the final output space, such as class scores in classification tasks. This is achieved through a weighted sum of the inputs followed by the addition of a bias term. Mathematically, if x is input vector, W is the weight matrix, and b is the bias vector, the output y of the FC layer is:

$$y = Wx + b$$

The weights and biases are learnable parameters adjusted during training through backpropagation.

FC layers are crucial in the final stages of CNNs, transforming feature maps from convolutional and pooling layers into a format suitable for classification or regression. After extracting and condensing spatial features, the FC layer integrates these features to produce a coherent output, such as an image label.

While FC layers are powerful in representing complex functions, they come with a high number of parameters, increasing computational complexity and the risk of overfitting, especially with limited training data. Regularization techniques, such as dropout, are used to reduce overfitting by randomly deactivating a portion of neurons during training process, promoting the development of robust features.

In summary, fully connected layers are essential in neural networks, translating high-level features into meaningful outputs. Despite their computational demands, their ability to capture intricate relationships makes them a vital component in deep learning architectures.

In this layer neurons are full connected to all neurons in both the preceding and succeeding layers, like in a standard fully connected neural network (FCNN). This full connectivity allows it to be computed using matrix multiplication followed by adding a bias term. The representation between the input and output is mapped with assistance from the fully connected (FC) layer.

- **Non-Linearity Layers**

Non-linearity layers are often placed immediately after convolutional layers to introduce non-linearity into the activation maps. This is because convolution is a linear operation where as pictures are by nature non-linear. Typical non-linear functions consist of:

- **Sigmoid**: The sigmoid function, squashes a real-valued number into the range [0, 1]. Mainly used in binary classification tasks.

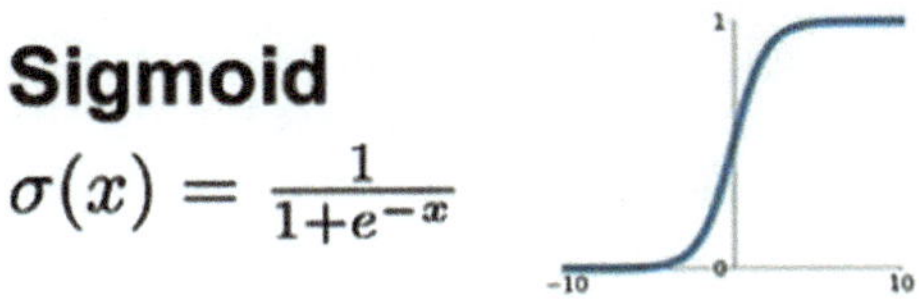

- **Tanh**: It outputs values in the range [−1, 1], with zero-centered.

- **ReLU (Rectified Linear Unit)**: The ReLU function, $f(\kappa) = \max(0,\kappa)$, sets all negative values to zero. It is more reliable than sigmoid and tanh, accelerating convergence by up to six times.

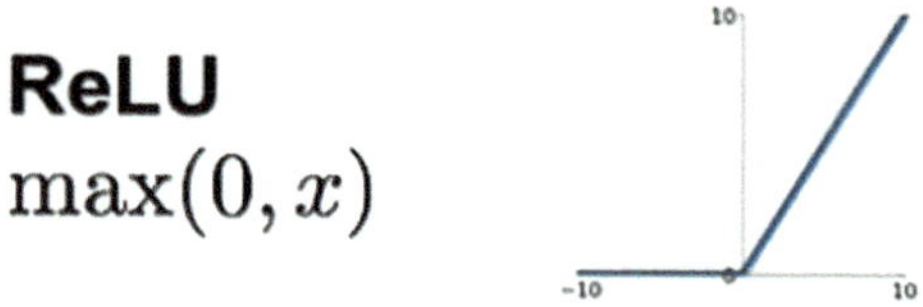

- **Softmax**: Softmax normalizes a network's output into a probability distribution across predicted output classes, often used as final activation function in a neural network. It is used for multi-class classification tasks.

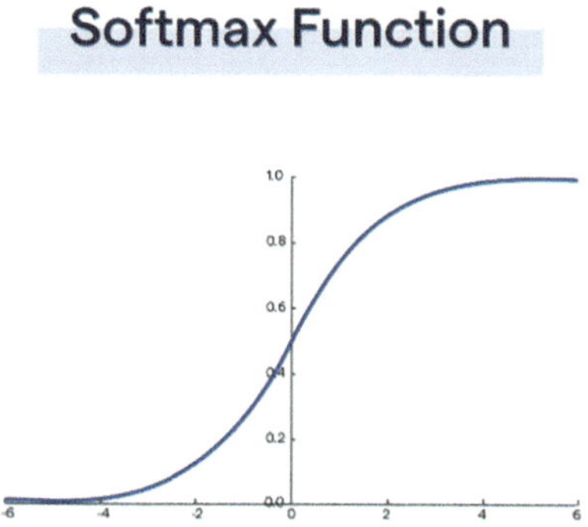

3.4 *About the Dataset*

The model is trained using a skin disease related dataset. The dataset compromises 9 different types of skin diseases. The dataset has been divided into two segments, i.e. one is training set and the other is test set. The images are in JPEG format, compromising 3 channels, i.e. RGB. The resolutions differ from one image to another image, and from category to category, but generally these are not of extremely high resolution.

The categories are,

1. Actinic keratosis,
2. Atopic Dermatitis,
3. Benign keratosis,
4. Dermatofibroma,
5. Melanocytic nevus,
6. Melanoma,
7. Squamous cell carcinoma,
8. Tinea Ringworm Candidiasis and
9. Vascular lesion.

After analyzing the face, the system will classify the image into the above classes.

3.5 NLP

Natural Language Processing is a subfield of artificial intelligence which emphasizes the interaction between computers and human language. It creates algorithms and models which can be used to interpret, and generate human language by a computer. Machine translation, sentiment analysis, speech recognition, and text summarization are some of the key uses of NLP. Advanced models like GPT-4 and BERT have significantly improved NLP by capturing the nuances and context of language. Overall, NLP is essential for making human–computer interactions more natural and efficient, enhancing various technological applications [5, 7].

3.6 Generative AI

Generative AI (Gen AI) represents a significant leap in artificial intelligence, focusing on creating models that generate new, original content. Unlike traditional AI, which relies on predefined rules, generative AI models like GPT-4 are trained on extensive datasets to understand and replicate patterns in a human-like manner. This allows them to perform tasks such as composing essays, creating art, and engaging in meaningful conversations. The core strength of generative AI is its ability to comprehend context and produce responses that are relevant and coherent.

In our project, we leverage generative AI to develop a simple chatbot that enhances user interaction by providing accurate and timely information. This chatbot assists users with questions about our website and offers information on skin problems. Utilizing advanced natural language processing (NLP) techniques, the chatbot comprehends and processes user queries effectively, ensuring a natural and intuitive interaction. Users receive quick, accurate responses whether they seek website navigation help or information about skin issues based on symptoms.

The generative AI-powered chatbot benefits users by being available 24/7, providing consistent and reliable information, and efficiently handling routine queries. This constant availability enhances user satisfaction, while the chatbot's consistency builds trust. By addressing common queries and offering basic advice, the chatbot educates users, promoting better skin health awareness and preventive care.

The application of generative AI in our chatbot illustrates the broader impact of this technology. By automating routine interactions, generative AI frees up human resources for more complex tasks, improving operational efficiency and user experience. As AI evolves, generative AI models will enable even more sophisticated and personalized user interactions, driving innovation across various fields.

4 Methodology

First we have to find a suitable Dataset. We have used Dermnet dataset. Which includes 9 types if skin diseases. After that the images are loaded in the system the images are preprocessed (Fig. 7).

Step 1: Next the CNN model architecture is defined. It is a sequential model.
Step 2: Then the model is compiled.
Step 3: Now to start the training process we fit the model with the declared hyper parameters. This can be simply done by using the keras model.fit() method and passing the parameters.
Step 4: After the model is trained the model will be evaluated.
Step 5: Classification.

5 Implementation

Following steps need to follow:

i. **Model Creation**:

- First we have to find a suitable Dataset. We have used a dataset. Which includes 9 types of skin diseases.
- After that the images are loaded in the system and the images are preprocessed.
- Next the CNN model architecture is defined. It is a sequential model. Then the model is compiled.

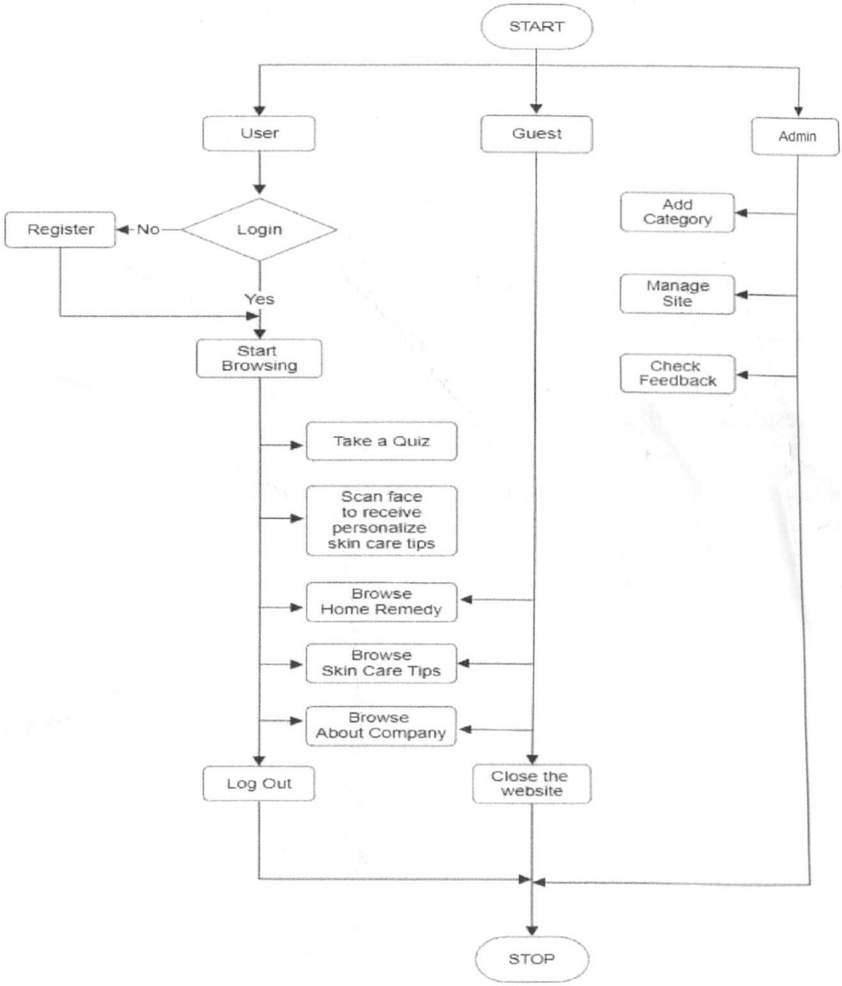

Fig. 7 Work flow diagram

- Next to train the model we fit the model with the declared hyperparameters. This can be simply done by using the keras model.fit() method and passing the parameters.
- After the model is trained the model will be evaluated.

ii. **Front end Creation**:

- The structure is created using HTML.
- Then is designed using CSS and bootstrap.
- And javascript is also used.

iii. **Back end creation:**

- To save the user log in details SQL database is used.
- Flask is used as the server scripting language.
- We have used SQLAlchemy to create the database.
- The CNN model will run in the backend.
- The cnn model is saved using model.save('model_name.hdf5') function. model_name.hdf5' is used to predict output later.
- The model will be deployed using flask in the website.
- To use the saved model in our website we'll use Flask, a web framework written in Python.

6 Result and Discussion

The successful development of the skin disease prediction model marks a significant milestone in the "**Implementation of Skin Disease Identification and Analysis Through Deep Learning**". The model, trained to identify 9 types of skin diseases, demonstrates robust capabilities in accurately diagnosing conditions such as melanoma, Actinic Keratosis, Atopic Dermatitis, Vascular Lesion etc. The model's predictions are currently accessible through backend processing, showcasing its effectiveness in analyzing user-uploaded images and providing relevant diagnostic information.

The frontend of the system has been meticulously crafted, allowing users to log in, log out, and sign up, facilitating a seamless and secure user experience. Although the model is yet to be fully integrated into the website, its predictive capabilities are validated independently.

Moving forward, the next steps involve connecting the trained model to the web interface, enabling real-time predictions for users. This integration will enhance the user experience by providing immediate feedback and personalized recommendations based on the identified skin conditions. The achievement of a functional backend, coupled with a user-friendly frontend, positions the project as a promising tool in the realm of personalized skincare diagnostics. The ongoing developments signify a significant stride toward the project's overarching goal of delivering accurate, accessible, and personalized skin health insights to users.

Training and Testing data load:

See Figs. 8, 9 and 10.

```
tranning_data=[]
def creating_tranning_data():
    for catagory in catagories:
        path= os.path.join(Data_Dir , catagory)
        lables= catagories.index(catagory)
        for img in os.listdir(path):
            img_arr = cv2.imread(os.path.join(path,img))
            rgb_img= cv2.cvtColor(img_arr, cv2.COLOR_BGR2RGB)
            new_array=cv2.resize(rgb_img , (Size,Size))
            tranning_data.append([new_array , lables])
creating_tranning_data()
```

```
Test_Data_Dir="C:/Users/DELL/Downloads/archive (5)/test"
testing_data=[]
def creating_testing_data():
    for category in catagories:
        path= os.path.join(Test_Data_Dir , catagory)
        test_lables= catagories.index(catagory)
        for img in os.listdir(path):
            test_img_arr = cv2.imread(os.path.join(path,img))
            test_rgb_img= cv2.cvtColor(test_img_arr, cv2.COLOR_BGR2RGB)
            test_new_array=cv2.resize(test_rgb_img , (Size,Size))
            testing_data.append([test_new_array , test_lables])
creating_testing_data()
```

Fig. 8 Some snapshot of the coding part

```
model = Sequential()
model.add(Conv2D(32, (3,3)  , input_shape=(Size,Size,3)))
model.add( Activation ("relu"))
model.add(MaxPooling2D(pool_size=(2,2)))
model.add(Conv2D(64, (3,3) , activation ='relu'))
model.add(MaxPooling2D(pool_size=(2,2)))
model.add(Flatten())
model.add(Dense(140 , activation='relu'))
model.add(Dense(180 , activation='relu'))
model.add(Dropout(0.3))
model.add(Dense(9 , activation='softmax'))
```

```
model.compile(optimizer='adam',loss='sparse_categorical_crossentropy',metrics=['accuracy'])
history=model.fit(X,Y,epochs=20)
```

Fig. 9 Some snapshot of the model training

Model Training:

```
20/20 [==============================] - 27s 1s/step - loss: 3.2553 - accuracy: 0.1308 - val_loss: 2.1246 - val_accuracy: 0.214
3
Epoch 2/15
20/20 [==============================] - 17s 878ms/step - loss: 2.0309 - accuracy: 0.2488 - val_loss: 1.9046 - val_accuracy: 0.
2286
Epoch 3/15
20/20 [==============================] - 16s 791ms/step - loss: 1.7490 - accuracy: 0.3397 - val_loss: 1.5882 - val_accuracy: 0.
3000
Epoch 4/15
20/20 [==============================] - 14s 701ms/step - loss: 1.5703 - accuracy: 0.4083 - val_loss: 1.4494 - val_accuracy: 0.
4714
Epoch 5/15
20/20 [==============================] - 15s 732ms/step - loss: 1.2785 - accuracy: 0.5407 - val_loss: 1.4549 - val_accuracy: 0.
4571
Epoch 6/15
20/20 [==============================] - 14s 677ms/step - loss: 1.1686 - accuracy: 0.5710 - val_loss: 1.2491 - val_accuracy: 0.
5286
Epoch 7/15
20/20 [==============================] - 20s 991ms/step - loss: 1.0660 - accuracy: 0.5837 - val_loss: 1.1306 - val_accuracy: 0.
6143
Epoch 8/15
20/20 [==============================] - 19s 943ms/step - loss: 0.9096 - accuracy: 0.6236 - val_loss: 1.1243 - val_accuracy: 0.
5857
Epoch 9/15
20/20 [==============================] - 18s 879ms/step - loss: 0.7722 - accuracy: 0.7145 - val_loss: 1.1308 - val_accuracy: 0.
5286
Epoch 10/15
20/20 [==============================] - 15s 740ms/step - loss: 0.7828 - accuracy: 0.6858 - val_loss: 1.0803 - val_accuracy: 0.
5714
```

```
model.summary()

Model: "sequential"
_________________________________________________________________
 Layer (type)                Output Shape              Param #
=================================================================
 conv2d (Conv2D)             (None, 178, 178, 32)      896

 activation (Activation)     (None, 178, 178, 32)      0

 max_pooling2d (MaxPooling2  (None, 89, 89, 32)        0
 D)

 conv2d_1 (Conv2D)           (None, 87, 87, 64)        18496

 max_pooling2d_1 (MaxPoolin  (None, 43, 43, 64)        0
 g2D)

 flatten (Flatten)           (None, 118336)            0

 dense (Dense)               (None, 140)               16567180

 dense_1 (Dense)             (None, 180)               25380

 dropout (Dropout)           (None, 180)               0

 dense_2 (Dense)             (None, 23)                4163

=================================================================
Total params: 16616115 (63.39 MB)
Trainable params: 16616115 (63.39 MB)
Non-trainable params: 0 (0.00 Byte)
_________________________________________________________________
```

Fig. 10 Some snapshot of the model summary

Model validation and evaluation:

See Fig. 11.

Confusion Matrix:

See Fig. 12.

Prediction:

See Figs. 13 and 14.

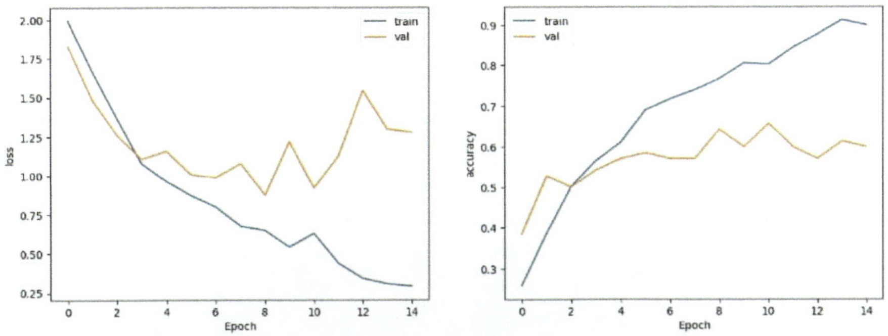

Fig. 11 Loss and accuracy graph

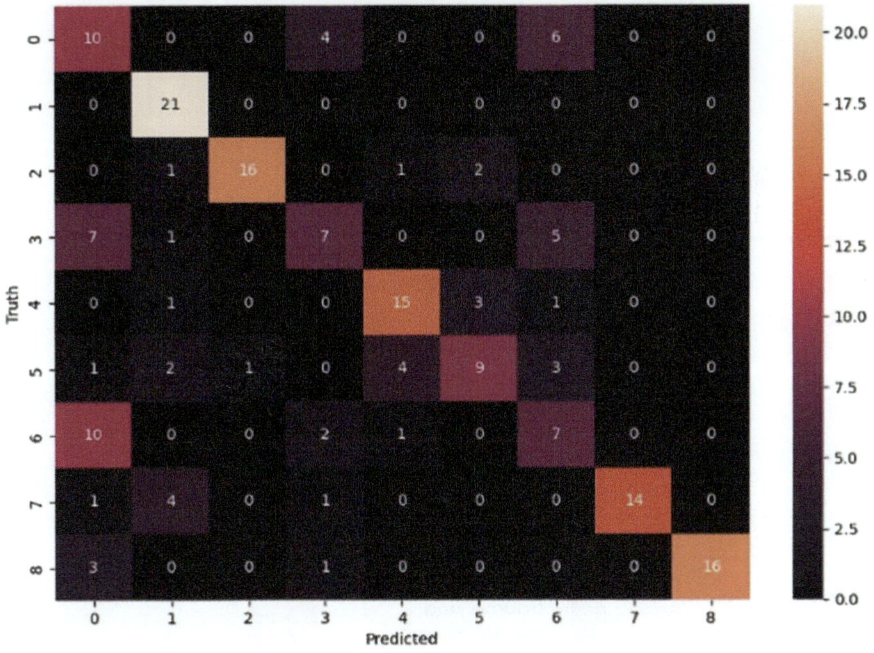

Fig. 12 Confusion matrix

7 Future Scope

The "Skin Care Analysis System using Deep Learning" has laid a robust foundation for future enhancements, extending its capabilities beyond current features. In future we will use data augmentation to overcome the model over fitting problem and also we will add this model in our website. The envisioned future scope encompasses two innovative modules designed to further empower users in understanding their

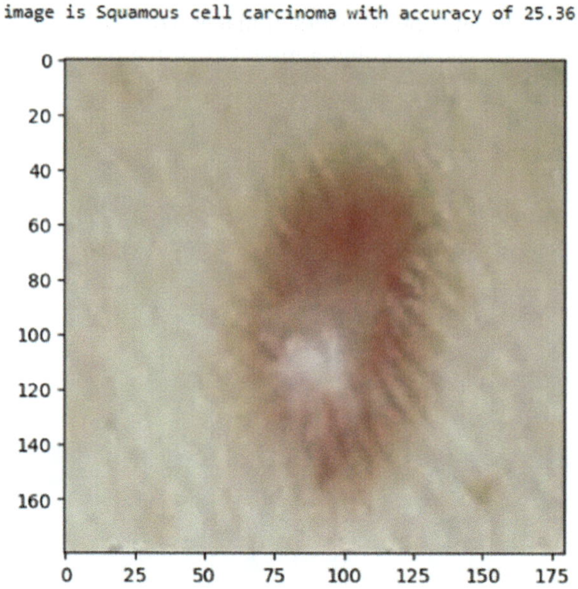

Fig. 13 Some snapshot of classification

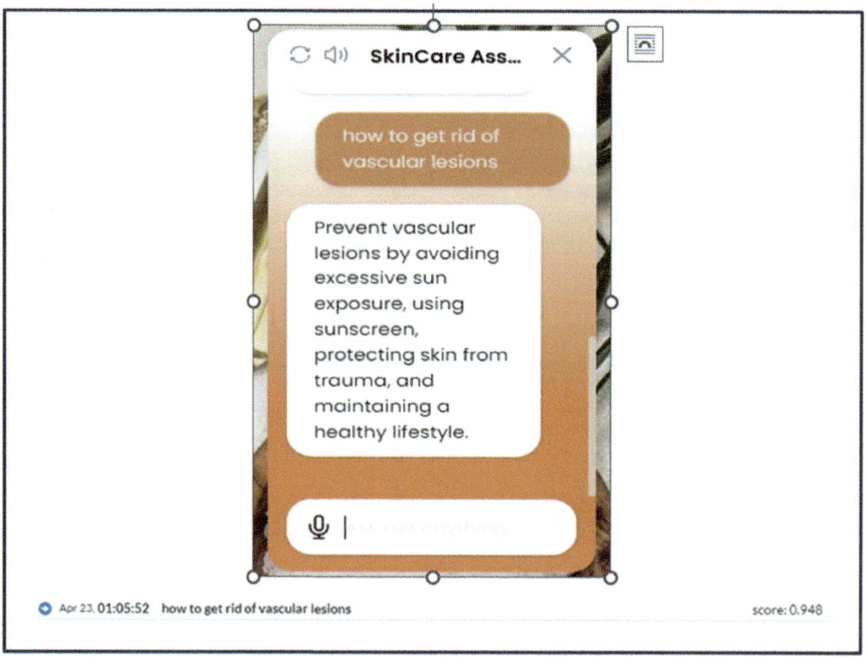

Fig. 14 Some snapshot of the skincare analysis chatbot

skin changes over time and optimizing their makeup routines. A pioneering addition to the system involves implementing a module capable of discerning the temporal evolution of an individual's skin. Users will be prompted to upload a current selfie along with a reference selfie taken six months prior. Leveraging advanced deep learning techniques, the system will analyze subtle variations in skin conditions, identifying changes such as pigmentation, texture, and aging markers. This module aims to provide users with valuable insights into the long term effects of their skincare routines, lifestyle changes, and environmental factors on their skin health. The second module focuses on revolutionizing the makeup industry by introducing an intelligent Foundation Match Recommendation system. Users will upload a high-resolution selfie, and the system, employing sophisticated color analysis and facial mapping algorithms, will accurately determine the ideal foundation shade for the individual's unique skin tone and undertones. This feature aims to streamline the makeup selection process, ensuring users achieve a seamless and natural look that complements their distinct complexion. The proposed future enhancements signify an exciting direction for the Skin Care Analysis System, demonstrating its potential to evolve into a comprehensive platform for personalized skincare and beauty guidance. By incorporating modules that analyze skin changes over time and recommend tailored foundation shades, the system continues to align with user-centric beauty solutions. As we embark on the next phase of development, these forward-looking modules exemplify the commitment to advancing technology's role in individualized skincare and beauty experiences. The integration of these features will not only empower users in their skincare and makeup choices but also position the system as a trailblazer in the intersection of technology and beauty.

8 Conclusion

In conclusion, the "Skin Care Analysis System using Deep Learning" project has successfully integrated advanced technologies to provide a comprehensive and personalized solution for skin care analysis. By leveraging Convolutional Neural Networks (CNN) implemented with Keras, the system accurately identifies common skin issues such as acne and wrinkles based on user-uploaded facial images and survey responses. The user-friendly web interface enhances accessibility, allowing individuals to receive tailored product recommendations aligned with their unique skin needs. Furthermore, the innovative inclusion of a bruise detection feature adds an extra layer of functionality to the system, allowing users to identify and understand various types of skin discolorations. This feature enhances the project's utility by providing users with information about potential skin conditions associated with bruises. The methodology employed in this project encompasses comprehensive data collection, meticulous preprocessing, effective model training, and rigorous testing. The system's deployment on a scalable hosting platform ensures widespread accessibility, and a well-defined plan for regular maintenance and updates safeguards its adaptability to evolving skin care trends. As we conclude this project, it is gratifying to see the successful integration of deep learning into the realm of personalized skin care analysis, offering users an informed and tailored approach to their skincare routine. The additional bruise detection feature enhances the system's versatility, contributing to its potential impact on users' overall skin health awareness. This project not only advances the capabilities of technology in the beauty and skincare domain but also holds promise for future developments in healthcare applications related to dermatological diagnostics.

References

1. Hawkes, J.E., Truong, A., Meyer, L.J.: Genetic predisposition to melanoma. Semin. Oncol. **43**(5), 591–597 (2016). https://doi.org/10.1053/j.seminoncol.2016.08.003. Epub 2016 Sep 1 PMID: 27899192
2. Knight, A., Karapetyan, L., Kirkwood, J.M.: Immunotherapy in Melanoma: recent advances and future directions. Cancers **15**(4), 1106 (2023). https://doi.org/10.3390/cancers15041106
3. Fabbrocini, G., Triassi, M., Mauriello, M.C., Torre, G., Annunziata, M.C., De Vita, V., Pastore, F., D'Arco, V., Monfrecola, G.: Epidemiology of skin cancer: role of some environmental factors. Cancers **2**(4), 1980–1989 (2010). https://doi.org/10.3390/cancers2041980
4. Spallone, G., Botti, E., Costanzo, A.: Targeted therapy in nonmelanoma skin cancers. Cancers **3**(2), 2255–2273 (2011). https://doi.org/10.3390/cancers3022255
5. Theobald, O.: Machine Learning for Absolute Beginners: A Plain English Introduction, 2nd edn
6. Kelleher, J.D., Namee, B.M., D'Arcy, A.: Fundamentals of Machine Learning for Predictive Data Analytics: Algorithms, Worked Examples, and Case Studies
7. Weidman, S.: Deep Learning from Scratch: Building with Python from First Principles, vol. 3

GPSR Compliance

The European Union's (EU) General Product Safety Regulation (GPSR) is a set of rules that requires consumer products to be safe and our obligations to ensure this.

If you have any concerns about our products, you can contact us on ProductSafety@springernature.com

In case Publisher is established outside the EU, the EU authorized representative is:

Springer Nature Customer Service Center GmbH
Europaplatz 3
69115 Heidelberg, Germany

Batch number: 10641589

Printed by Printforce, the Netherlands